Ecological Studies

Analysis and Synthesis

Volume 49

Forest Ecosystems in Industrial Regions

Studies on the Cycling of Energy Nutrients and Pollutants in the Niepołomice Forest Southern Poland

Edited by
W. Grodziński, J. Weiner, and P. F. Maycock

With 116 Figures

Springer-Verlag
Berlin Heidelberg New York Tokyo 1984

Prof. Dr. WŁADYSŁAW GRODZIŃSKI
Dr. JANUARY WEINER
Department of Animal Ecology, Jagiellonian University,
Institute of Environmental Biology, Krupnicza 50, 30-060 Kraków, Poland

Prof. Dr. PAUL F. MAYCOCK
Department of Botany, University of Toronto, Erindale College,
Mississauga, Ontario, Canada L5L, 1C6

ISBN-13: 978-3-642-69804-0 e-ISBN-13: 978-3-642-69802-6
DOI: 10.1007/978-3-642-69802-6

Forest ecosystems in industrial regions. (Ecological studies ; v. 49) Bibliography: p. Includes index. 1. Forest ecology – Poland – Niepolomice Forest. 2. Pollution – Environmental aspects – Poland – Niepolomice Forest. 3. Niepolomice Forest (Poland) I. Grodzinski, W., II. Weiner, J. (January), 1947– . III. Maycock, Paul F., 1930– . IV. Series. QH162.F67 1984 574.5'2642'0943 84-10522

Softcover reprint of the hardcover 1st edition 1984

Typesetting, printing, and binding: Brühlsche Universitätsdruckerei, Giessen
2131/3130-543210

Preface

A number of well-known forests have been the object of studies by ecological teams. Every ecologist is familiar with the names of forests such as Whytham Woods in England (Elton 1966), Solling in West Germany (Ellenberg 1971), Hubbard Brook in New Hampshire, U.S.A. (Likens et al. 1977, Bormann and Likens 1979), and Jadraös in Central Sweden (Persson 1980). The number of such team studies grew rapidly during the period of the International Biological Programme (Reichle 1981). Each of these forests represented aspects of local and general interest.

The Niepołomice Forest in Southern Poland (Fig. 1) does not possess any spectacular features and it may be regarded as a typical forest of the lowlands of continental Europe. Situated in the Vistula River Valley, 20 km east of Cracow (50°07′ N, 20°23′ E) it consists of two major sections: an extensive pine forest of the Pino-Quercetum type, and a smaller, deciduous, oak-hornbeam-

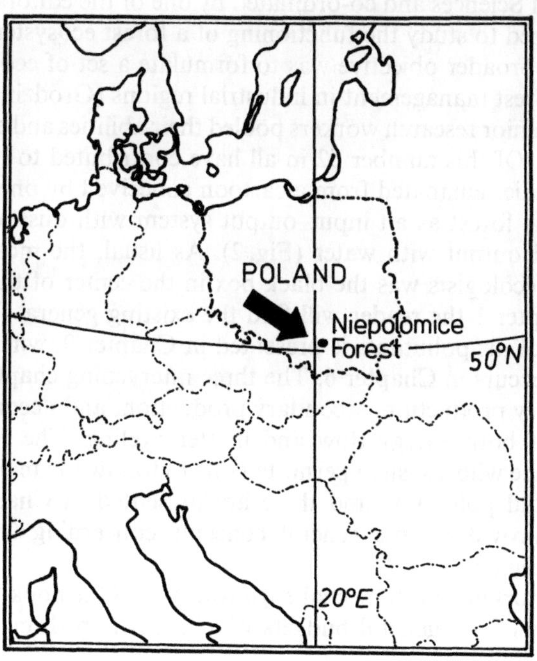

Fig. 1. The location of the Niepołomice Forest in Southern Poland

Tilio Carpinetum forest. This huge forested area (11,000 ha) owes its survival in one of the most densely populated parts of Europe to a long period of protection as a royal hunting ground. The period of royal protection ended abruptly at the turn of 20th century under the devastating blows of two world wars followed by perhaps an even greater danger, the effects of modern developments in heavy industry. During the last three decades the forest has been exposed to long-term air pollution, chiefly sulfur dioxide, heavy metals and fluoride emitted from coal combustion mainly from metallurgical industry in the region (Grodziński and Yorks 1981). Such chronic exposure to airborne pollution at moderate levels seems to have become a typical situation for temperate zone forests of Europe and North America.

The Niepołomice Forest has traditionally been a study area for ecologists from the nearby academic center of Cracow. The first studies of forest habitats were begun by Professors E. Chodzicki and B. Pawłowski as early as the mid 1940's. Comprehensive studies on the productivity of deciduous forests were initiated by Professor A. Medwecka-Kornaś in the late 1960's and continued well into the 1970's. Later, a group under the direction of the late Professor S. Myczkowski studied the effects of industrial emission on pine forests. More than 200 papers and reports have been published on subjects relating to the Niepołomice Forest, 167 of which are listed in the bibliography covering the 1946–1978 period (Banasik 1978b).

This book originated from a team project funded from 1976 by the Polish Academy of Sciences and co-ordinated by one of the editors (W.G.). The project was organized to study the functioning of a forest ecosystem under air polution stress. The broader objective was to formulate a set of ecological recommendations for forest management in industrial regions (Grodziński 1978, Rich 1979). Nearly 80 senior research workers pooled their abilities and experience to work on the project. Of this number 52 in all have contributed to this book.

Our studies emanated from a cartoon conceived by one of the editors (J.W.) showing the forest as an input–output system with dust and aerosol as inputs and a total output with water (Fig. 2). As usual, the most exciting aspect for inquisitive ecologists was the black box in the center of the sketch.

In Chapter 1 the reader will find the existing general features of the forest. The input of air pollutants is presented in Chapter 2, whereas the release from the forest occurs in Chapter 6. The three intervening chapters (3, 4, and 5) deal with primary production, secondary production, and decomposition considered in terms of both energy flow and matter cycling. The relationships of these aspects to the whole system permitted one to follow the main transfers of energy, nutrients and pollutants and these are presented in Chapter 7. The Epilogue (Chap. 8) provides some general concepts concerning forest management in industrial regions.

The editors and authors challenge you to read the book through. If not, begin with Chapter 7 on general budgets which, it is hoped, may send you browsing among others.

This book has had a rather long gestation period although parturition was smooth. Most of the editorial work was done by Dr. J. Weiner. In the final phases of the editorial work Professor P. F. Maycock, Visiting Professor at the Institute

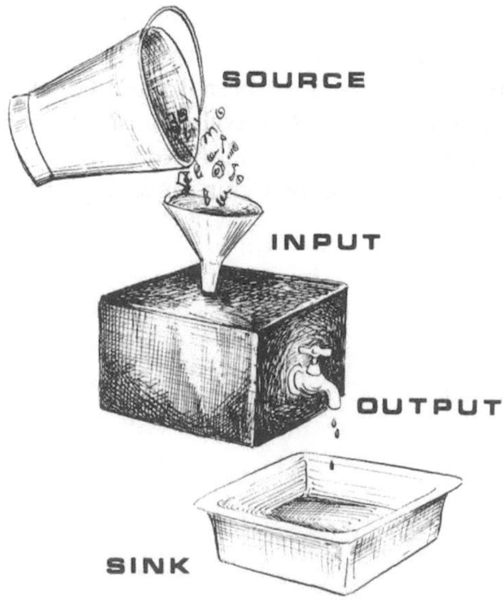

Fig. 2. An input-outout black box model of an ecosystem

of Botany, Polish Academy of Sciences in Cracow, on leave from Erindale College of the University of Toronto, Canada, contributed immensely to the final shape of the book. In the course of preparation the editors worked with the authors mainly via chapter editors. Without their indispensable efforts this book would have never been completed.

The concept of the book developed gradually at a series of annual meetings of the group as a whole. The project co-ordinator (W. Grodziński) tried desperately to keep this multidisciplinary team of 80 scientists together, promoting a common language among ecologists, forest scientists, dendrologists, hydrologists and geochemists, soil biologists and even engineers and economists, and attempted to ensure an influx of funds.

This project has recently been extended for another 5 years and new field experiments have been introduced (e.g., application of industrial dust and fertilizers to the forest floor). Some preliminary reports have already been published (Greszta et al. 1982a, b). Our book has been preceded by a more practical report entitled *Ecological principles for re-building the region of the Niepolomice Forest* (Grodziński and Lesiński 1978) with practical recommendations on the gradual re-structuring of forest stands, amelioration, aerial fertilization and wildlife management. Some of these have already been implemented.

We all are very grateful to the Institute of Ecology, Polish Academy of Sciences, for supporting this project. Special thanks are due to Professor R. Z. Klekowski, at the time Director of the Institute, and current chief coordinator of the ecological program MR.II/15, for his long-standing interest in our studies. The assistance we have received from the authorities of the City of Cracow and the Forest Administration has been most helpful. Among the many persons involved we would like to mention Mr. T. Irzabek, M.Sc., Manager of the Niepołomice Forest District. We appreciate very greatly the patient and stimulating co-operation of the Publishers, Springer Verlag, especially that of Dr. D. Czeschlik. Chapters have been translated by Dr. A. Hoffman (1–6), Dr. R. Tertil (7), and Prof. M. Giertych (8). Ms. F. Skolimowska, M.Sc. typed the complete text several times. To all these people and to many who have not been mentioned here we express our sincere gratitude for their invaluable contributions.

August 1984 W. GRODZIŃSKI

Contents

1. The Niepołomice Forest: Existing Features. Editor: K. GRODZIŃSKA. (With 18 Figures) 1

1.1 Flora and Vegetation of the Niepołomice Forest. By A. ĆWIKOWA, A. GRABOWSKI, A. J. LESIŃSKI, and S. MYCZKOWSKI 2
1.1.1 Tree Layer 2
1.1.2 Flora 3
1.1.3 Forest Communities 6

1.2 Soils. By B. ADAMCZYK 11
1.3 Climate. By J. KLEIN 16
1.3.1 Solar Radiation and Insolation 16
1.3.2 Air Circulation 17
1.3.3 Thermal Relations 19
1.3.4 Atmospheric Precipitation 20
1.3.5 Air Humidity 22
1.3.6 Final Remarks 22
1.4 History of the Niepołomice Forest. By M. ROKOSZ 24
1.4.1 Forest Management 24
1.4.2 Water Relationships 26
1.4.3 Wildlife 26
1.4.4 Concluding Remarks 27
1.5 Bioindication of Environmental Deterioration. By K. GRODZIŃSKA . 27

2. Transport and Input of Air Pollutants in the Niepołomice Forest Area. Editor: A. MANECKI. (With 28 Figures) 35
2.1 Introduction. By A. MANECKI 36
2.2 Emission of Industrial Air Pollutants. By E. GARŚCIA, A. MANECKI, and K. P. ZAJĄC 36
2.2.1 Emission by Distant Industrial Complexes 37
2.2.2 Emission by Adjacent Industrial Installations 37
2.3 Evaluation of Aerosol Pollution Transport Using Remote-Sensing Methods. By J. WALCZEWSKI 40
2.4 Meteorological Aspects of Air Pollutant Spread in Forest Areas. By J. LEWIŃSKA 44

2.5 Concentration, Transformation, and Deposition of Sulfur Compounds. By S. KASINA, J. KWIEK, and J. LEWIŃSKA 45
2.5.1 Concentration of Sulfur Dioxide and Sulfate Aerosol 47
2.5.2 Estimation of the Dry Deposition of Sulfur 51
2.5.3 Wet Deposition of Sulfur 51
2.5.4 Precipitation Acidity 54
2.5.5 Input of Sulfur to the Substrate 56
2.5.6 Particulate Matter 56

2.6 Input and Mineral Composition of Dust to the Substrate. By A. MANECKI and A. SKOWROŃSKI 58
2.6.1 Amounts of Dust Fall 59
2.6.2 Highly Soluble Particulate Components 61
2.6.3 Slightly Soluble Particulate Components 61
2.6.4 Chemical Composition of the Dust Fall 64
2.6.5 Geochemical Characteristics of the Substrate 67

2.7 Conclusion 68

3. Primary Production in the Niepołomice Forest Ecosystems. Editor: R. RIEGER. (With 10 Figures) 69

3.1 Introduction. By R. RIEGER 70
3.2 Growing Stock and Increment of Tree Stands. By R. RIEGER, S. GRABCZYŃSKI, S. ORZEŁ, and J. RAIMER 70
3.2.1 The Characteristics of Tree Stands of the Niepołomice Forest . 70
3.2.2 Dendrometric Characteristics of Tree Stands in Selected Areas . 71
3.2.3 Productivity of Tree Stands 75

3.3 Photosynthetic Productivity. By M. CZARNOWSKI and W. MACZEK . 78
3.3.1 Seasonal Changes in the Photosynthetic Activity of Leaves in Areas with Variable Air Pollution 79
3.3.2 Seasonal Photosynthetic Production and Respiration in Leaves of Selected Tree Species 80
3.3.3 Productivity and Accumulation of Solar Energy in Oak-Hornbeam Forests 81
3.3.4 Seasonal Changes in the Rate of Photosynthesis and Respiration of Pine Shoots 83
3.3.5 Effects of Industrial Air Pollution on the Morphology and Abundance of Pine Needles 84
3.3.6 Potential Photosynthetic Production of Long Pine Shoots . . . 85
3.3.7 Estimation of Photosynthetic Production of Pine Forests . . . 86

3.4 Productivity of the Herb Layer. By J. BANASIK and B. ZEMANEK . 87
3.4.1 Study Areas 87
3.4.2 Methods 88
3.4.3 Production of the Herb Layer in Forest Communities 88
3.4.4 Discussion 89

3.5 Productivity of Nonforest Vegetation. By Z. DENISIUK 91
3.5.1 Nonforest Communities 91
3.5.2 Productivity of Nonforest Communities 92

3.6 The Concentration of Nutrients and Pollutants in Plant Materials in the Niepołomice Forest. By K. GRODZIŃSKA 95
3.7 Conclusions 98

4. Energy and Matter Flow Through Consumers in the Niepołomice Forest Ecosystem. Editor: A. GÓRECKI. (With 12 Figures) 101

4.1 Introduction. By W. GRODZIŃSKI 102
4.2 Ecology, Energetics and the Significance of Phytophagous Insects in Deciduous and Coniferous Forests. By Z. WITKOWSKI, and K. BORUNSIEWICS 103
4.2.1 Sampling Plots 103
4.2.2 Species Composition, Bioenergetics, and Biomass Structure . . 104
4.2.3 Nutrients (Nitrogen, Phosphorus, Potassium, and Calcium) in Insect Bodies and Excrement 108
4.2.4 Phytophagous Insects in the Ecosystems of the Niepołomice Forest 110
4.2.5 Effects of Industrial Pollution on the Insect Fauna. By K. BORUSIEWICZ 112

4.3 Energy and Matter Flow Through Homeothermic Consumers 113
4.3.1 Energy and Matter Flow Through Rodents. By A. GÓRECKI . 113
4.3.2 Energy and Matter Flow Through Ungulates. By B. BOBECK, and K. PERZANOWSKI 121
4.3.3 Energy and Matter Flow Through Bird Populations. By Z. GŁOWACIŃSKI, J. KOZŁOWSKI, and J. WEINER 125
4.3.4 Energy and Matter Flow Through Carnivores. By W. GRODZIŃSKI and J. WASILEWSKI 132

4.4 Flow of Heavy Metals Through Selected Homeotherm Consumers. By K. SAWICKA-KAPUSTA and J. KOZŁOWSKI 139
4.5 Impact of Consumers on the Forest Ecosystems of the Niepołomice Forest. By A. GÓRECKI 143

5. Decomposition in the Pine Forests on Niepołomice. By J. ZIELIŃSKI. (With 8 Figures) 149

5.1 Introduction 150
5.2 Study Areas 151
5.3 Matter Dynamics and Balance in the Litter 153
5.4 Turnover Rate of Matter in Forest Ecosystems 161
5.5 Decomposition Rate of Dead Organic Matter 163

6. The Outflow of Macro- and Microelements from the Niepołomice Forest Watersheds. Editor: M Reczyńska-Dutka. (With 17 Figures) 167

6.1 The Watershed and Typical Streams of the Niepołomice Forest. By K. Pasternak and J. Szlaga 168

6.1.1 Watershed of Selected Streams 168
6.1.2 Analysis of Water Flow in Traczówka Creek 170
6.1.3 Average Annual Flow in Traczówka and Śmigne Creeks . . . 174

6.2 Effect of Drainage Systems on the Hydrological Regime of the Niepołomice Forest. By A. Lewiński 175

6.3 Hydrobiologic Characteristics of Selected Streams. By T. Bednarz, K. Trela, and J. Zygmuntowa 176

6.3.1 Bacteriological Characteristics of the Streams 176
6.3.2 Concentration of Dissolved Proteins and Free Amino Acids in the Streams 179
6.3.3 Variation in the Trophic Value of Water 180
6.3.4 Characteristics of Phytoplankton 181
6.3.5 Hydrobiological Characteristics of the Streams 182

6.4 Outflow of Macroelements from the Forest Watershed Affected by Industrial Air Pollution. By S. Wróbel and M. Bombówna 182

6.4.1 Characteristics of the Creeks 183
6.4.2 Physicochemical Characteristics of the Water 183
6.4.3 Ionic Composition of the Water 185
6.4.4 Organic Matter and Nitrogen and Phosphorus Compounds in the Water 187
6.4.5 Chemical Erosion by the Streams 190
6.4.6 Mechanisms of Macroelement Run-Off from the Traczówka Watershed 192

6.5 Outflow and Accumulation of Heavy Metals in Streams of the Noepołomice Forest. By K. Pasternak and M. Reczyńska-Dutka 193

6.5.1 Seasonal Variation in Water Flow in the Southern Forest Complex 194
6.5.2 Concentration of Heavy Metals 195
6.5.3 Seasonal Variation in the Daily Outflow of Heavy Metals . . . 196
6.5.4 Spatial Variation in the Outflow of Heavy Metals from the Southern Forest Complex 197
6.5.5 Outflow and Accumulation of Heavy Metals in Streams 198
6.5.6 Transport and Accumulation of Heavy Metals in the Niepołomice Forest 199

6.6 Conclusion. By K. Pasternak and M. Reczyńska-Dutka 202

7. Energy, Nutrient, and Pollutant Budgets of the Forest Ecosystems. By J. Weiner and W. Grodziński. (With 20 Figures) 203

7.1 Introduction 204

7.2 Biomass Production, Nutrient Uptake, and Pollutant Accumulation by Plants 205
7.2.1 Data Sources 205
7.2.2 Energy Budget of Primary Production 207
7.2.3 Standing Crop of Biomass, Energy, and Chemical Elements . 207
7.2.4 Nutrient Uptake by Forest Vegetation 208
7.2.5 Heavy Metal Accumulation 212
7.2.6 Standing Crop, Production, and Turnover Rates of Biomass, Energy, Nutrients, and Pollutants 212

7.3 Energy and Matter Flow Through Consumer Populations 213
7.3.1 Energy and Nutrient Intake by Consumers 213
7.3.2 Intake and Accumulation of Pollutants by Consumers 215

7.4 Chemical Budgets in the Forested Ecosystems 216
7.5 Discussion 218
7.5.1 Primary Production 218
7.5.2 Nutrient Budgets 219
7.5.3 Pollutant Budgets 220
7.5.4 The General Pattern of Energy, Nutrient, and Pollutant Behavior in Forest Ecosystems 222
7.5.5 Disruptions in Elemental Cycles and the Future of the Niepołomice Forest 223

7.6 Appendix 226

8. Concepts of Forest Management in Industrial Regions.
Editors: S. Białobok and J. Fabijanowski. (With 1 Figure) 231

8.1 Forest Functions in Industrial Regions. By J. Fabijanowski and J. A. Lesiński 232
8.2 Stand Conformity with Site. By B. Adamczyk, J. A. Lesiński, and S. Białobok 234
8.3 Site Formation. By B. Adamczyk 237
8.4 Selection of Trees and Shrubs for Forest Restructuring in Industrial Regions. By S. Białobok, J. Oleksyn, and L. Rachwał 239
8.4.1 Tolerance of Trees and Shrubs to Industrial Fumes 240
8.4.2 Principles of Seed and Nursery Work 243
8.5 The Niepołomice Forest as an Example of a Forest Affected by Industrial Stress. By S. Białobok, J. Fabijanowski, J. Greszta, and J. A. Lesiński 244

References 246

Subject Index 269

Contributors

ADAMCZYK, BOLESŁAW, Department of Soil Science, Agricultural University, Al. 29 Listopada 48, 31-425 Kraków

BANASIK, JULITA, Nature Protection and Natural Resources Research Centre, Polish Academy of Sciences, ul. Lubicz 46, 31-512 Kraków

BEDNARZ, TERESA, Institute of Fresh Water Biology, Polish Academy of Sciences, Hydrobiological Laboratory, ul. Sławkowska 17, 31-016 Kraków

BIAŁOBOK, STEFAN, Institute of Dendrology, Polish Academy of Sciences, Department of Tree Resistance, ul. Parkowa 5, 62-035 Kórnik

BOBEK, BOGUSŁAW, Department of Animal Ecology, Institute of Environmental Biology, Jagiellonian University, ul. Karasia 6, 30-060 Kraków

BOMBÓWNA, MARIA, Institute of Fresh Water Biology, Polish Academy of Sciences, Ecosystem Studies Laboratory, ul. Sławkowska 17, 31-016 Kraków

BORUSIEWICZ, KRYSTYNA, Nature Protection and Natural Resources Research Centre, Polish Academy of Sciences, ul. Lubicz 46, 31-512 Kraków

CZARNOWSKI, MARIAN, Department of Plant Physiology, Polish Academy of Sciences, ul. Sławkowska 17, 31-016 Kraków

ĆWIKOWA, ALIĆJA, Department of Forest Ecology, Agricultural University, Al. 29 Listopada 48, 31-425 Kraków

DENISIUK, ZYGMUNT, Nature Protection and Natural Resources Research Centre, Polish Academy of Sciences, ul. Lubicz 46, 31-512 Kraków

FABIJANOWSKI, JERZY, Department of Silviculture, Agricultural University, Al. 29 Listpada 48, 31-425 Kraków

GARŚCIA, EDWARD, Institute of Environmental Protection, Academy of Mining and Metallurgy, Al. Mickiewicza 30, 30-059 Kraków

GŁOWACIŃSKI, ZBIGNIEW, Nature Protection and Natural Resources Research Centre, Polish Academy of Sciences, ul. Lubicz 46, 31-512 Kraków

GÓRECKI, ANDRZEJ, Department of Animal Ecology, Institute of Environmental Biology, Jagiellonian University, ul. Karasia 6, 30-060 Kraków

GRABCZYŃSKI, STANISŁAW, Department of Forest Mensuration, Agricultural University, Al. 29 Listopada 48, 31-425 Kraków

GRABOWSKI, ANDRZEJ, Department of Forest Ecology, Agricultural University, Al. 29 Listopada 48, 31-425 Kraków

GRESZTA, JAN, Institute of Botany, Polish Academy of Sciences, Department of Ecology and Plant Geography, ul. Lubicz 46, 31-512 Kraków

GRODZIŃSKA, KRYSTYNA, Institute of Botany, Polish Academy of Sciences, Department of Ecology and Plant Geography, ul. Lubicz 46, 31-512 Kraków

GRODZIŃSKI, WŁADYSŁAW, Department of Animal Ecology, Institute of Environmental Biology, Jagiellonian University, ul. Karasia 6, 30-060 Kraków and Institute of Fresh Water Biology, Polish Academy of Sciences, Ecosystem Studies Laboratory, ul. Sławkowska 17, 31-016 Kraków

KASINA, STANISŁAW, Research Institute for Environmental Development, Cracow Branch, ul. Salwatorska 14, 30-109 Kraków

KOZŁOWSKI, JAN, Department of Animal Ecology, Institute of Environmental Biology, Jagiellonian University, ul. Karasia 6, 30-060 Kraków

KLEIN, JAN, Nature Protection and Natural Resources Research Centre, Polish Academy of Sciences, ul. Lubicz 46, 31-512 Kraków

KWIEK, JÓZEF, Research Institute for Environmental Development, Cracow Branch, ul. Salwatorska 14, 30-109 Kraków

LESIŃSKI, JERZY, A., Department of Forest Ecology, Agricultural University, Al. 29 Listopada 48, 31-425 Kraków

LEWIŃSKI, ALEKSANDER, Nature Protection and Natural Resources Research Centre, Polish Academy of Sciences, ul. Lubicz 46, 31-512 Kraków

LEWIŃSKA, JANINA, Research Institute for Environmental Development, Cracow Branch, ul. Salwatorska 14, 30-109 Kraków

MACZEK, WIESŁAW, Department of Plant Physiology, Polish Academy of Sciences, ul. Sławkowska 17, 31-016 Kraków

MANECKI, ANDRZEJ, Department of Mineralogy and Geochemistry, Institute of Geology and Mineral Deposits, Academy of Mining and Metallurgy, Al. Mickiewicza 30, 30-059 Kraków

MAYCOCK, PAUL, F., Department of Botany, University of Toronto, Erindale College, Mississauga, Ontario, Canada L5L, 1C6

MYCZKOWSKI, STEFAN, Department of Forest Ecology, Agricultural University, Al. 29 Listopada 48, 31-425 Kraków

OLEKSYN, JACEK, Institute of Dendrology, Polish Academy of Sciences, Department of Tree Resistance, ul. Parkowa 5, 62-035 Kórnik

ORZEŁ, STANISŁAW, Department of Forest Mensuration, Agricultural University, Al. 29 Listopada 48, 31-425 Kraków

PASTERNAK, KAZIMIERZ, Institute of Fresh Water Biology, Polish Academy of Sciences, ul. Sławkowska 17, 31-016 Kraków

PERZANOWSKI, KAJETAN, Department of Animal Ecology, Institute of Environmental Biology, Jagiellonian University, ul. Karasia 6, 30-060 Kraków

RACHWAŁ, LESŁAW, Institute of Dendrology, Polish Academy of Sciences, Department of Tree Resistance, ul. Parkowa 5, 62-035 Kórnik

RAIMER, JERZY, Department of Forest Mensuration, Agricultural University, Al. 29 Listopada 48, 31-425 Kraków

RECZYŃSKA-DUTKA, MARTA, Institute of Fresh Water Biology, Polish Academy of Sciences, Ecosystem Studies Laboratory, ul. Sławkowska 17, 31-016 Kraków

RIEGER, ROMAN, Department of Forest Mensuration, Agricultural University, Al. 29 Listopada 48, 31-425 Kraków

ROKOSZ, MIECZYSŁAW, Institute of History, Jagiellonian University, ul. Gołębia 13, 31-007 Kraków

SAWICKA-KAPUSTA, KATARZYNA, Department of Animal Ecology, Institute of Environmental Biology, Jagiellonian University, ul. Karasia 6, 30-060 Kraków

SKOWROŃSKI, ANDRZEJ, Department of Mineralogy and Geochemistry, Institute of Geology and Mineral Deposits, Academy of Mining and Metallurgy, Al. Mickiewicza 30, 30-059 Kraków

SZLAGA, JAN, Institute of Meteorology and Water Management, ul. Borowego 14, 30-107 Kraków

TRELA, KRYSTYNA, Institute of Fresh Water Biology, Polish Academy of Sciences, Microbiological Laboratory, ul. Sławkowska 17, 31-016 Kraków

WALCZEWSKI, JACEK, Division of Space Meteorology, Institute of Meteorology and Water Management, ul. Borowego 14, 30-107 Kraków

WASILEWSKI, JACEK, Department of Zoopsychology and Ethology of Animals, Institute of Environmental Biology, Jagiellonian University, ul. Karasia 6, 30-060 Kraków

WEINER, JANUARY, Department of Animal Ecology, Institute of Environmental Biology, Jagiellonian University, ul. Karasia 6, 30-060 Kraków

WITKOWSKI, ZBIGNIEW, Nature Protection and Natural Resources Research Centre, Polish Academy of Sciences, ul. Lubicz 46, 31-512 Kraków

WRÓBEL, STANISLAW, Nature Protection and Natural Resources Research Centre, Polish Academy of Sciences, ul. Lubicz 46, 31-512 Kraków

ZAJĄC, KONRAD, P., Institute of Environmental Protection, Academy of Mining and Metallurgy, Al. Mickiewicza 30, 30-059 Kraków

ZEMANEK, BOGDAN, Botanic Garden, Institute of Botany, Jagiellonian University, ul. Kopernika 27, 31-501 Kraków

ZIELIŃSKI, JAN, Department of Animal Ecology, Institute of Environmental Biology, Jagiellonian University, ul. Karasia 6, 30-060 Kraków

ZYGMUNTOWA, JADWIGA, Institute of Fresh Water Biology, Polish Academy of Sciences, Ecosystem Studies Laboratory, ul. Sławkowska 17, 31-016 Kraków

Chapter 1

The Niepołomice Forest: Existing Features

Chapter Editor: K. GRODZIŃSKA

Contributors:

B. ADAMCZYK, A. ĆWIKOWA, A. GRABOWSKI, K. GRODZIŃSKA, J. KLEIN, A. J. LESIŃSKI, S. MYCZKOWSKI, and M. ROKOSZ

1.1 Flora and Vegetation of the Niepołomice Forest

A. Ćwikowa, A. Grabowski, A. J. Lesiński, and S. Myczkowski

The Niepołomice Forest is a lowland complex approximately 110 km^2, situated 10–35 km east of Cracow and the great steel complex, and between the forks of the Vistula and Raba rivers. It is crossed by a creek, the Drwinka. The northern third totals 19.8 km^2 and the Grobla, Grobelczyk, and Koło sections contribute 15.0, 2.6, and 2.2 km^2, respectively. The southern part is a total of 88.7 km^2. The northern part is confined to a flood plain terrace of the Vistula, and is characterized by high ground water levels. The southern part occurs on a higher terrace and under much less humid conditions. Those differences in habitat conditions are reflected in the prevalence of mixed oak-pine and mesic oak-hornbeam forest in the southern part, while moist oak-hornbeam and alder forest as well as alder and elm forest predominate in the northern forest areas.

The forest has been wastefully exploited since the 16th century (Smólski 1965, Dziewolski 1976, Rokosz, in press), and in the past 25 years it has been greatly affected by air pollutants emitted by industrial installations in the Cracow region (Grodzińska et al. 1978, Grodziński and Yorks 1981, Grabowski 1981 a, Grodzińska 1982 a, b). This natural forest complex has thus undergone considerable changes (Myczkowski and Lesiński 1976). Tree stands and plant communities, as well as the flora, have thus been affected.

1.1.1 Tree Layer

The present state of the tree layer has resulted from both habitat conditions and human activities in the past. The early period of wasteful exploitation involved highgrading of oak, ash, and elm for timber, mine supports, charcoal, etc. This produced a negative selection and consequently resulted in a decrease in quality of natural regeneration. Natural regeneration was considerably hampered as well by pasturing in the forest up to the end of World War II. Cutting, combined with artificial regeneration with exotic species, has resulted in the development of vast pine monocultures, often in unsuitable habitats (Myczkowski and Lesiński, in press). Natural habitat conditions have also been affected by repeated drainage (Smólski 1965). The forest management plan has had disastrous effects on the health of the forest stands. There were mass outbreaks of forest insects, such as *Panolis flammea, Acantholyda nemoralis, Neodiprion pini, Ocneria monacha, Dendrolimus pini,* and *Bupalus piniarus,* during the last century (Smólski 1965). Weakened stands of pine monoculture are often subsequently blown down by wind or severely damaged by snow (Myczkowski and Lesiński, in press).

Industrial air pollution has recently resulted in a considerable increase in the production of dead wood, particularly from pine. Furthermore, secondary insect pests which lower the commercial value of timber have appeared, and are typical of forests confined to industrial areas. These are: *Blastophagus piniperda*, and *B. minor*, *Criophalus rusticus*, *Rhagium inquisitor*, and *Rh. mordax*, *Pissodes piniphilus*, *Spondylis buprestoides*, *Leiopus nebulosus*, *Saperda scalaris*, *Plagionotus arcuatus*, and others (Starzyk 1975, Starzyk et Starzyk 1981, Myczkowski and Lesiński 1976). Physiological pests, mostly *Tortrix viridana*, have become much more common in deciduous oak stands than previously (Witkowski and Langer 1978). The present unsatisfactory state of the Niepołomice Forest is to a considerable extent due also to excessive cutting, some during World War II.

Consequently the tree stands include three major tree species: pine (*Pinus silvestris*) – 71% of the area and 72% of standing timber; pedunculate and sessile oak (*Quercus robur* and *Q. petraea*) – 17% of area and 19% of standing timber; and black alder – 9% of area and 8% of standing timber. Pine stands of variable age are confined almost exclusively to the southern part of the forest. Oak stands occur mostly in the northern sections, and alder occurs in small scattered patches in the Grobla section and in the southern part near the so-called Błoto (Fig. 1.1).

There are no larger older fir, spruce, or lime. Old beech, ash, maple, and sycamore maple occur very infrequently, and the last few old elm are dying from infestations of the fungus *Ceratocystis ulmi* and the beetle *Scotylus* sp. All these tree species were once quite common in the Niepołomice Forest (Trela 1938, Myczkowski 1954, Szponder 1955, Smólski 1965, Ćwikowa 1974).

The age structure of the tree stands is far from satisfactory. Most areas are covered with stands from 21 to 50 years old, while older stands of 51–100 years are of minor occurrence (36% of area).

Most tree stands are single-storied. Some oak stands in the northern sections and at the southern margin have, however, a sparse lower layer including hornbeam and lime. In places, these stands show a considerable undergrowth of hawthorn, spindle-tree, and elder. Older pine stands, particularly those recently affected by a mass outbreak of *Acantholyda nemoralis*, possess a low undergrowth of oak regeneration. Oak, hornbeam, lime, and beech were planted in certain older pine stands in attempts to restructure the forest. Generally, these plantations are in good health, except for the oak type. Fir were also planted in some pine plantations but they have been intensively browsed by game. Black alder, as well as stands confined to habitats favorable for alder and elm forest, often support bird cherry in the undergrowth; buckthorn is also a common component.

1.1.2 Flora

The Niepołomice Forest comprises a part of the lowland Sandomierz basin geobotanical region (Szafer 1977). The continental climate influencing this region results in the absence of certain species typical of milder Atlantic climatic conditions. There are also certain xerophytic species, as well as several Carpathian species which have ingressed from the adjacent mountains.

The earliest contribution to the knowledge of the flora of the forest is that of Berdau (1859). Valuable accounts of the flora are given by Krupa (1876), Raciborski (1884), Żmuda (1920), and Pawłowski (1932). Some data can also be found in a paper by Łoziński (1935). Studies by Dubiel (1971, 1973) and the present authors permit an analysis of the changes in the flora over the past 120 years.

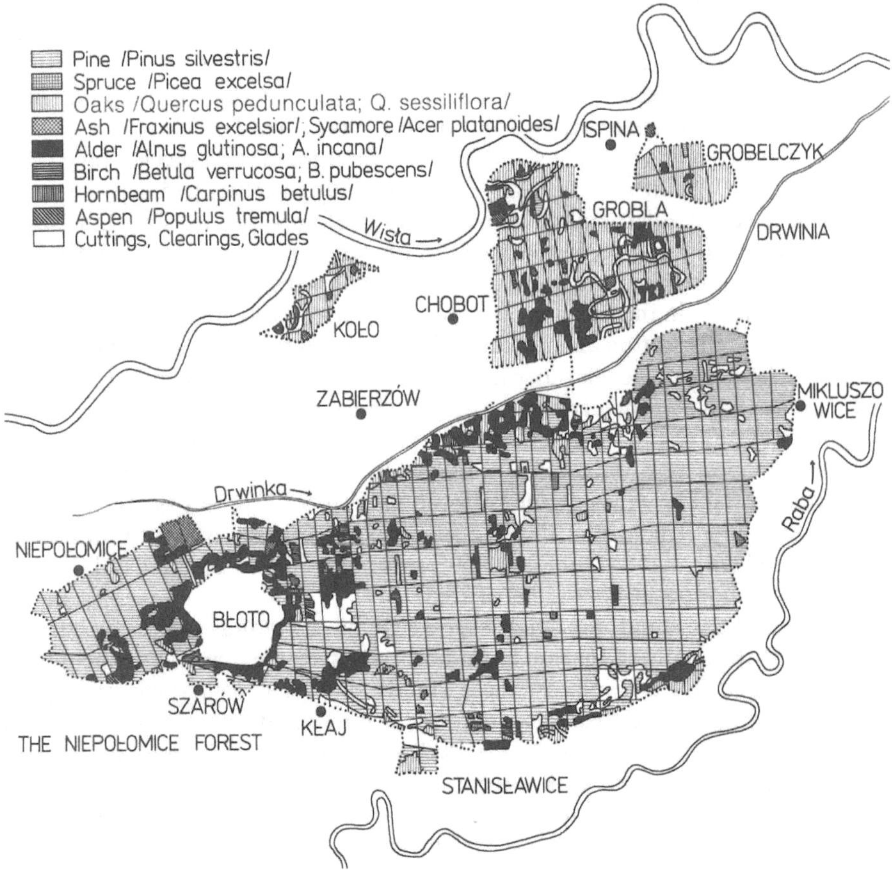

Fig. 1.1. Forests of the Niepołomice Forest, classified according to dominant tree species

The forest management mentioned above associated with changes in the hydrological pattern has effected either total extermination, or at least considerable decrease in the distribution of various plant species. These changes affected not only trees, but also herbaceous plants and low shrubs. For instance, there is little doubt that soil hydrological disturbances resulted in the impoverishment of the herb layer in swamp pine forest; *Drosera rotundifolia, Saxifraga hirculus, Lycopodium inundatum,* and other species have disappeared from these forests, while *Comarum palustre, Oxycoccus quadripetalus, Andromeda polifolia,* and *Ledum palustre,* formerly abundant, are now rare.

The forest has been converted into a network of drainage trenches, which contributes to the disappearance of such species as *Dryopteris cristata, Calla palustris,* and *Calamagrostis lanceolata* from alder forests, as well as to the restriction of species such as *Circaea alpina, C. intermedia, Carex remota,* and *Climacium dendroides,* characteristic of fertile and humid habitats and common in alder and elm forests in the past. *Cucubalus baccifer* and *Saponaria officinalis* decreased significantly in response to the Vistula river having been dyked and isolated from the Niepołomice Forest, and the riverside having been deforested. The oak-hornbeam forests have undergone an impoverishment in plant species not only due to a general decrease in soil moisture, but also to massive plantations of pine and spruce, which have caused an increase in acidity and hence the disappearance of some stenotopic species. *Daphne mezereum* and *Evonymus verrucosa* are now rare in the undergrowth of the oak-hornbeam forests. This is also true for various species that occurred frequently in the herb layer in the past, including *Lilium martagon, Corydalis solida, Galanthus nivalis, Platanthera bifolia, Veratrum lobelianum, Phaegopteris dryopteris, Ph. polypodioides.*

Potentilla alba, Carex montana, and *Lathyrus niger* were formerly abundant, but are now absent from dune hills in the forest. Furthermore, other species with similar ecological requirements to those mentioned above, *Cytisus ratisbonensis, Lycopodium clavatum, Polygonatum odoratum, Anthericum ramosum,* and even *Convallaria majalis,* occur now very infrequently on dune slopes. *Blechnum spicant, Pirola uniflora, Chimaphila umbellata, Monotropa hipopitys,* and *Homogyne alpina,* all characteristic of pine forests, are known to have become extinct here during the last century, whereas *Lycopodium annotinum* has undergone a considerable decline in abundance. Plant species attributable to the montane element in the flora, namely *Dryopteris oreopteris, Senecio fuchsi, Epilobium montanum, Arabis halleri,* and others, are now among the rarest species in the forest.

Plant species typical of the xerothermic oak forest association (Potentillo albae-Quercetum), as well as those characteristic of pine forests and montane species, were confined to the most favorable microhabitats in the forest. Forest management has degraded these microhabitats and caused, in turn, the disappearance of these species. Clear-cuttings have been planted with pines and occasionally spruce, which has temporarily produced a situation favorable for plants adapted to high acidity, especially the blueberry (*Vaccinium myrtillus*). This shrub is now rapidly declining, perhaps in response to decreases in the contribution of pine to the tree stands and in density. Concurrently, an increase in sedge (*Carex brizoides*), moorgrass (*Molinia arundinacea*), bracken (*Pteridium aquilinum*), and raspberry (*Rubus hirtus*) has been noted. Various synantropic species have immigrated recently into the Niepołomice Forest, as for example *Poa annua, Impatiens parviflora*, and *Chelidonium majus*. When detailing changes in the composition of the plant components of the ecosystems of the forest, one must also recall the observations of Kiszka (1977, 1981), who described the harmful effects of industrial air pollution on the lichen flora.

These changes in the composition of the plant components of the ecosystems of the Niepołomice Forest are indicative of the general impoverishment of the communities, and may even be responsible for the occurrence of transitional types.

1.1.3 Forest Communities

The detailed studies and descriptions of the forest communities are due mainly to the research efforts of this research group (Ćwikowa and Lesiński 1981). In addition, all previously published phytosociological data on the northern section of the forest (Ferchmin 1976, Ferchmin and Medwecka-Kornaś 1976, Denisiuk and Medwecka-Kronaś 1976) has also been considered. The map of the existing forest communities is a result of the present research with the exception of the northern Grobla section which is slightly modified and simplified from the map by Denisiuk et al. (1977).

Six associations and three transitional communities are recognized (Table 1.1) and their contributions to total area are given in Table 1.2.

Mixed oak-pine forests (Pino-Quercetum) comprise more than half of the area (51.8%) and are confined to the higher, rendzina terrace of the Vistula River (Trafas 1975, Suliński 1978) in the southern sections (Fig. 1.2). The forests of this association include *Pinus silvestris* with minor amounts of *Quercus robur* and *Betula verrucosa*, and rarely *Alnus glutinosa* and *Betula pubescens*. The sparse undergrowth comprises exclusively *Rhamnus frangula* and naturally regenerated poor-quality *Quercus robur* and *Pinus silvestris*. The herb layer is almost totally composed of *Vaccinium myrtillus, V. vitis-idaea, Trientalis europaea, Pteridium aquilinum*, and *Melampyrum vulgatum*.

Table 1.1. Phytosociological classification of forest communities in the Niepołomice Forest

Class	Order	Alliance	Association	Subassociation	Variant
Alnetea glutinosae Br.-Bl. et Tx. 1943	Alnetalia glutinosae Tx. 1937	Alnion glutinosae Malc. 1929/ Meijer Dress 1936	Carici elongatae-Alnetum Koch 1926		
Querco-Fagetea Br.-Bl. et Vlieg. 1937	Fagetalia silvaticae Pawł. 1928	Alno-Padion Knopp 1942 emend. Medw.-Kornaś 1959	Circaeo-Alnetum Oberd. 1953 Ficario-Ulmetum campestris Knapp 1942		
		Carpinion betuli Oberd. 1953	Tilio-Carpinetum Traczyk 1962	Stachyetosum Traczyk 1962 Typicum Traczyk 1962	
Vaccinio-Piceetea Br.-Bl. 1939	Vaccinio-Piceetalia Br.-Bl. 1939	Pino-Quercion Medw.-Kornaś 1959	Pino-Quercetum Kozl. 1925 em. Matuszkiewicz et Polakowska 1955		Moist Mesic Dry
		Dicrano-Pinion Libb. 1933	Vaccinio uliginosi-Pinetum Kleist 1929 Kobendza 1933		

Table 1.2. Percentage contribution of different forest communities to the total area of the Niepołomice Forest

Forest communities		Percent
Community transitional between alder forest and mixed oak-pine forest	Carici elongatae-Alnetum/Pino-Quercetum	3.9
Alder forest	Carici elongatae-Alnetum	0.4
Lowland alder forest	Circaeo-Alnetum	1.1
Community transitional between alder and elm lowland forests and moist oak-hornbeam forest	Circaeo-Alnetum/Ficario-Ulmetum/ Tilio-Carpinetum stachyetosum	4.3
Moist oak-hornbeam forest	Tilio-Carpinetum stachyetosum	12.7
Mesic oak-hornbeam forest	Tilio-Carpinetum typicum	0.2
Community transitional between oak-hornbeam forest and mixed oak-pine forest	Tilio-Carpinetum/Pino Quercetum	24.6
Mixed oak-pine forest–dry variant		0.1
Mixed oak-pine forest–mesic variant	Pino-Quercetum	7.1
Mixed oak-pine forest–moist variant		44.6
Swamp pine forest	Vaccinio uliginosi-Pinetum	0.7

There are three phases of the association. A dry variant, confined to the highest dune hills, is distinct in the occurrence of *Calamagrostis epigeios, Calluna vulgaris, Cytisus ratisbonensis,* and the lichens *Cetraria islandica* and *Cladonia rangiferina*. The most common variant is moist, dominated by *Molinia arundinacea* accompanied by *Sphagnum cuspidatum, S. fuscum,* and *S. formosum*. A mesic variant occurs on the slopes of dune hills in a belt (Izmaiłow 1975) at the southwestern edge of the southern forest section. It lacks any distinctive species and may be recognized merely by the infrequency of *Molinia arundinacea* and *Sphagnum* spp. as well as by the presence of species typical of dry microhabitats. The Pino-Quercetum association develops under conditions of a moderately fertile, variously humid habitat.

Moist oak-hornbeam forests (Tilio-Carpinetum stachyetosum) are rather well preserved and quite widespread in the complex. They account for 12.7% of the total area, especially on the rendzina terrace in the northern sections (Bzowski 1973) and also in island patches in the southern part (Fig. 1.2). These tree stands are species-poor, and include mostly *Quercus robur* accompanied by single individuals of *Tilia cordata* and *Alnus glutinosa*. The lower layer as a rule is composed of *Carpinus betulus*. The shrub undergrowth is well developed, comprising mostly *Crataegus oxyacantha, Evonymus europaea,* and *Cornus sanguinea*. The herb layer includes species characteristic of the association as well as of higher phytosociological units including *Aegopodium podagraria, Impatiens noli-tangere,* and *Anemone nemorosa*. There are also some plant species typical of moist habitats, e.g., *Stellaria nemorum, Stachys silvatica, Circaea lutetiana*. The moist oak-hornbeam forests develop on fertile and moist sites.

Mesic oak-hornbeam forests (Tilio-Carpinetum typicum) cover only 0.2% of the total area (Fig. 1.2), as a few patches in the Grobla section and in the southern area. When compared to the above-mentioned subassociation, these stands differ in *Quercus robur*, being frequently associated with single individuals of *Pinus sil-*

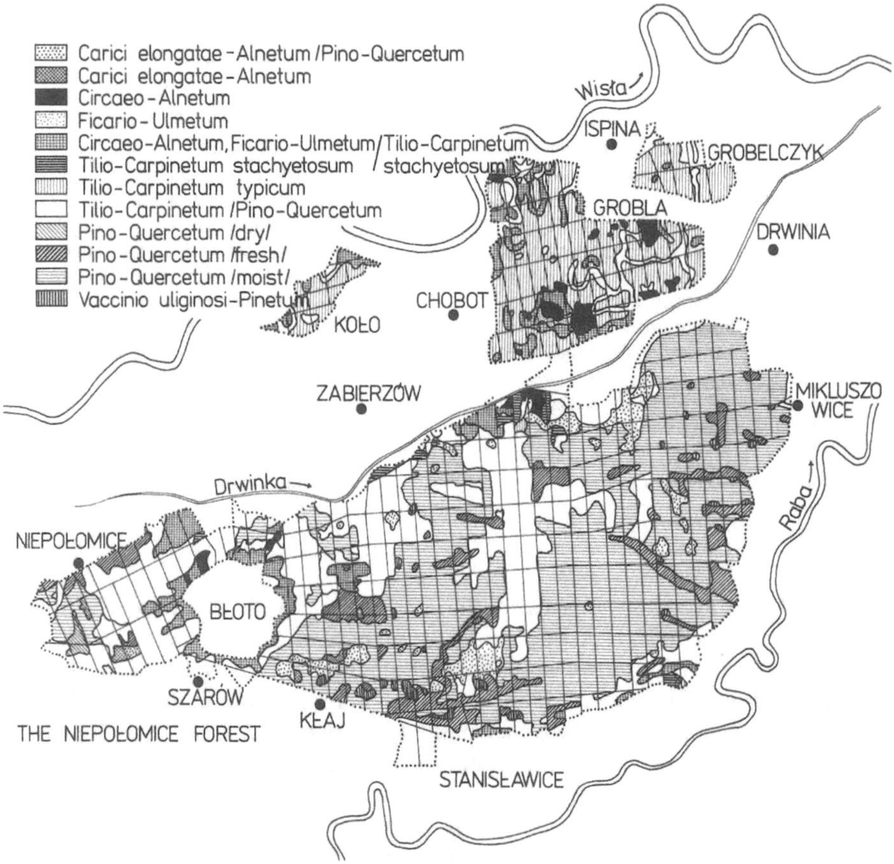

Fig. 1.2. Distribution of forest communities in the Niepołomice Forest

vestris, Populus tremula, and *Betula verrucosa*. The undergrowth is somewhat poor as a rule, composed mainly of *Crataegus oxyacantha, C. monogyna,* and *Sambucus nigra*. There are no species of the alliance Alno-Padion in the herb layer, while species characteristic of the association, such as *Stellaria holostea* and *Galium schultesi,* commonly are accompanied by *Moehringia trinervia, Poa nemoralis, Polygonatum multiflorum, Galeobdolon luteum, Geum urbanum,* and *Scrophularia nodosa,* and rarely by *Majanthemum bifolium*. Artificial planting of pine has modified the environment and community of Tilio-Carpinetum typicum and given it the character of Pino-Quercetum. Tilio-Carpinetum typicum can be identified as a moist variant to this association. Mesic oak-hornbeam forests occur mainly in fertile, mesic sites.

A transitional successional community between oak-hornbeam forest (Tilio-Carpinetum) and mixed oak-pine forests (Pino-Quercetum) accounts for 24.6% of the total area, particularly in the southern but also in the northern areas (Fig. 1.2). Undoubtedly, it has developed due to artificial regeneration of clear-cut areas with pine, under fertility levels excessive for this species. The stands are composed almost exclusively of *Pinus silvestris* associated with minor amounts of *Quercus robur, Q. sessilis, Alnus glutinosa, Betula verrucosa, Populus tremula,* and other species. The undergrowth is poor as a rule, composed mainly of *Frangula alnus* sometimes with *Corylus avellana* often including naturally regenerated *Quercus robur* and planted *Tilia cordata, Carpinus betulus, Fagus silvatica, Larix decidua, Acer platanoides, A. pseudoplatanus,* and others. The herb layer is dominated by *Carex brizoides* accompanied by species of the class Vaccinio-Piceetea (*Majanthemum bifolium, Vaccinium myrtillus, Pteridium aquilinum, Calamagrostis arundinacea*), as well as others of Querco-Fagetea (*Rubus hirtus, Moehringia trinervia, Polygonatum multiflorum, Anemone nemorosa, Viola silvestris, Carex silvatica, Lathyrus vernus*). *Luzula pilosa* commonly occurs in this community. The community is found on fertile to moderately fertile mesic to moist sites.

Three types of wet lowland forest dominated by alder, alder and elm occupy only a small and steadily diminishing area.

Alder forests (Carici elongatae-Alnetum) account for less than 0.4% of the area, and in general are confined to depressions with high ground water. They occur mostly in the southern Grobla section (Denisiuk et al. 1977). The stands include mainly *Alnus glutinosa* associated locally with *Betula pubescens. Salix cinerea* and *Frangula alnus* are among the characteristic shrubs. The herb layer is dominated by *Carex elongata, Lycopus europaeus, Dryopteris thelypteris,* and *Calamagrostis canescens.* Plant species characteristic of the classes Phragmitetea and Potametea are also common. The irregular structure of alder woods is typically found in those forests (Ferchmin 1976).

Communities transitional between alder forests (Carici elongatae-Alnetum) and mixed oak-pine forests (Pino-Quercetum) are characteristic of flat, poorly defined watershed divides between the southern tributaries of the Drwinka, and their wide valleys in the southern part of the forest (Fig. 1.2). No doubt, these communities originated due to drainage, as well as to the planting of pine and spruce. The stands consist either of pine and spruce often with minor amounts of black alder, or exclusively of black alder. The undergrowth is poorly developed, composed almost only of *Frangula alnus,* whereas the herb layer is dominated by *Carex brizoides.* The latter species usually occurs in dense masses causing considerable impoverishment of the original layer. It is often accompanied by some species of the classes Phragmitetea, Molinio-Arrhenatheretea, and Vaccinio-Piceetea, which push through the sedge layer or occur in small patches close to the trees where the density of *Carex brizoides* is decreased. The irregular structure is poorly preserved, and stagnant water occurs at the surface only occasionally. This transitional community develops in the degraded habitats typical of alder forests.

Alder and elm lowland forests (Circaeo-Alnetum) account for 1.1% of the area, being confined almost exclusively to the Grobla section and scattered patches in the southern part (Fig. 1.2). They occur only on flooded terraces, under conditions of high fertility (Karkanis 1973; cf. Chap. 1.2). These habitats used to be periodically inundated some time ago, but now are supplied only from atmospheric precipitation and thawing snow (Denisiuk and Medwecka-Kornaś 1976). The association Circaeo-Alnetum is species-rich and multilayered. The tree stands consist of *Alnus glutinosa,* sometimes accompanied by *Alnus incana* and

Quercus robur. The undergrowth is abundant in places, including *Padus avium, Ribes schlechtendalii, Crataegus oxyacantha, Evonymus europaea,* and *Frangula alnus*. Saplings of *Alnus glutinosa* are also common. The herb layer is rich, often many-layered, consisting of species characteristic of the alliance Alno-Padion, commonly associated with nitrophilous species such as *Rumex obtusifolius* and *Urtica dioica*.

Lowland elm forests (Ficario-Ulmetum) contribute only a minor proportion (0.3%) to the total area but there is little doubt that they were more widespread in the past. This association is almost always confined to the southeastern part of the Grobla section and to the Gibiel Reserve (Fig. 1.2). It is certainly transitional between the association Circaeo-Alnetum and the subassociation Tilio-Carpinetum stachyetosum (Ćwikowa 1974), but it is distinct because of the occurrence of *Ulmus campestris* and *U. laevis* which here attain their optima. The association Ficario-Ulmetum differs from Circaeo-Alnetum in the total absence of species representative of the class Alnetea glutinosae (*Lycopus europaeus, Carex elongata*), and from Tilio-Carpinetum stachyetosum in the absence of representatives of the alliance Carpinion betuli (*Stellaria holostea, Ranunculus cassubicus*). Locally differentiating species of the elm forests include *Gagea lutea, Cardamine impatients,* and *Veronica montana*. The tree stands consist of dying elms, and the undergrowth is equally species-rich in the association Circaeo-Alnetum.

A community transitional between alder-elm forests and moist oak-hornbeam forests appears more commonly in the Niepołomice Forest. It accounts for 4.3% of the total area and is characteristic of the northern part of the Grobla section (Ferchmin and Medwecka-Kornaś 1976, Denisiuk et al. 1977) and of the margins of the "Błoto" clearing. The transitional nature of this community consists in the tree stands containing *Alnus glutinosa, Ulmus laevis,* and *U. campestris*, as well as *Quercus robur, Tilia cordata,* and *Carpinus betulus*; the undergrowth including *Padus avium* and *Ribes schlechtendalii* along with *Sambuccus nigra*; and the herb layer containing plant species characteristic of the alliance Alno-Padion and higher phytosociological units. The occurrence of masses of *Carex brizoides* is notable, because this species is typical of habitats with abnormal water flux. The community is presumed to have developed relative to the limits set by flood dikes on the Vistula river.

Small patches of the association Vaccinio uliginosi-Pinetum occur in deflationary niches (Izmaiłow 1975) at the foot of dunes, under swampy conditions. They account for a mere 0.7% of the total area. The tree stands are composed of *Pinus silvestris* with minor amounts of *Betula pubescens, B. verrucosa, Alnus glutinosa,* and *Quercus robur*. The undergrowth is dominated by *Frangula alnus* associated with *Salix cinerea* and *S. aurita*. The distinctive herb layer includes *Vaccinium uliginosum, Oxycoccus quadripetalus, Ledum palustre, Eriophorum vaginatum, Molinia arundinacea, M. coerulea,* and also *Osmunda regalis* locally. The moss layer is abundant, composed mostly of *Sphagnum* spp.

When the actual distribution of forest communities is compared to the available habitat conditions, and is conditioned by forest remnants in recognizable habitats, it is clear that the proportion of disturbed communities amounts to 80% (Myczkowski and Lesiński, in press; cf. Chap. 7).

Phytosociological research on the existing vegetation of the Niepołomice Forest and a recognition of the possible directions of ecological succession may facilitate the prediction of future affects of industrial pollutants on these ecosystems. Moreover, such predictions may permit one to suggest what sort of remedial action on the composition and structure of the tree stands should be undertaken in order to attempt to lessen sensitivity to air pollution (Fabijanowski and Lesiński 1978).

1.2 Soils

B. ADAMCZYK

In spite of intense research by botanists and the long period that has passed since drainage work was first undertaken 150 years ago, the soils have never been studied in detail. General information on soils is given by Łoziński (1935), Chodzicki (1947), Karkanis (1973, 1974), and recent mapping (Gruszczyk et al. 1976, Gruszczyk 1981). Detailed pedological research was started in 1977 (Adamczyk and Maciaszek 1977, Adamczylk et al. 1978).

There is great variability in texture, moisture-oxygen relationships, pH reaction, nutrient content, biological activity, and exchange capacity in soils in the area. Soils that are stable and relatively resistant to secondary disturbances occur less commonly than those weakly buffered and sensitive to industrial pollution and lowering of the ground water.

The main determinant of the ecological properties of the soils, and by implication, the resulting habitat conditions in the Niepołomice Forest area, is the rock substrate (Łoziński 1935, Chodzicki 1947, Adamczyk et al. 1978). Topography, climate, and other soil-forming factors are of minor significance.

At least three main soil–habitat complexes can be distinguished. Those of the deciduous forest, the mixed forest, and the coniferous forest (Fig. 1.3). The deciduous forest complex is confined to the northern part between the Vistula river and Drwinka creek. The other two complexes occur in the southern part, the mixed forest in the west, and the coniferous in the east (Fig. 1.3).

In the deciduous forest complex (Fig. 1.3) including the Koło, Grobla, and Grobelczyk sections, the substrate is represented by fertile alluvia of the Vistula river (Doktorowicz-Hrebnicki and Różycki 1949, Świdziński 1952). These are cohesive, very fine sands and heavy loams. The bottom layers commonly include sandy or sandy-loam intercalations (Fig. 1.4). On such substrates fertile mesotrophic-eutrophic soils, resistant to disturbance, have originated. These are so-called old warp soils, often partly transformed into brown, gleyed, pseudogley, or gley types. A negative feature of these soils is high cohesiveness and poor permeability. Consequently, the bottom and often the top layers are in a state of permanent or temporary anaerobiosis (Fig. 1.4, horizons g, G). In the deciduous forest complex fertile habitats favorable for moist deciduous forests are predominant and habitats typical of lowland alder and alder and elm forests are of minor occurrence (Gruszczyk et al. 1976). These habitats support moist oak-hornbeam forest (Tilio-Carpinetum stachyetosum) and alder and elm lowland forests (Circaeo-Alnetum and Ficario-Ulmetum) (cf. Sect. 1.1).

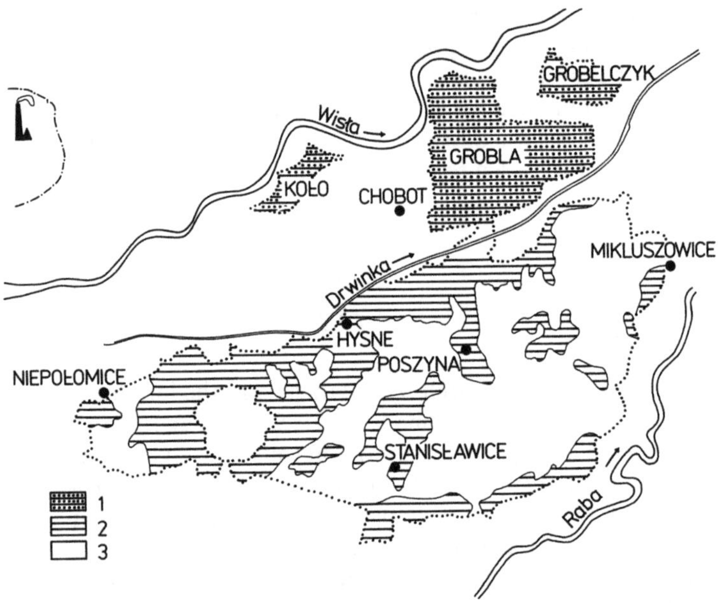

Fig. 1.3. Main soil-habitat complexes in the Niepołomice Forest: *1* deciduous forest; *2* mixed forest; *3* coniferous forest

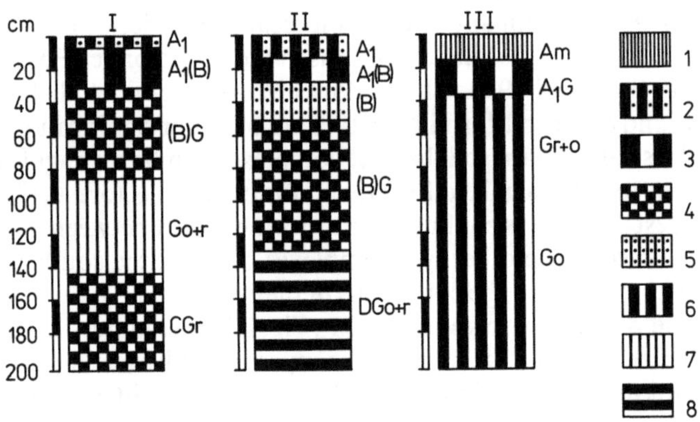

Fig. 1.4. Main soil types in the deciduous forest complex of the Niepołomice Forest (forest communities named according to Ćwikowa and Lesiński 1981, cf. Sect. 1.1): *I* humic brown, gleyed soil with community transitional between Ficario-Ulmetum and Tilio-Carpinetum stachyetosum; *II* brown, weakly leached, gleyed soil with Tilio-Carpinetum stachyetosum; *III* muck-gley soil with Ficario-Ulmetum; *1* mulch-humus layer; *2* mull-humus layer; *3* transitional humus layer; *4* very fine sand; *5* very fine clay sand; *6* heavy loam; *7* loamy sand; *8* less-than-loamy sand

The mixed forest complex is representative of poorer soils and habitat conditions (Fig. 1.3). The substrate is Pleistocene fluvioglacial sands (Starkel 1972), often underlain by or interlayered with sediments more promising for soil formation, namely sandy-loam, marly, or loamy-clay sediments (Fig. 1.5). The nature of these interlayers and their depth below the ground level determine the nutrient

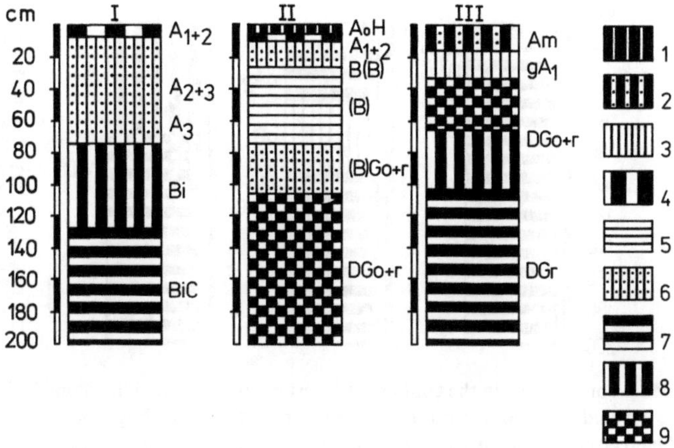

Fig. 1.5. Main soil variants of the mixed forest complex of the Niepołomice Forest: *I* grey-brown podzolic soil with mesic variant of community transitional between Tilio-Carpinetum and Pino-Quercetum; *II* podzolized brown gleyed soil with moist variant of community transitional between Tilio-Carpinetum and Pino-Quercetum; *III* muck-gley, mesotrophic-eutrophic soil with community transitional between Circaeo-Alnetum and Tilio-Carpinetum stachyetosum; *1* mor-humus layer; *2* muck-humus layer; *3* hydromull-humus layer; *4* transitional humus layer (accumulation-eluvial horizon); *5* loose sand; *6* less-than-loamy sand; *7* light loamy sand; *8* loamy sand or light, sandy loam; *9* heavy loam

content and moisture–oxygen relationship in the soil. Thus, the mixed forest complex includes grey-brown (lessive) podzolic soils (Fig. 1.5, I) favorable for mesic to moist mixed forests and mesic mixed forests; brown, podzolized, gleyed soils (Fig. 1.5, II) favorable for moist mixed forests; and muck-gley, mesotrophic soils (Fig. 1.5, III) favorable for moist deciduous or alder forests (cf. Adamczyk and Maciaszek 1977). At present, the habitats of the mixed forest complex support strongly modified communities, transitional between Tilio-Carpinetum and Pino-Quercetum, between Circaeo-Alnetum and Ficario-Ulmetum and Tilio-Carpinetum stachyetosum, and between Carici elongatae-Alnetum and Pino-Quercetum (cf. Sect. 1.1). Two upper layers of the soils are sandy as a rule (Fig. 1.5) and, under conditions of low ground water, are subject to soil and habitat degradation. Podzolization (acidification and deposition of raw humus A_0F+H) is often initiated under such conditions and muck-gley soils often undergo an accelerated mineralization of organic matter.

The coniferous forest complex shows the poorest soil and habitat conditions (Fig. 1.3). The parent rock is represented almost exclusively by Pleistocene sands. In spite of considerable soil moisture, more or less homogenous sands attain great thickness in this complex (Fig. 1.6). It includes mainly wet podzolic, semi-bog, and stagnogley soils (Fig. 1.6, II) producing a habitat typical of moist mixed oak-pine forests (Pino-Quercetum, moist variant). Oligotrophic peat soils are unimportant, supporting (Fig. 1.6, III) patchy communities of swamp pine forests (Vaccinio uliginosi-Pinetum). Locally, podzolic soils (Fig. 1.6, I) occur at the tops of dunes (Ismaiłow 1975), becoming suitable habitat for mesic mixed oak-pine forests (Pino-Quercetum, mesic variant). Sandy-loam and loamy intercalations

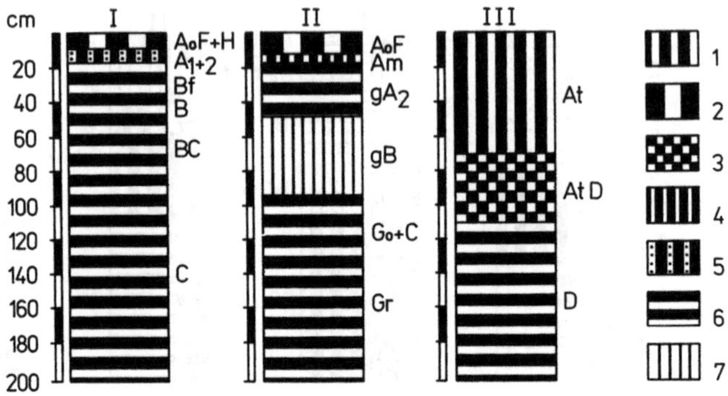

Fig. 1.6. Main soil variants in the coniferous forest complex of the Niepołomice Forest: *I* podzolic soil on a dune covered with mesic variant of Pino-Quercetum; *II* half-bog, podzolic soil with moist variant of Pino-Quercetum; *III* oligotrophic peat soil with Vaccinio uliginosi-Pinetum; *1* peat layer; *2* mor or hydromor-humus layer; *3* peaty-sand layer; *4* muck-humus layer; *5* transitional humus layer (accumulation-eluvial horizon); *6* loose sand; *7* less-than-loamy sand

occur in this part of the Niepołomice Forest at a considerable depth, and their contribution to the soil formation consists mostly in an increase in soil moisture, and only slightly in controlling nutrient content.

There is variation in the proportions of the main macro- and microelements among the soils representative of the three soil–habitat complexes. The deciduous forest complex includes soils rich in magnesium, moderately rich in nitrogen, calcium, and potassium, and poor only in phosphorus (Table 1.3). The ratio of organic carbon to total nitrogen averages 10.6. Almost 90% of the nitrogen available in minerals is accounted for by nitrate N, and some 10% by ammonia. Thus, the C:N ratio and nitrate-ammonia are, as in fertile habitats, favorable for deciduous forests (Ellenberg 1964, Brożek 1979).

The coniferous forest complex includes soils deficient in potassium, phosphorus, magnesium, and calcium, as well as in copper and molybdenum among other microelements (Adamczyk and Maciaszek 1977). However, these soils are sufficiently rich in manganese and zinc, and moderately rich in nitrogen. Some 20–25% of the mineral nitrogen is accounted for by nitrate N, and the remainder by ammonia. The ratio of organic carbon to total nitrogen ranges from 22:1 to 36:1, averaging 26:1 (Table 1.3). Soils of the mixed forest complex are more widely variable in their chemical characteristics than those discussed above; however, they show more affinity to the coniferous forest complex than to the deciduous.

Acid and very acid soils prevail in the coniferous and mixed forest complexes. Their pH in water ranges from 3.5 to 4.5 as a rule, but there are also some patches of soils with pH ranging from 5.0 to 6.5 and even to 7.0. In the deciduous forest complex, soils with pH from 5.0 to 6.0 are prevalent (Table 1.3).

The ecological properties and environmental resistance of soils and habitats are to a considerable extent dependent upon exchange capacity.

The exchange capacity depends mostly upon the amounts of humus and clay minerals in the soil. Average values approximate 100 me per sample 100 g of raw

Table 1.3. Soil properties in the major soil-habitat complexes of the Niepołomice Forest

Soil Properties	Soil habitat complexes	Depth in cm 0–10 1	0–10 2	0–10 3	20–70 1	20–70 2	20–70 3	100–150 1	100–150 2	100–150 3
1. Percent clay minerals		Humus layer			26.9	8.0	2.6	19.4	9.5	2.5
2. pH in	H_2O	4.9	4.7	3.6	5.0	4.5	3.8	5.8	4.8	4.0
	KCl	3.7	3.5	2.7	3.7	3.7	3.1	4.5	4.0	3.7
3. Exchange capacity in me per 100 g of soil		34.5	18.9	31.2	24.1	11.5	3.3	27.1	9.5	2.3
4. Nutrient content in mg per 100 g of soil	K_2O	25.7	8.5	12.1	8.3	2.9	1.3	8.9	3.0	2.3
	P_2O_5	4.8	5.0	5.8	0.4	0.8	1.2	1.4	1.8	1.0
assimilable	MgO	31.9	12.5	10.3	31.2	12.2	1.4	40.5	13.1	0.7
exchangeable	CaO	222.5	164.0	120.0	180.1	160.0	16.3	294.0	143.4	19.0
5. C:N		10.6	17.5	25.7	n.m.	n.m.	n.m.	n.m.	n.m.	n.m.
6. Percent total nitrogen		0.43	1.17	1.41	n.m.	n.m.	n.m.	n.m.	n.m.	n.m.
7. Percent mineral nitrogen	$N–NH_4$	8.0	29.0	78.0	n.m.	n.m.	n.m.	n.m.	n.m.	n.m.
	$N–NO_3$	92.0	71.0	22.0						

1 Deciduous forest – Tilio-Carpinetum stachyetosum and Tilio-Carpinetum typicum with *Aegopodium podagraria* (Ferchmin and Medwecka-Kornaś 1976; Denisiuk et al. 1977; Ćwikowa and Lesiński 1981)
2 Mixed forest complex (patches I and II-b or intermediate in soil properties) – forest communities similar to either Tilio-Carpinetum, or moist variant of Pino-Quercetum, or Circaeo-Alnetum (Ćwikowa and Lesiński 1981)
3 Coniferous forest complex – moist variant of Pino-Quercetum and less commonly with mesic variant of Pino-Quercetum (Ćwikowa and Lesiński 1981)
n.m. Not measured

humus and peat horizons (horizons A_0F, A_0H, At). 15–40 me in mull and moder humus-bearing horizons (A_1, Am) and in loamy deposits, and only 0.7 to 2.0 me in loose to less-than-loamy sands (Figs. 1.4–1.6). This indicates that organic matter is the only soil and habitat stabilizer in sandy soils and hence, mineralization should be hindered.

As judged from the composition of humus and the proportions of various forms of heavy metals and sulfur in the soils (Adamczyk et al. 1978), it appears that they have already undergone partial transformation into industrial soils. When compared to control samples taken from Białowieża National Park, the Niepołomice soils are richer in bitumens, heavy metals, and sulfur. The horizons A_0F+H in the mixed and coniferous forest complexes contain an average of 3–4 times more Mn, Zn, and Cu in assimilable forms and the horizons A_1 and A_m in the deciduous forest complex contain 2–3 times more Mn and Zn and up to 6 times more Cu than do the controls. Sulfate ions are only insignificantly absorbed by soil (3–6 times more are passed to the ground water) but nevertheless, deeper horizons (G_0+r, DG_0+r) (Figs. 1.4–1.6) of the sandy soils show 3.5 times more sulfur per 1% of silt and clay than do the control samples. As measured in an acetate-phosphoric extract, the sandy and loamy soils are respectively 4.5 and 2 times richer in sulfate ions than are the controls.

The habitat potential of these soils and some remarks on soil fertilization and amelioration of moisture relationship will be discussed in the chapter on restructuring of the forest stands.

1.3 Climate

J. Klein

This account of climatic conditions in the Niepołomice Forest region is prepared from the literature (Romer 1949, Hess 1969, Klein 1978) and employing observations from 1977–1978.

Climatological stations were established in the northern (Ispina) and southern (Szarów) sections. Total solar radiation was measured with a Robitsch-Fuess pyranograph 58dc; insolation with a Campbell-Stokes heliograph; air temperature with an Assmann psychrometer, a thermograph, and extremum thermometers; and atmospheric precipitation with a Hellmann rain gauge. Phytoclimatic conditions were investigated in four areas, one in each of the following tree stands: oak-hornbeam forest, dry mixed oak-pine forest, mesic mixed oak-pine forest, and moist mixed oak-pine forest. In each area, the infiltration of atmospheric precipitation was measured with 10 modified rain gauge totalizers, and the light in the forest understory with a Lange-Mensing luxometer, the photocelis having been placed at every rain gauge.

The climate of the forest is representative of the submontane lowlands and valleys (Romer 1949). The characteristics include a long vegetation period (223 days), total annual precipitation exceeding 700 mm, and mean annual temperature approximating 8 °C (Hess 1969).

1.3.1 Solar Radiation and Insolation

The total annual solar radiation amounts to 237.4 $kJcm^{-2}$, with the maximum energy influx in July (41.1 $kJcm^{-2}$) and the minimum in December (−2.1 $kJcm^{-2}$) (Mackiewicz 1954). As measured in 1977–1978, the total annual radiation approximates very closely the multiannual average, even though the monthly observations differ from the average values (Table 1.4).

The number of days with solar radiation exceeding arbitrarily set critical values is informative of the monthly and seasonal distributions. As indicated in Table 1.5, these distributions were different in 1977 and 1978, having been more favorable in the previous year.

There is a relationship between the inflow of solar energy to ground level and insolation. Insolation of the area is shown in Fig. 1.7. The maximum (20 h) occurs in summer, in the morning, prior to the formation of convection clouds. In

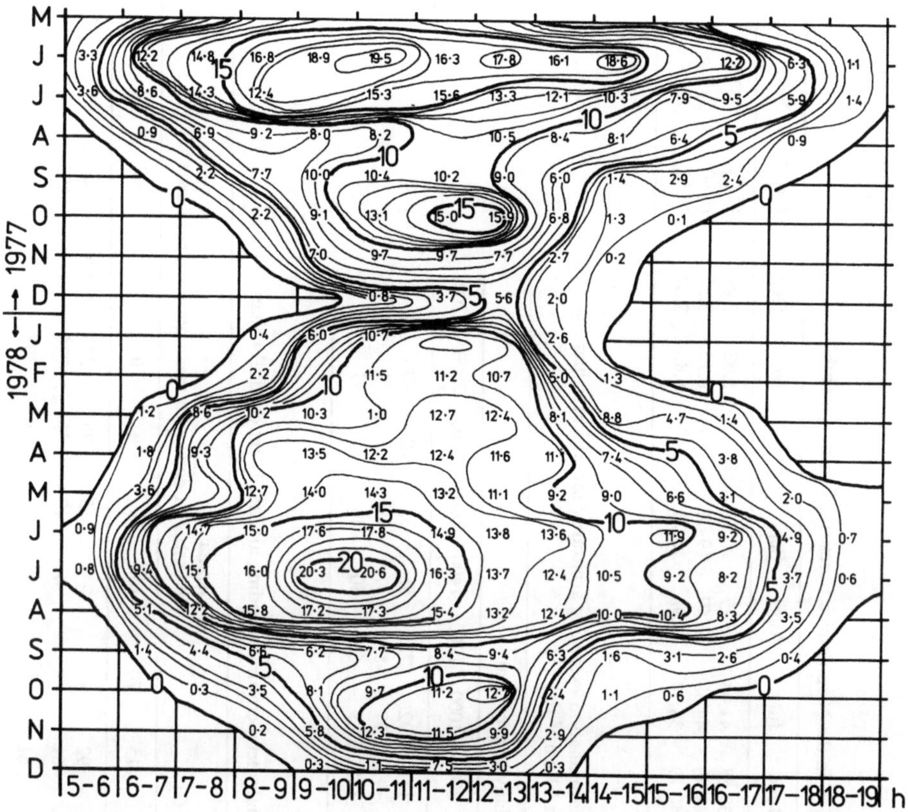

Fig. 1.7. Insolation in the Niepołomice Forest (in monthly sunny hours in specific time intervals)

autumn, maximum insolation approximates 12–15 h, mostly at noon. In December, maximum insolation is only 5 h.

The amount of light reaching the understories of various forest communities depends upon insolation and tree canopies and the undergrowth. As measured at the peak of the vegetation period in selected forest areas, the lowest proportion of light reaches the understory of the oak-hornbeam forest (10.2%), this value being close to one previously obtained from studies on light distribution in similar forests here (Klein 1977). The greatest proportion of light reaches the understory of the dry oak-pine forest (24.2%), while the figures obtained for the moist and mesic oak-pine forests are 18.2% and 16.8% respectively.

1.3.2 Air Circulation

Anticyclonic situations prevail in the area (52% days per year). Various air masses affect this location. The weather is most commonly influenced by humid, polar-marine air masses (60.2%); dry polar-continental air masses account for 25%, arctic for 7%, and tropical for 7.8% (Niedźwiedź 1969).

Table 1.4. Total monthly solar radiation in the Niepołomice Forest (in kJ cm^{-2} $month^{-1}$)

	Jan.	Feb.	March	Apr.	May	June	July	Aug.	Sept.	Oct.	Nov.	Dec.	May–Dec.	Year (total)
After Mackiewicz (1954)	2.93	6.28	15.91	24.79	34.33	37.68	41.03	35.17	20.93	11.30	5.02	2.09	187.55	237.37
1977	–	–	–	–	38.52	49.40	44.38	33.49	19.68	15.07	6.70	3.35	210.60	210.60
1978	7.54	11.30	19.68	23.45	32.66	40.19	38.10	30.98	16.74	10.47	5.86	2.51	177.52	239.48

Table 1.5. Number of days with total radiation exceeding arbitrarily set critical values (in kJ cm^{-2} day^{-1})

Month	Number of days with total daily sum of heat											
	0	200	400	600	800	1000	1200	1400	1600	1800	2000	2200
May–Dec.	245	196	156	134	115	92	72	56	33	18	10	1
May–Dec.	245	184	139	114	93	77	57	37	19	7	1	–
Year (total)	365	276	207	150	115	89	66	40	19	7	1	–

Table 1.6. Number of days with mean daily temperature exceeding arbitrarily set crtitical values

	Szarów Number of days with mean daily temperature (°C)								Ispina Number of days with mean daily temperature (°C)							
	–10	–5	0	0	5	10	15	20	–10	–5	0	0	5	10	15	20
May–Dec. 1977	–	4	24	221	197	142	92	11	–	5	25	220	196	146	91	12
May–Dec. 1978	4	10	20	225	193	150	61	8	4	10	21	224	191	153	64	7
Total 1978	6	24	56	309	225	155	61	8	6	25	58	307	221	158	64	7
Average for 1966–1970	10	29	75	290	222	167	97	21	10	29	75	290	222	167	97	21

Atmospheric fronts occur here 42% days per year, especially in winter and summer. Cold fronts account for 45% of this, warm fronts for 34.8%, occlusion for 17.2%, and stationary for 3.2%.

Winds blow mostly from the western (NW – 12.9%, W – 16.6%, SW – 12.8%) and eastern quadrants (SE – 4.2%, E – 13.6%, NE – 6.2%), whereas those from the northern and southern occur infrequently (5% and 7.2%, respectively). Windstills are a common phenomenon (22.5%).

1.3.3 Thermal Relations

Thermal relations in the area display local "oceanic" characteristics, the autumn (September to November) temperatures being higher than the spring (March to May) (Romer 1949, Zinkiewicz 1962), and the Kerner thermodromic index is high (3.9).

The mean annual temperature is 7.8 °C, with the average maximum 12.2 °C, and the average minimum 3.4 °C. The monthly and 10-day temperature distribution in 1977–1978 differs considerably from the average or 1966–1970, even though the mean annual temperature (7.5 °C) is maintained (Fig. 1.8).

Minimum and maximum temperatures and perhaps their amplitudes at ground level are crucial for herbaceous plants and small animals. As shown in Fig. 1.8, maximum temperatures are always higher in the southern part of the forest than in the northern, whereas minimum temperatures are almost the same at both stations, being a little lower in the south than in the north in the cool half-year. Consequently, the amplitudes are also greater in the south, especially in summer.

The number of days with average temperature exceeding arbitrarily set critical values is characteristic of the annual distribution of heat in an area (Hess 1965); it also permits recognition of thermal seasons.

Thermal seasons designate time spans in which temperature is maintained continuously within a characteristic range. They are established from yearly averages and hence may be incompatible with the absolute number of days with any given mean temperature (cf. Table 1.6). As computed by Klein (1978), the thermal winter ($T < 0$ °C) takes 86 days in the Niepołomice Forest area, early spring ($T = 0°–5$ °C) 34 days, thermal spring ($T = 5°–10$ °C) 27 days, early summer ($T = 10°–15$ °C) 31 days, thermal summer ($T > 15$ °C) 97 days, late summer ($T = 10°–15$ °C) and thermal autumn ($T = 10°–5$ °C) 36 days each, and the early winter ($T = 5°–0$ °C) only 18 days.

Periods other than winter account for 279 days, the vegetation period lasts 227 days, the period of intense plant growth 164, and the maturation period 97 days.

Thermal relations in an area can also be described by the number of so-called characteristic days. In this area, there are 26 very frosty days ($T_{min} < -10$ °C) per annum, 51 frosty days ($T_{min} < 0$ °C), 63 slightly frosty days ($T_{min} < 0$ °C and $T_{max} < 0$ °C), and 47 hot days ($T_{max} > 25$ °C). The frostless period is also an important ecoclimatologic characteristic (Baumgartner 1968 a, b). It averages 168 days but it lasted only 134 and 157 days in 1977 and 1978 respectively.

Fig. 1.8. Ten-day average temperature 2 m above ground level, and minimum and maximum temperature 5 cm above ground level in the Niepołomice Forest in 1977–1978; *1* average temperature at Ispina as measured in 1966–1970; *2* average temperatures at Szarów; *3* average temperatures at Ispina; *4* maximum temperatures at Szarów; *5* maximum temperatures at Ispina; *6* minimum temperatures at Szarów; *7* minimum temperatures at Ispina

1.3.4 Atmospheric Precipitation

The area is within a zone of moderately high precipitation which provides plants with sufficient amounts of water (Gumiński 1951). The average annual precipitation amounts to 729.5 mm (Table 1.7). Compared to May–December 1977, the total precipitation was much higher during May–December 1978, the differences amounting to 151.6 mm at Szarów and 150.8 mm at Ispina.

There is variation in the spatial distribution of precipitation (Table 1.8). Total precipitation is higher in the southern part of the forest (Szarów) than in the northern (Ispina). The former station is located at the margin of the Wieliczka Uplands, some 30 m higher than the latter.

Summer precipitation accounts for 42.1% of the total annual amount and the spring for 22.9%. Autumn precipitation accounts for 77.2% of that of the spring. Permanent snow cover is maintained for 62 days per annum on the average.

Precipitation is lower in fields than in the forest, as a rule (Geiger 1961, Tomanek 1963). Total annual precipitation measured at Szarów and Ispina, mar-

Table 1.7. Total monthly precipitation (h) and number of days with rainfall or snowfall in the Niepołomice Forest in 1977–1978, and mean values for 1966–1970

Month	Szarów				Ispina				1977–1970			
	h	0.1	1.0	10.0	h	0.1	1.0	10.0	h	0.1	1.0	10.0
1977												
May	43.4	7	5	3	43.2	7	4	2	88.2	12	9	2
June	23.8	6	5	–	30.5	6	5	–	98.4	12	10	4
July	113.1	13	11	4	86.1	13	12	4	114.8	12	11	4
Aug.	104.3	16	12	4	111.0	17	11	5	93.7	13	10	4
Sept.	99.0	13	11	4	85.8	14	14	3	39.1	8	6	1
Oct.	9.4	5	2	–	16.5	6	2	1	42.0	11	8	1
Nov.	53.7	18	12	2	53.8	19	12	2	48.1	14	8	1
Dec.	26.9	10	8	1	24.2	11	8	1	42.4	18	10	–
May–Dec.	473.6	88	66	18	451.1	93	68	18	566.7	100	72	17
1978												
Jan.	13.8	7	5	–	12.8	7	4	–	42.8	17	13	–
Feb.	26.3	8	4	–	26.5	7	3	–	41.2	15	9	–
March	23.2	13	7	–	23.3	12	6	–	30.7	14	8	–
April	56.0	13	9	2	53.0	13	7	2	48.1	12	9	1
May	136.9	20	18	3	135.2	19	17	3	88.2	12	9	2
June	74.0	13	10	4	73.9	14	13	2	98.4	12	10	4
July	56.9	14	10	2	54.8	13	10	1	114.8	12	11	4
Aug.	154.6	14	12	6	151.8	15	13	4	93.7	13	10	4
Sept.	82.9	21	16	4	74.9	20	13	4	39.1	8	6	1
Oct.	42.8	13	10	–	39.6	13	10	–	42.0	11	8	1
Nov.	45.7	4	4	2	40.2	4	4	2	48.1	14	8	1
Dec.	31.4	18	7	–	31.5	18	8	–	42.4	18	10	–
May–Dec.	625.2	117	87	21	601.9	116	92	14	566.7	100	72	17
Year (total)	744.5	158	112	23	717.5	155	112	16	729.5	158	109	18

ginal sites, was 641.7 and 593.7 mm, respectively. In contrast, precipitation was 675.3 and 622.2 mm at Poszynka and in forest division No. 44, respectively (Table 1.8), the latter two stations being located in open areas within the forest. One may conclude that the precipitation is 3.2 to 10.2% higher in the forest than in the fields.

Infiltration rate of atmospheric precipitation to the forest understory depends upon the nature of the forest association and the closeness of tree canopies and undergrowth. The lowest proportion of rainfall (66.6% on average) passes through the canopies of the dry oak-pine forest, and the highest (73%) through canopies of the mesic oak-hornbeam (Fig. 1.9). These measurements are obviously also dependent upon the developmental stage of leaves. Total precipitation at crowns of trees was almost equal in April and July 1978; however, the understory of the leafless oak-hornbeam forest was reached by 77% of the precipitation, while only 70% passed through the fully developed canopies.

Infiltration rate increases relative to the total monthly precipitation in the forest (Fig. 1.10). Interception by tree canopies depends upon the nature of the tree

Table 1.8. Atmospheric precipitation in the Niepołomice Forest from July 11 to November 2, 1977, and March 31 to November 3, 1978, as measured by rain gauge totalizers

	Szarów	Poszynka	Forest district No. 44	Ispina
1977				
July 11–Aug. 6	98.5	77.7	91.9	71.2
Aug. 7–Sept. 1	95.7	136.2	131.6	93.2
Sept. 2–Oct. 1	91.6	92.4	78.5	68.7
Oct. 2–Nov. 2	15.0	22.2	17.5	13.5
Total	298.1	328.5	319.5	246.6
Average	74.5	82.1	79.9	61.6
1978				
March 31–May 2	69.7	69.7	65.2	56.0
May 3–June 1	146.7	179.5	138.4	120.0
June 2–July 1	74.0	72.3	72.4	75.1
July 2–Aug. 1	57.1	64.5	57.9	58.7
Aug. 2–Sept. 1	155.0	160.2	160.8	158.7
Sept. 2–Oct. 3	97.4	91.6	90.8	90.2
Oct. 4–Nov. 3	41.8	37.5	36.7	35.0
Total	641.7	675.3	622.2	593.7
Average	91.7	96.5	88.9	84.8

stand, closeness of the canopies, etc. In oak-hornbeam forests interception of 70% is achieved with a 40 mm monthly precipitation at the tops of the trees. In moist oak-pine forests the same interception is achieved with 60 mm, whereas in mesic oak-pine and in dry oak-pine forests with only 80 mm and 100 mm respectively.

1.3.5 Air Humidity

The mean annual relative air humidity is high (85%). The monthly means are slightly variable but never fall below 80%. This is because of high substrate moisture, a considerable number of dead river channels and oxbows, a dense network of drainage ditches with stagnant water, and the ameliorating affects of the forest complex. Fogs and mists occur here three times more frequently than in Cracow.

1.3.6 Final Remarks

For the past 25 years, the Niepołomice Forest has been affected by dust and gaseous air pollutants emitted by the nearby steelworks. There are no climatological data collected in this area prior to that period. Hence, it is not possible to recognize any trends in local climatic conditions. One may only refer to the literature dealing with this problem in other industrial areas.

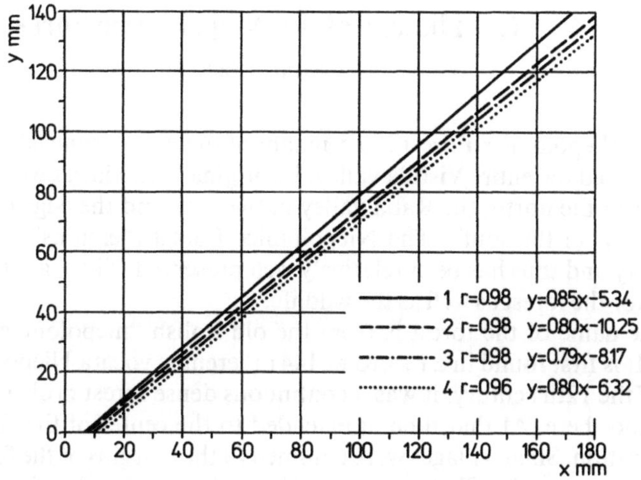

Fig. 1.9. Total rainfall reaching the understory (y) and canopy (x) in various forest communities in the Niepołomice Forest; *1* mesic oak-hornbeam forest; *2* dry variant of mixed oak-pine forest; *3* mesic variant of mixed oak-pine forest; *4* moist variant of mixed oak-pine forest

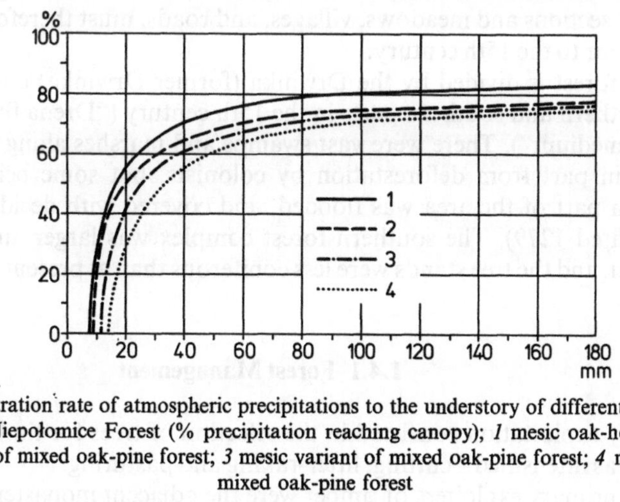

Fig. 1.10. Infiltration rate of atmospheric precipitations to the understory of different forest communities in the Niepołomice Forest (% precipitation reaching canopy); *1* mesic oak-hornbeam forest; *2* dry variant of mixed oak-pine forest; *3* mesic variant of mixed oak-pine forest; *4* moist variant of mixed oak-pine forest

As estimated by Munzar (1977), there is a 5–10% increase in cloudiness, 100% increase in the number of misty days in winter and a 30% increase in summer, and a 10% increase in number of days with rainfall or snowfall in the regions of Ostrava and Most, Czechoslovakia. The increase in cloud and mist in the vicinity of Cracow cause a 15–30% decrease in total solar radiation (Hess 1969). As measured by Kasina et al. (1978), direct solar radiation is decreased at the rate of 80% in the proximity of the steelworks, and by 40% at the western margin of the forest when the wind is from the northwest. The anthropogenic sources of energy cause an increase in heat received by the area at the rate of a few to a dozen Kcal $cm^{-2}y^{-1}$ (Budyko 1974) and consequently, an increase in local mean annual temperature at the rate of 0.5°–1.0 °C (Oke 1973).

1.4 History of the Niepołomice Forest

M. Rokosz

The Niepołomice Forest is a remnant of the great primeval lowland that long ago covered the entire Vistula valley. Its original boundaries were set by the Vistula river in the north, the Raba valley in the east, and the edge of the Carpathian Highlands in the south. The Niepołomice Forest ("eremus") was always royal property and thus has been relatively well preserved. There are thus historical records for the forest as well as the wildlife.

The name of the forest is from the old Polish "niepołomne" – unbreakable trees. It is first found in a record of 1441 ("eremus vocata Niepołomska"). To the end of the 12th century, it was a continuous dense forest broken only by swamps here and there. At that time, it extended to the center of the Cracow urban agglomeration. Small villages were confined to the margins of the forest.

Anthropogenic affects started in the 14th century when intense colonization under German law occurred. Colonization in this area was continued throughout the 14th and 15th centuries, but declined later when the forest became a royal property to maintain game. The development pattern, including the distribution of forest sections and meadows, villages, and roads, must therefore have been initiated prior to the 15th century.

The forest is divided by the Drwinka (former Drwinka) creek which split it into northern and southern parts in the 15th century ("Dręna fluvium dividit silval per medium"). There were vast swamps and marshes along the riverside, resulting in part from deforestation by colonists, but some being natural. The northern part of the area was flooded, and covered with deciduous forests (records dated 1279). The southern forest complex was larger and drier than the northern, and the tree stands were less coniferous than at present.

1.4.1 Forest Management

Early exploitation consisted in the selection of trees for timber; later, exploitation was intensified – cutting, litter-raking and pasturing.

The primary exploiters of timber were the adjacent monasteries including the Benedictine convent at Staniatki, the Premonstratensian monastery at Hebdów, and the rectors of the church at Niepołomice. As documented in historical records, the Benedictines cut 15 four-horse wagons of timber a week from the beginning of May until the end of September, and 70 wagons a week for the remainder of the year during the reign of Sigismundus III Vasa (16th century). A total of 5,437 wagons of timber were taken from the forest in 1736, with the share of the Benedictines amounting to 2,485. Beginning in the middle of the 13th century, considerable amounts of timber were taken for the royal salt mines at Wieliczka and Bochnia. It was used for props, as well as for barrels and firewood. New colonists and the homeless victims of fire were also allowed to cut trees. Furthermore, timber was taken for repairs to royal buildings even if it had to be transported to distant locations.

Owing to persistent and excessive exploitation of timber, the forest was subjected to devastation even though it had long ago been afforded protection. Forty five foresters were released from mandatory service to the landlords in order to maintain the forest and wildlife in good health. Beginning in the 17th century, when the Polish kings took up residence in Warsaw, the forest was left entirely at the mercy of leaseholders and land agents, and the decline became serious. As evidenced by inspection reports dated 1710, 1730, 1736, and 1744, the southern part was the most affected by this wasteful exploitation. The inspectors recorded various clear-cuts, partial-cut, and open areas. In their opinion the Grobla section was the best preserved, where the tree stands included mainly oak, alder, willow, and lime, and scattered pine.

At the beginning of the 19th century, the forest was divided into parcels and exploited only as clear-cuts, open stands and partial cutting having been dispensed with. The cutovers were planted with pines but the plantations were heavily grazed. Nevertheless, an inspection report of 1858 provides evidence that there were at that time only 326 ha of partial-cut and open stands and swamp of 1,250 ha previously forested. Because of the nature of the forest management, of clear-cuts regenerated chiefly with pine, pine comprised 89% of the total forested area. In 1852, a forest bill was promulgated that prohibited further felling and recommended regeneration of existing clear-cut areas.

There was no forest management plan at the beginning of the 20th century. In 1915, the first was presented for the Grobla, Niepołomice, and Damienice forest sections. Later plans (1932/33 to 1941/42) set regulations for variable cutting age, and extent of fellings. These management goals fell short of achievement due to mass outbreaks of insect pests. At the beginning of the 20th century, the pine forests of the south were hit by an outbreak of the pine noctuide, and in the 1920's, by the sawfly. Infested trees were cut and this resulted in the formation of sawfly-stricken stands of variable density and age structure. These weakened forests were heavily damaged by wind- and snowbreaks. Tree stands in the northern part of the forest were more resistant, being less monospecific. Cuttings in that area were regenerated with oak (60%), pine, lime, alder, ash, and elm.

The forest was seriously devastated during World War II when the Germans harvested 885,000 m^3 of timber. Several 300-year-old, healthy oaks were then felled in the Koło section.

In 1945, the forest suffered considerable damage. A new management plan was prepared for the period 1945/46 to 1954/55. The northern part was projected as oak management, and the southern as pine. New fellings were recommended to be followed by regeneration with diverse tree species (pine, oak, beech, hornbeam, ash, and alder) in the north, and with pine and oak in the south.

In the late 1950's, the largest steel complex in Poland was located between Cracow and Niepołomice. The forest came under the strong impact of industrial air pollution emitted by the entire Cracow urban-industrial agglomeration (cf. Chap. 2).

According to the forest management plan for the period 1957/58 to 1966/67, forests of economic and conservation value were recognized here. Recently in 1979 the entire forest complex has been slated for conservation.

1.4.2 Water Relationships

Water relationships have long undergone repeated changes. In the northern part of the forest these were induced mainly by shifts in the channels of the Vistula River. At one time the valley of the Drwinka served as such a channel (Trafas 1975). The southern part was drained by tributaries to the Raba River, and the water relationships have been kept more or less constant.

Until the end of the 18th century, water relationships were controlled naturally, by flood waters in the north and atmospheric precipitation in the south. Drainage was first undertaken after the disastrous inundation of 1813. However, these early drainage installations were soon abandoned. In the middle of the 19th century, the channel of the Drwinka was dredged and a network of drainage ditches leading to the Vistula and the Raba was excavated in both sections of the forest. At the same time, stagnant water was drained from meadow depressions, and oxbows located in the northern sections. A system of high dikes was developed along the Vistula to prevent flood waters from moving far into the forest, and flooded areas were planted to willows. In fact, the regulated Vistula channel is less affected by floods, even though the incidence of floods has considerably increased during the last century due to rapid deforestation of the Carpathians. Flooding occurred only every 4.2 yr in the 19th century, but every 2.8 yr in the present century.

Additional drainage was undertaken in the period between the World Wars. The channel of the Drwinka was once again dredged and dikes were constructed along its lower portion. A system of drainage channels was then excavated in the Grobla section. More recently, in the 1970's, extensive regulation of the Drwinka and other drainage installations were completed.

1.4.3 Wildlife

The forest was a royal hunting preserve for over 400 years. Casimir the Great had a castle built at Niepołomice and a road leading to the forest, now known as the Royal Road. Later in the 15th century, Ladislaus Jagiełło, who hunted here at least on 50 occasions during his 44-year reign, built a hunting lodge with a stable and baths at Poszyna. His descendants, as well as the kings of the Vasa and Saxon dynasties, also often hunted in the forest.

Wildlife was abundant but somewhat poor in species. The oldest historical records indicate that the European bison, the auroch, and the bear were absent from the forest even in the Middle Ages. However, there was an abundance of European elk, deer, wild boar, fox, and wolves. Pigs were fattened in the oakwoods. Beaver were frequently found in meanders of the Vistula River. The abundance of wildlife is well documented by records of the royal hunt. For example, King Ladislaus IV killed 50 deer during a 4-day hunt in July 1644; King Augustus II Saxon killed 3 elk, 17 deer, 23 boar, 12 wolves, 2 lynx, 32 fox, 88 roe deer, and innumerable hare.

The wildlife became impoverished slowly but surely. Elk disappeared at the end of th 18th century, and the wolf at the beginning of the 19th. A wolf killed in 1979 came undoubtedly from some distance away, possibly the montane forests. This impoverishment of the wildlife was caused mostly by poaching. Until the 16th century poaching in royal forests called for the death penalty. This was mitigated under the reign of King Sigismundus Augustus at the end of the 16th century, but in the time since the number of poachers has continuously increased. The wildlife decreased considerably due to the hunting of deserters during World War I, as well as during World War II. After both World Wars, the wildlife rapidly returned to abundance (cf. Chap. 4).

1.4.4 Concluding Remarks

Five stages may be distinguished in the history of man's influence on the natural environment of the Niepołomice Forest. Until the middle of the 13th century the effects of human activities were insignificant. From then to the beginning of the 16th century the forest decreased in area because of colonization and adjacent road developments. From the 16th to 18th centuries, the forest, and especially the southern part, was intensively exploited, and various cuttings, open stands, and partial cuts were produced. From the 19th to the middle of present century, the forest has been intensively exploited by clear-cuttings planted subsequently to pine, which resulted in mass outbreaks of insect pests and extensive snow damage. Beginning in the 1950's deterioration due to industrial air pollution has affected the forest ecosystems and attempts have been made to restructure tree stands into more resistant mixed forests.

1.5 Bioindication of Environmental Deterioration

K. GRODZIŃSKA

Pollution of an environment is usually estimated by the application of physicochemical methods aimed at measuring the concentration of toxic elements and their compounds in the air. However, it may be also estimated by the use of various bioindicators (Grodziński and Yorks 1981). Such methods are straightforward and not costly. Moreover, they are advantageous because the organisms display the affects of the cumulative action of toxic pollutants.

There are various plant indicators which have been used (Grodzińska 1978a, b, Hawksworth and Rose 1970, Kiszka 1977, de Witt 1976) and animal indicators (Jop 1979, Nováková and Paukert 1973), as well. Pollution can be estimated from changes in the geographical distribution of various organisms, as well as from changes in morphological, cytological, chemical, physiological, or biochemical characteristics.

The deterioration of the environment of the forest was estimated from changes in (1) the distribution of lichens, the most sensitive group of plants to air pollution; (2) concentration of pollutants in selected plant species; (3) characteristics of needles, twigs, and crowns of the Scotch pine; (4) photosynthetic efficiency

Table 1.9. Cover of pine trunks with lichens and algae in the Niepołomice Forest, relative to distance from air pollution emitters

Distance	Coverage coefficient	
km	Lichens	Algae
20.5	1826	1700
22.5	2970	1201
25.5	5138	808
28.0	5409	532
31.5	6323	520
35.0	5404	790

and timber production in the ecosystem; and (5) characteristics of stag antlers in the roe deer.

As evidenced by lichenological research (Kiszka 1964, 1977, 1981), species richness of the lichen flora has decreased at the rate of approximately 10% over the last 20 years, with terricolous types having been much more seriously affected (23.3%) than epiphytic species (6%). It is notable that the lichen flora has been more heavily affected in the northern part of the forest, along the Vistula valley (26.7% decrease in species richness), than in the southern part (5.3%). Those lichen species that are most sensitive to air pollution (*Usnea* sp., *Parmeliopsis aleurites, Parmelia furfuracea, Cetraria islandica*) disappeared from the northern part of the forest as well as from the western area of the southern forest section, while their frequencies in the remaining areas have considerably decreased. There is also much less coverage to tree trunks with lichens, as they become replaced by more resistent algae (Table 1.9). Many lichen populations have changed habitats, and have abandoned coniferous trees with strongly acid bark and shifted to deciduous trees with less acidified bark. Commonly, they have displayed a change in thallus morphology, and an excessive development of soredia. Finally toxic-tolerant lichen species have appeared, such as *Lecanora conizaeoides* or *L. varia*. The pattern of pollution intensity zones is recognized in the area relative to the distribution of the lichen flora (Fig. 1.7). The entire northern part of the forest and the western and eastern margin of the southern part comprise a zone of very severe to severe environmental deterioration, and only the central part of the southern forest section is less heavily affected by air pollution.

The content of total sulfur in organisms strongly increases in areas affected by emissions of sulfur dioxide (Keller 1973, Lihnell 1968, Turunen and Visapää 1972). Concentration of sulfur in pine needles ranges from 1,287 to 2,418 ppm in the forest (Grodzińska 1977) (Fig. 1.12), increasing with needle age, i.e., with the time span of the pollutant action, and decreasing with distance from the point of emission. Conversely, the concentration is lower (1,200–1,250 ppm) and constant with age in needles taken from control areas.

When affected by sulfur dioxide emitted by industrial installations, tree bark undergoes acidification (Grodzińska 1971, 1979, 1982a, b, Skye 1968, Zdanowska 1976). In the Niepołomice Forest, the bark of lime (*Tilia cordata*), oak (*Quer-*

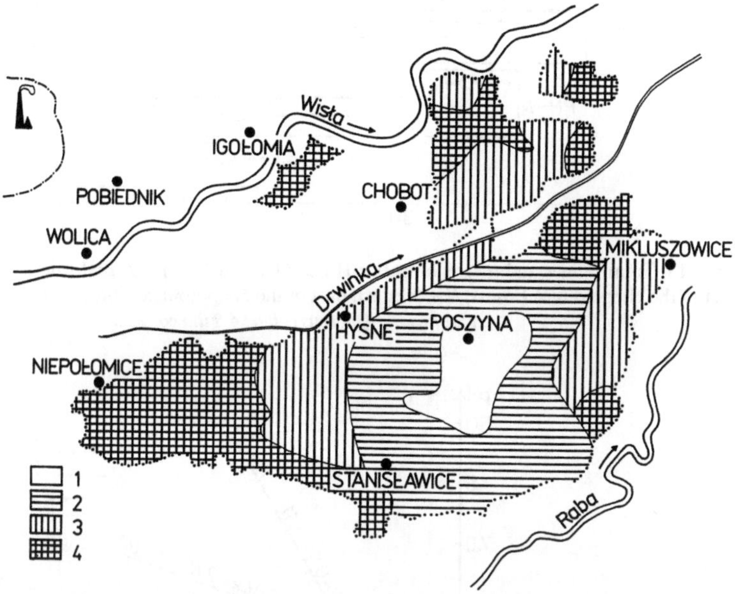

Fig. 1.11. Lichen zones in the Niepołomice Forest. (After Kiszka 1977); *1* slight damage; *2* moderate damage; *3* severe damage; *4* very severe damage

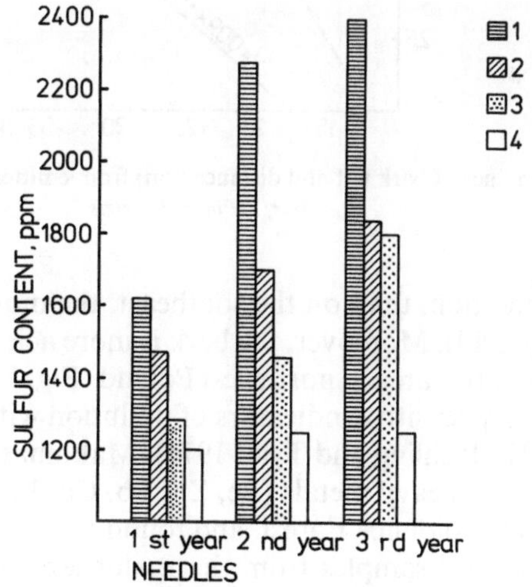

Fig. 1.12. Concentration of total sulfur in needles (related to dry weight) of the Scotch pine (*Pinus silvestris*) (Grodzińska 1977); *1–3* Niepołomice Forest, localities at varying distances from emitters; *4* control area Białowieża Forest

cus robur), black alder (*Alnus glutinosa*), and pine (*Pinus silvestris*) shows an acid reaction, ranging from pH 2.2 to 4.7 (Fig. 1.13). The most acid reaction is shown by pine bark, and the least by lime. There are differences in bark pH between the western and eastern margins of the forest. In all investigated tree species the bark is more acidified on the western side, in proximity to the ironworks and the

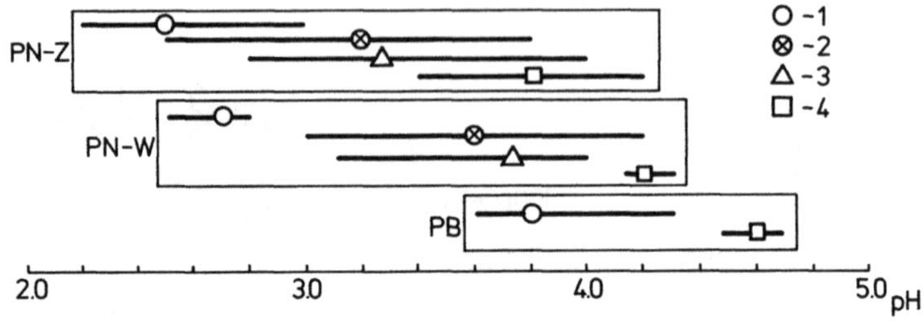

Fig. 1.13. Total ranges and means of bark pH (Grodzińska 1977): *PB* Białowieża Forest; *NE* eastern part of the Niepołomice Forest; *NW* western part of the Niepołomice Forest; *1 Pinus silvestris; 2 Alnus glutinosa; 3 Quercus robur; 4 Tilia cordata*

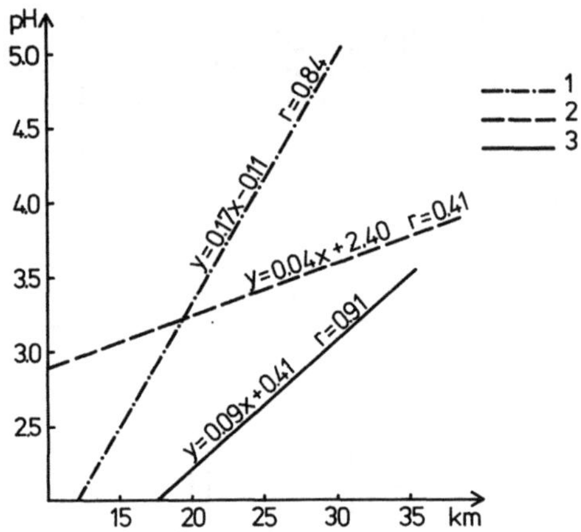

Fig. 1.14. Regression lines of bark pH and distance (km) from emitters; *1 Tilia cordata; 2 Quercus robur; 3 Pinus silvestris*

Cracow agglomeration, than on the southeast, the most distant from the source of emission (Fig. 1.14). Moreover, the bark is more acidified at all locations in the forest than in a control area in northwest Poland (Fig. 1.13).

Mosses are very sensitive indicators of pollution with heavy metals (Grodzińska 1978 b, 1982 b, Rühling and Tyler 1971). Makomaska (1978) investigated the concentration of six heavy metals (Fe, Zn, Pb, Cd, Mn) in the moss *Pleurozium schreberi* in the Niepołomice Forest, and found that it is higher by a factor of 1.3 to 4.8 than in control samples from the Białowieża Forest (Table 1.10). Based upon the global index of pollution (GI), defined to compare the increased concentration of five heavy metals (Cd, Pb, Cu, Zn, Fe) in the moss in the Białowieża Forest, Makomaska (1978) distinguished five pollution zones at Niepołomice (Fig. 1.15). The western part is most severely affected and the eastern margin somewhat less. The least affected by industrial air pollution is the central part of the forest. The zone of severe to very severe deterioration of the environment (Global Index = 17–22) accounts for 65.6% of the total area of the forest.

Coniferous trees undergo considerable morphological change in areas influenced by industrial emission. This is why characteristics of their needles, an-

Table 1.10. Heavy metal accumulation in *Pleurozium schreberi* moss. (After Makomaska, 1978)

Element/ ppm. dr. wt.	Niepołomice Forest (N.F.)	Białowieża Forest (B.F.)	Pollution index N.F./B.F.
Cd	2.4	0.5	4.87
Cu	15.3	11.6	1.30
Pb	88.0	20.0	4.40
Zn	270.0	86.0	3.14
Fe	1384.0	286.0	4.84

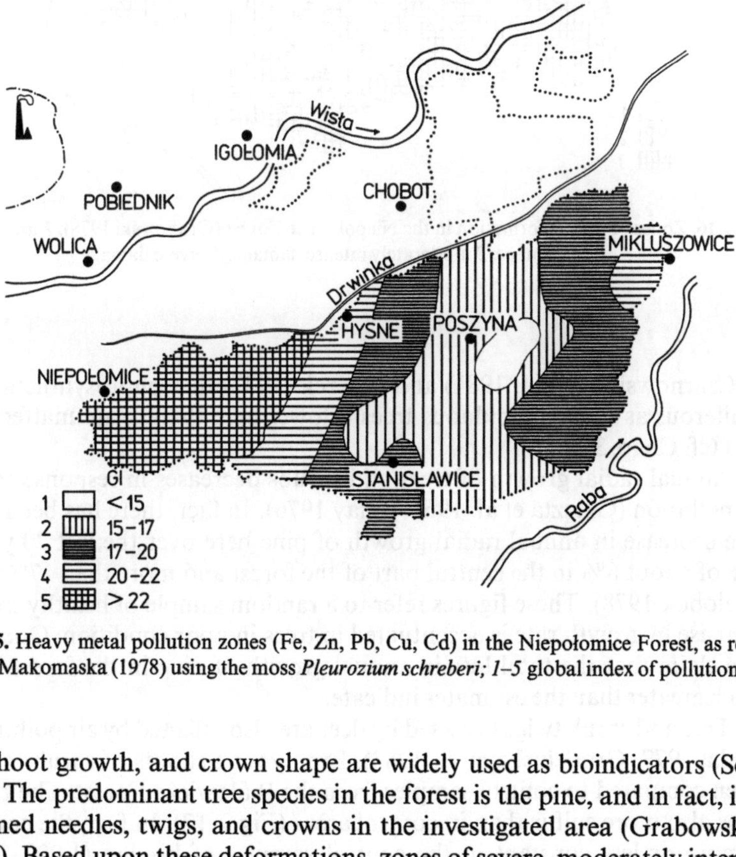

Fig. 1.15. Heavy metal pollution zones (Fe, Zn, Pb, Cu, Cd) in the Niepołomice Forest, as recognized by Makomaska (1978) using the moss *Pleurozium schreberi; 1–5* global index of pollution (*GI*)

nual shoot growth, and crown shape are widely used as bioindicators (Schubert 1977). The predominant tree species in the forest is the pine, and in fact, it shows deformed needles, twigs, and crowns in the investigated area (Grabowski 1978, 1981 b). Based upon these deformations, zones of severe, moderately intense, and moderately typical damage are recognized (Fig. 1.16). The zone of severe damage accounts for 47% of the total area, being confined mostly to the western part of the eastern margin of the southern section. Moderately intense damage occurs over 47% of the total area, while moderately typical damage is confined to as little as 6%. There are no pine free of damage in the mixed oak-pine forests.

Air pollution by gaseous and dust pollutants can be estimated also in relation to the intensity of photosynthetic activity in the ecosystem. It was demonstrated

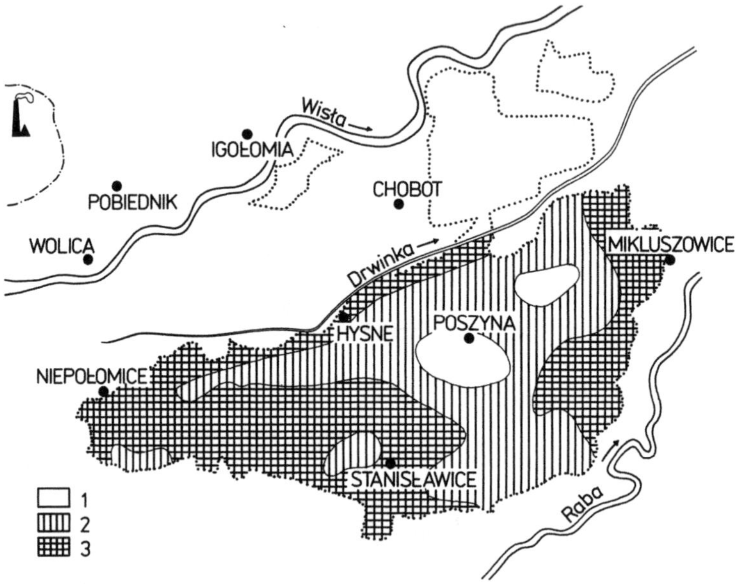

Fig. 1.16. Zones of pine deformation in the Niepołomice Forest (Grabowski 1978): *1* moderately typical damage; *2* moderately intense damage; *3* severe damage

by Czarnowski (1977a, 1981) and Maczek (1977) that photosynthetic activity of coniferous as well as deciduous trees is lower here than in an unaffected control area (cf. Chap. 3.5).

Annual radial growth of trees and shrubs decreases in response to industrial air pollution (Greszta et al. 1969, Pillay 1976). In fact, there has been a considerable decrease in annual radial growth of pine here over the last 20 years, at the rate of about 6% in the central part of the forest and more than 17% in the west (Białobok 1978). These figures refer to a random sample of healthy trees, but the decrease in growth rate is accentuated in trees in poor condition. One may therefore claim that the total loss in annual growth of pine in the forest is probably much greater than the estimates indicate.

Tree and shrub twigs browsed by deer are also affected by air pollution (Grodzińska 1977, Grodzińska et al. 1983). Even young shoots of shrubs are acidified when compared to control samples from the Białowieża Forest. The browse contains also more sulfur than in control areas (Fig. 1.17). In fact, the concentration is more or less constant in the control samples of browse (1,000–1,100 ppm), whereas the browse as the herbs of the forest is enriched in sulfur at the rate of 20–70%.

This pollution of browse and herbs with sulfur compounds and heavy metals (Grodzińska 1977, 1978c, Grodzińska et al. 1983) has resulted in an increase in their concentration in deer antlers (Sawicka-Kapusta 1978, 1979). When compared to the Białowieża Forest control sample, deer antlers from the Niepołomice Forest show a 30% increase in sulfur, and a 10% increase in heavy metals. The

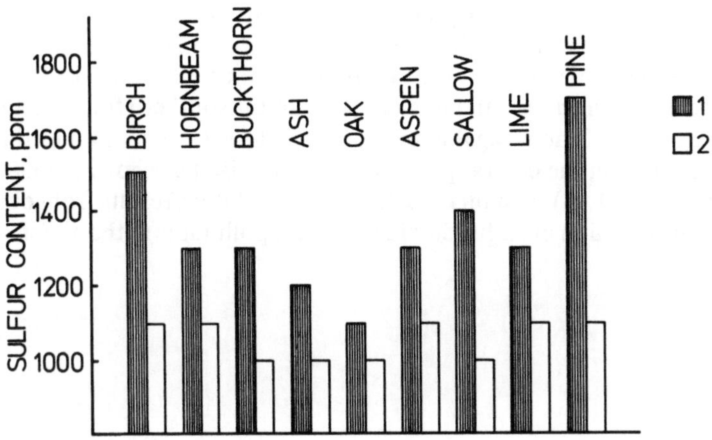

Fig. 1.17. Concentration of total sulfur in browse (Grodzińska 1977): *1* Niepołomice Forest; *2* Białowieża Forest

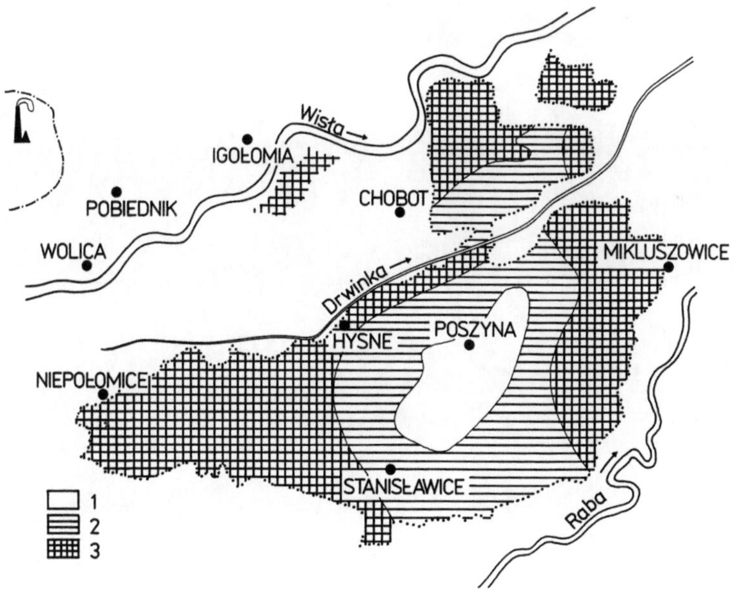

Fig. 1.18. Zones of forest deterioration, as recognized in the Niepołomice Forest after Kiszka (1977), Grabowski (1978), and Makomaska (1978); *1* slight deterioration; *2* moderate deterioration; *3* severe deterioration

pollution of food may thus provide an explanation for a drop in quality of antlers in the Niepołomice Forest (Jop 1979). In addition, in the last 20 years there has been a 32% decrease in the average weight of antlers, a 29% decrease in antler length, and a 25–30% decrease in trophy quality as estimated from international CIC tables.

The bioindicator data presented above are mutually consistent and demonstrate that the natural environment of the Niepołomice Forest has already undergone extensive deterioration from industrial air pollution. The most strongly affected are the northern and western parts of the southern forest section, which are nearest to the Cracow agglomeration, and the eastern margin of the forest. Only a small area in the central part may be recognized as showing only slight deterioration (Fig. 1.18). A similar pattern of air pollution results from observations on the concentration of sulfur dioxide and dust pollutants in the air (cf. Chap. 2).

Chapter 2

Transport and Input of Air Pollutants in the Niepołomice Forest Area

Chapter Editor: A. MANECKI

Contributors:
E. GARŚCIA, S. KASINA, J. KWIEK, J. LEWIŃSKA, A. MANECKI
A. SKOWROŃSKI, J. WALCZEWSKI, and K. P. ZAJĄC

2.1 Introduction

A. MANECKI

The following model representing the introduction of dust and gaseous air pollutants in the Niepołomice Forest ecosystem has been accepted in the present study:

emission → transport and transformation → input to the substrate

To evaluate this model the amounts of dust and gaseous pollutants emitted by the urban-industrial agglomeration of Cracow were determined, as well as the grain size, chemical, and mineralogical characteristics of the dust. The pollutant concentration, as well as the meteorological parameters in the near-ground air layer, were determined from direct measurements. These data were supplemented with photos of shifting pollutant plumes.

The sulfur budget in the atmosphere-ground system was computed using a "box model". The input of sulfur to substrate was estimated as the sum of both dry and wet deposition.

Dust pollutants were measured not only with standard methods recognizing infall in a given area, their grainsize and chemical composition, but also with mineralogical-geochemical methods. In this way, mineral composition, origin (natural vs. artificial), and amounts of components easily soluble in water were determined. In fact, the geochemistry of a natural environment is most affected by those components of industrial dust that have considerable water solubility. When the mineral composition of dust pollutants is known, it is then possible to estimate the contribution of specific factories.

The input of some heavy metals to substrate was also determined. This permits an estimation of the "pollutant accumulation" and "sorption capacity" of the ecosystem because the emission of those elements is known (cf. Chap. 6).

2.2 Emission of Industrial Air Pollutants

E. GARŚCIA, A. MANECKI, and K. P. ZAJĄC

The Niepołomice Forest ecosystem may be affected by dust and gaseous pollutants emitted by distant industrial agglomerations as well as by the nearby Cracow agglomeration.

2.2.1 Emission by Distant Industrial Complexes

The Upper Silesian Industrial District is 70 km west of the Forest, and winds from the west prevail in the area investigated. There are hundreds of modern as well as older factories in Upper Silesia, causing a considerable emission of air pollutants. The most deleterious are power stations fueled with coal with high sulfur content and consequently emitting large volumes of sulfur dioxide. Coking plants, iron and other metallurgical works, chemical and cement plants are also injurious to the natural environment. The Upper Silesian Industrial District, because of the high industrial concentration, ranks highest among pollutant emitters in Poland, its contribution amounting to 30% of total emission of dust and sulfur dioxide (G.U.S. – Chief Central Statistical Office). The location of the Upper Silesian industrial site relative to the Niepołomice Forest is shown in Fig. 2.1.

2.2.2 Emission by Adjacent Industrial Installations

The urban-industrial agglomeration of Cracow occurs some 15–30 km west of the margin of the forest. It includes more than a dozen great industrial works. The locations of the four largest are shown in Fig. 2.1.

The dust and gaseous pollutants emitted by that agglomeration were studied in detail in 1976–1978. Eleven major contributors to air pollution were selected relative to production type and size and smokestack heights were taken into account. These included an ironworks, aluminum foundry, power station, generating stations, cement plant, fireproofing material plant, inorganic chemical works, and others.

The total dust emission of each factory was computed relative to the concentration of dust in combustion gases and the rate of outflow, as well as the total

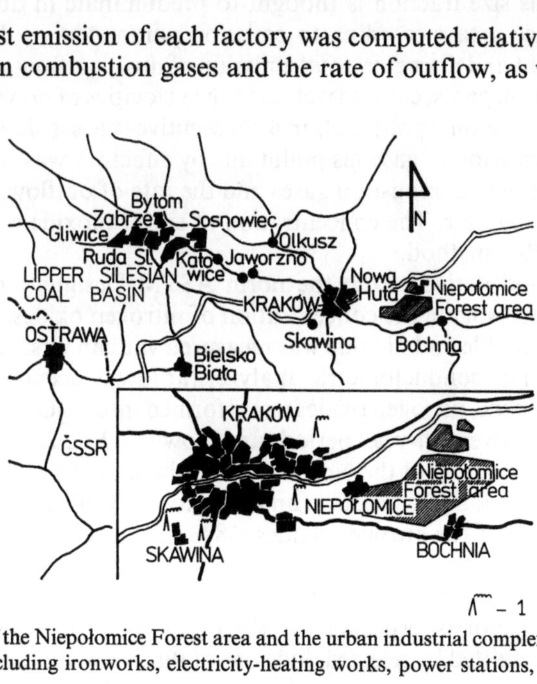

Fig. 2.1. Location of the Niepołomice Forest area and the urban industrial complex of Cracow *1* major industrial plants including ironworks, electricity-heating works, power stations, aluminum foundry

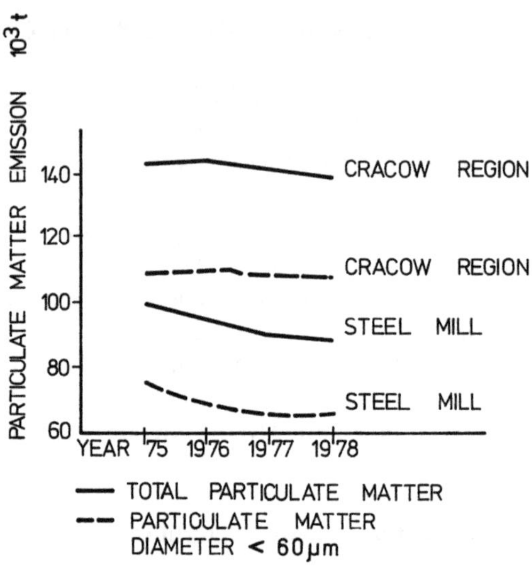

Fig. 2.2. Emission of dust by industrial works of the urban industrial agglomeration of Cracow. Emission by the ironworks in 1975–1978 included

activity span of the emitter. The rate of gas outflow was measured using a Prandtl tube and micromanometer. Concentration of dust in gas was measured with a combination of sensory probe, filter, gasometer, and vacuum pump.

Fine dust (grain size below 60 μm) was separated from other dust pollutants because this size fraction is thought to predominate in dust transported to the Niepołomice area, according to emission parameters in the Cracow agglomeration (height and diameter, total emission of dust, outflow speed and temperature of combustion gases, etc.), as well as to the velocities of prevailing, westerly winds. The emission of dust pollutants in 4 consecutive years is shown in Fig. 2.2.

Total emission of gaseous pollutants by a factory was computed from the gas concentration in combustion gases and the rate of outflow, and the total activity span of the emitter. The concentration of sulfur dioxide was measured applying the iodometric method.

A method derived from the norm ASTM-Designation P-1608-60 was employed to investigate the concentration of nitrogen oxides. The concentration of carbon monoxide and dioxide was measured with an Orsat analyzer (at high concentrations) or conductometric analyzer (at low concentrations). The emissions of sulfur dioxide, nitrogen oxides (transformed into N_2O_5), and carbon monoxide and dioxide, over a 4-year period, are shown in Fig. 2.3.

The contribution of the Nowa Huta steelworks to the total pollutant emission of the industrial agglomeration amounts to 64% of the dust, 44% of the sulfur dioxide, 46% of the nitrogen oxides, 68% of the carbon monoxide, and 60% of the carbon dioxide.

The grain size, mineral, and chemical composition of dust pollutants emitted by specific factories, especially by the ironworks, was also analyzed. The grain size of dust emitted by selected divisions of the ironworks is shown in Fig. 2.4. The

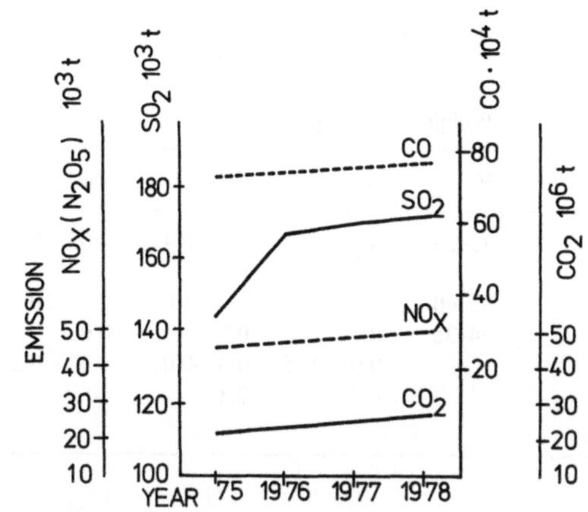

Fig. 2.3. Emission of sulfur dioxide, nitrogen oxides, and carbon monoxide and dioxide by industrial works of the urban industrial complex of Cracow 1975–1978

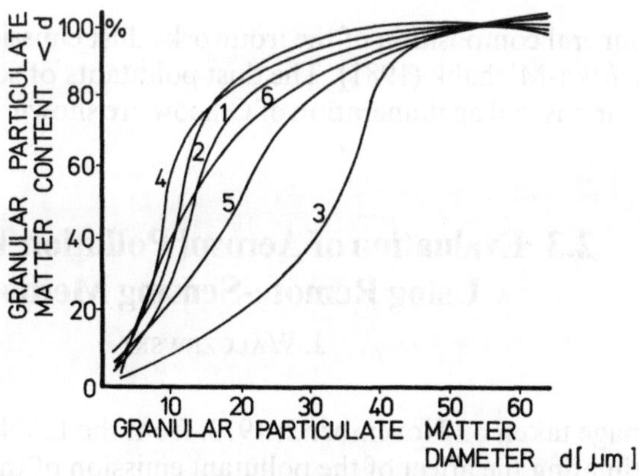

Fig. 2.4. Particle composition of dust emitted by various divisions of the steelworks. *1* Agglomerating Plant I and II; *2* Open Hearth Furnaces; *3* Blast Furnaces; *4* Refractory Works; *5* Coke Chemistry; *6* Power Plant

content of various metals in fine dust emitted by some divisions is provided in Table 2.1. Total amounts of the metals (lead, zinc, cadmium, and iron) in the dust emissions in 1976 are estimated at 130, 470, 3, and 14,000 t, respectively. Total emission of dust and gaseous pollutants is 15–20% higher during the cool half-year, the heating period, than during the warm half-year, the vegetative period.

The dust pollutants emitted vary in mineral composition in the respective factories. Dusts emitted by the ironworks division and the open-hearth and blast furnaces are composed mostly of iron oxides. Carbonate minerals predominate in those emitted by the coke oven-chemical factories and the fireproofing-material works. The dust produced by the power station is composed mostly of silica glass, quartz, and mullite, but carbonate minerals along with calcium oxide and silicates

Table 2.1. Percentages of selected metals in fine dust (60 μm in diameter) emitted by the steel works, Cracow

Emitted dust	Weight percentages					
	Fe	Mn	Pb	Zn	Cd	Cr
Agglomeration-division dust	33–38	0.3 –0.5	0.1 –0.4	0.05– 0.06	0.002–0.01	0.008
Steelworks dust	35–40	0.2 –0.9	0.3 –1.0	2 –10	0.006–0.04	0.01 –0.4
Blastfurnace dust	34–36	0.4	0.1	0.02	0.001	0.006
Volatile coal ash	2–5	0.03–0.05	0.01–0.07	0.02– 0.2	0.001	0.004–0.02
Dolomitic-calcareous dust	3	0.2	0.3	0.9	0.01	0.03

contribute mostly to those emitted by the cement plant. The dust pollutants of the inorganic-chemical works include calcium phosphate, carbonate minerals, gypsum, and quartz. Corundum and cryolite occur in the dust of the aluminum foundry.

The mineral composition of the ironworks dust emission was studied in detail by Wilczyńska-Michalik (1981). The dust pollutants of some industrial works of the urban-industrial agglomeration of Cracow are shown in Table 2.2.

2.3 Evaluation of Aerosol Pollution Transport Using Remote-Sensing Methods

J. Walczewski

An image taken on November 2, 1973, from the LANDSAT satellite may assist in estimating the affect of the pollutant emission of the Upper Silesian industrial agglomeration on the Cracow area. This District appears clearly with well-defined smoke plumes in a green-yellow channel and stretched parallel to the NW–SE wind direction. Walczewski (1976) demonstrated that assuming a wind velocity of 3–7 ms^{-1}, the aerosol component of the pollutant emission which can be recognized in a LANDSAT image may extend to a distance of up to 70–80 km (Fig. 2.5). This demonstrates that when the wind blows from directions ranging from the 270° to 300° azimuth, the aerosol component may well reach the Niepołomice Forest area; this is even more probable for the gaseous component.

The inorganic chemical works at Tarnów, some 45 km east of the forest, is also one of the emitters of detrimental air pollutants which may affect the ecosystem. Dworak (1978) demonstrated in a LANDSAT photograph of June 26, 1975, that, under conditions of a wind velocity of 4 ms^{-1} and southeasterly smoke propagation, the recognizable aerosol component extended some 35 km (Fig. 2.6). The topography is more favorable for westerly than easterly smoke production and thus the Niepołomice Forest may well be endangered by the Tarnów emitter. When the winds are easterly, the aerosol component of the Tarnów works

Table 2.2. Mineral composition of dust emitted by various industrial plants of the urban industrial agglomeration in Cracow

Iron work. (After Wilczyńska-Michalik, 1981)	
Anglesite	$PbSO_4$
Anhydrite	$CaSO_4$
Burkeite	$Na_6[Co_3(So_4)_2]$
Zincite	ZnO
Dolomite	$CaMg[CO_3]_2$
Epsomite	$MgSO_4 \cdot 7H_2O$
Gypsum	$CaSO_4 \cdot 2H_2O$
Glauberite	$CaNa_2[SO_4]_2$
Goslarite	$ZnSO_4 \cdot 7H_2O$
Halite	$NaCl$
Hexahydrite	$MgSO_4 \cdot 6H_2O$
Hematite	Fe_2O_3
Ilesite	Mn, Zn, Fe $SO_4 \cdot 4H_2O$
Illite	silicate
Calcite	$CaCO_3$
Kaolinite	silicate
Quartz	SiO_2
Langbeinite	$K_2Mg_2[SO_4]_3$
Larnite	silicate
Magnetite	Fe_3O_4
Mascagnite	$(NH_4)_2SO_4$
Melanterite	$FeSO_4 \cdot 7H_2O$
Mirabilite	$Na_2[SO_4] \cdot 10H_2O$
Mullite	silicate
Nitrocalite	KNO_3
Nitronatrite	$NaNO_3$
Olivine	silicate
Periclase	MgO
Salmiac	NH_4Cl
Feldspars	silicates
Sylvite	KCl
Syderite	$FeCO_3$
Spinels	$ZnOFe_2O_3$
Silica glass	
Tachhydrite	$CaCl_2MgCl_2 \cdot 12H_2O$
Trona	$Na_3H[CO_3]_2 \cdot 2H_2O$
Wüstite	FeO
A compound resembling	$ZnSO_4\ K_2SO_4 \cdot 6H_2O$
Aluminum foundry	
Cryolite	$Na_3[AlF_6]$
Chiolite	$Na_5[Al_3F_{14}]$
Corundum	Al_2O_3
Chemical works	
Apatite	$Ca_5[F(PO_4)_3]$
Calcite	$CaCO_3$
Quartz	SiO_2
Gypsum	$CaSO_4\ 2H_2O$
Power stations and electricity heating works	
Silica glass	
Mullite	silicate
Quartz	SiO_2

may settle on the eastern margin of the forest, while the gaseous component may extend even farther westward.

Ground photographs (Fig. 2.7) were taken with a camera with filters, from stations situated approximately 4 and 8 km distant from the main source of air pollution in the Forest area. The photos taken from the nearer station show smoke, while those taken from the other portray a dust-smoke layer underlying the breaking layer above the emitting source.

As recognized in airborne observations, the emission zone of the main source near the Niepołomice Forest area extends over an angle of 35° at a wind velocity of 4 ms^{-1}, and 45° at a velocity of 8 ms^{-1}. An image taken on June 26, 1975, from the LANDSAT satellite also indicates the source zone of smoke emitted in the investigated area. The zone is marked in the channel 4 (green-yellow) by a considerable change in undertone (Fig. 2.8). It extends over a distance of 25 km and at an angle of 35°. The wind velocity was 4 ms^{-1} at the time.

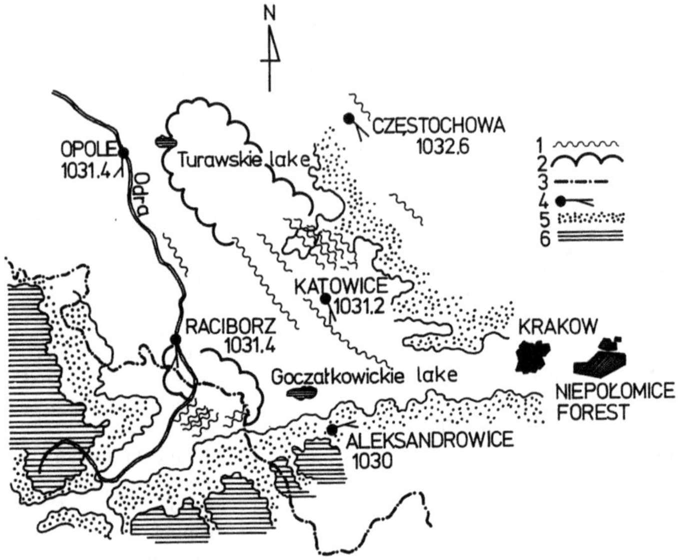

Fig. 2.5. Interpretation of image taken on November 2, 1973, from the LANDSAT satellite (Silesian region, S.W. Poland). Note position and extension of industrial smoke. *1* smoke plumes discernibly distinct; *2* approximate limits of an emission zone within an area influenced by several emittors; *3* Polish-Czechoslovakian state boundary; *4* meteorological stations and observed wind directions and velocities. *Numbers below station names* refer to atmospheric pressure (in mbar) measured simultaneously with wind observations; *5* 300 to 500 m a.s.l.; *6* more than 500 m a.s.l.

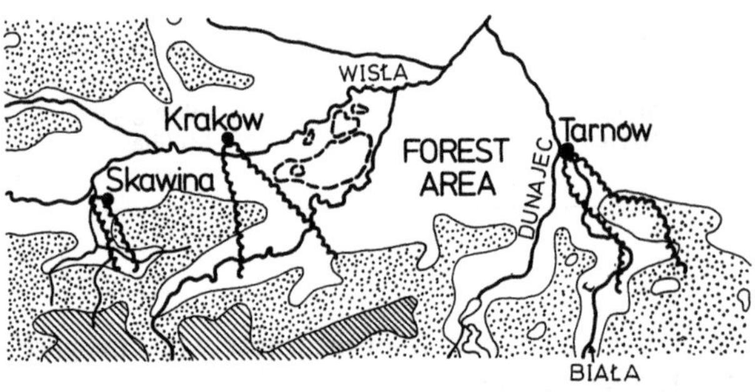

Fig. 2.6. Interpretation of image taken on June 26, 1975, from the LANDSAT satellite. Note extension of smoke plume emitted at Tarnów

The source zone of smoke, determined for the wind blowing from the 290° azimuth (WNW) from data presented above, is shown in Fig. 2.9. The Niepołomice Forest area is almost entirely within the source zone under such conditions. One is dealing here with a collective emitter of considerable spatial extension rather than with a single smoke plume, so that the aerosol density in the source zone will not fit the Gaussian distribution.

As judged from ground photos of the main emitters in the area, the aerosol emission is commonly split into two layers at the original source. A large group

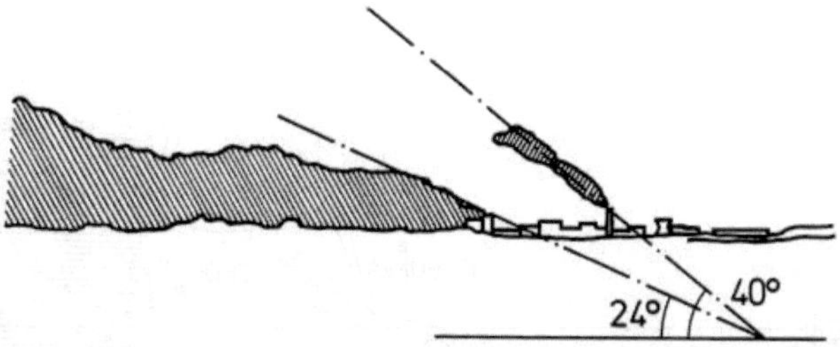

Fig. 2.7. Vertical distribution of smoke above an emitter, from a ground-based photo. Note separation of elevated layer

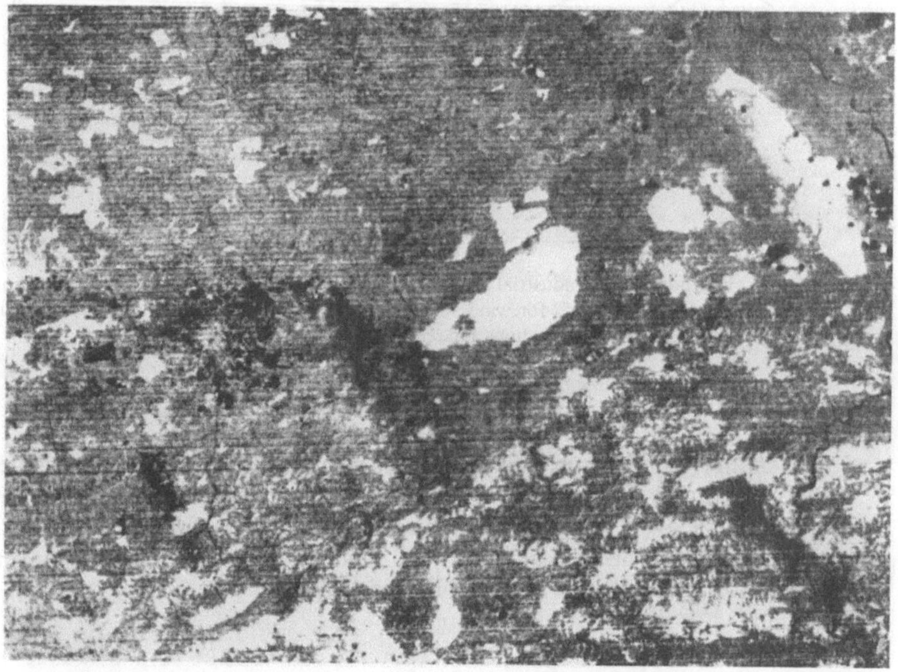

Fig. 2.8. Negative of image taken on June 26, 1975; from the LANDSAT satellite; channel MSS 4 (special range: green to yellow, 0.5–0.6 μm). Note light patch of the Niepołomice Forest near center of image, and dark, triangular smoke plume emitted west of the forest and extending southeastwards

of low smokestacks contributes to the bottom layer, reaching the surface of the ground, while the smoke and dust emitted by a smaller group of high smokestacks is spread much higher (Fig. 2.7). During flights over the emitter site, smoke, and combustion gases were detected at 1,000 m above the ground surface. The elevated layer is usually distinctly separated from the bottom layer. This may be partly due to the occurrence of low inversion layers between the low and high smokestack levels. In autumn and winter, a homogenous layer with high proportions of smoke and dust was often observed in ground photos taken in the vicinity

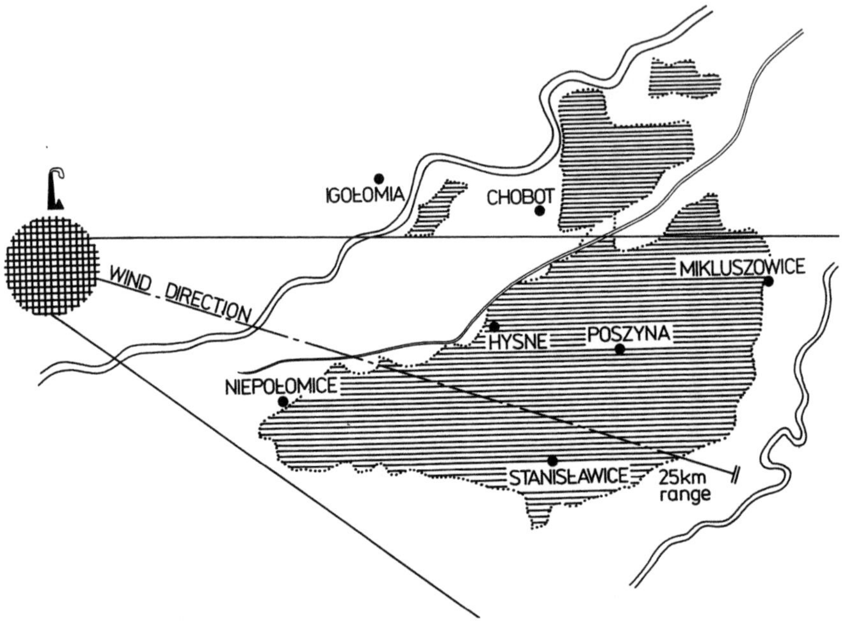

Fig. 2.9. Propagation zone of industrial smoke in the Niepołomice Forest area for wind direction 290° in azimuth. Gaping angle is 35° for wind velocity 4 m s^{-1}, as determined from airborne visual observations and remote-sensing techniques

of an emitter. This layer extended from the ground surface to 80–140 m in height. This observation is confirmed by the results of acoustic sounding of the atmosphere in Cracow. These studies have confirmed that the lowest inversion layers during winter are situated at a level of 150 m (Walczewski 1980). The inversion layers sometimes continue to hold for several days.

2.4 Meteorological Aspects of Air Pollutant Spread in Forest Areas

J. Lewińska

The affect of industrial air pollution upon a given area depends not only on the emission of air pollutants, but also to a considerable extent on the thermodynamics of the local atmosphere, which is influenced by a forest ecosystem in accordance with its size and the nature and spacing of trees.

Changes in air temperature depend upon the amounts of solar energy reaching the surface of a forest. The solar energy is to a considerable extent absorbed by the aerosol and water vapor under conditions of industrial air pollution.

The influx of radiant solar energy decreases at an average rate of 15–20% in urban industrial agglomerations, and the decrease in the intensity of ultraviolet radiation in winter may attain 30% (Lewińska 1977). In Cracow, the average decrease in total radiation is 20%, but it may reach 32% under conditions of low wind velocity and high concentrations of aerosol pollution (Lewińska and Rutkowski 1977), and even 50% when the air masses shift from Cracow toward the Niepołomice Forest (measurements under a smoke plume).

In spite of these changes in the influx of solar energy, considerable difference in insolation of the upper canopy surface and the forest interior results in maintenance of a positive thermal gradient which hampers a vertical exchange of air. As indicated by measuring air temperature at various levels, the forest interior shows a lower temperature than the canopy on an average 7 of every 10 days. This thermal stratification induces convection currents hindering air, and consequently also air pollutants, from moving downward.

A dense windward forest wall results in the piling up and whirling of air currents and a decrease in wind velocity. On the other hand, wind velocity increases considerably above the forest because of an increase in the density of air current lines and the development of turbulent flow due to roughness of the active surface. Finally, wind velocity decreases on the leeward side where air whirls develop owing to suction. As indicated by observations in the Niepołomice Forest area, the number of calms increases to 20% on the windward side of the forest relative to the adjacent fields. In turn, the wind velocity is, on the average, twice as high above the forest as in the interior, hence, the rate of air flow is also twice as high above the forest canopy.

These changes in wind velocity effected by a forest may result in the deposition of air pollutants, especially larger-sized fractions of dust, and is more intense on the windward leeward sides of the forest than in the interior. It is to be noted that because of regional orientation of winds the windward side of the Niepołomice Forest is the western margin, 50% of the time, and the eastern for 40%.

2.5 Concentration, Transformation, and Deposition of Sulfur Compounds

S. Kasina, J. Kwiek, and J. Lewińska

Continuous measurement of sulfur dioxide concentration in the Niepołomice Forest area began in 1967 (Morawska-Horawska 1970, Kasina 1971), and were intensified from 1975 to 1977. In the latter period, not only the concentration of sulfur dioxide was measured, but also the aerial sulfate aerosol, sulfate ions in the precipitation, and the acidity of precipitation.

The concentration of sulfur dioxide was measured using the West-Gaeke (1956) method and sampling over a 24-h period. To determine the vertical gradient of sulfur dioxide concentration in the forest, air sampling was carried out every 20 min, and the lead peroxide method was also used (Meas. of Air Poll., 1957) with monthly exposure of lead peroxide candles. The concentration of sulfate ions

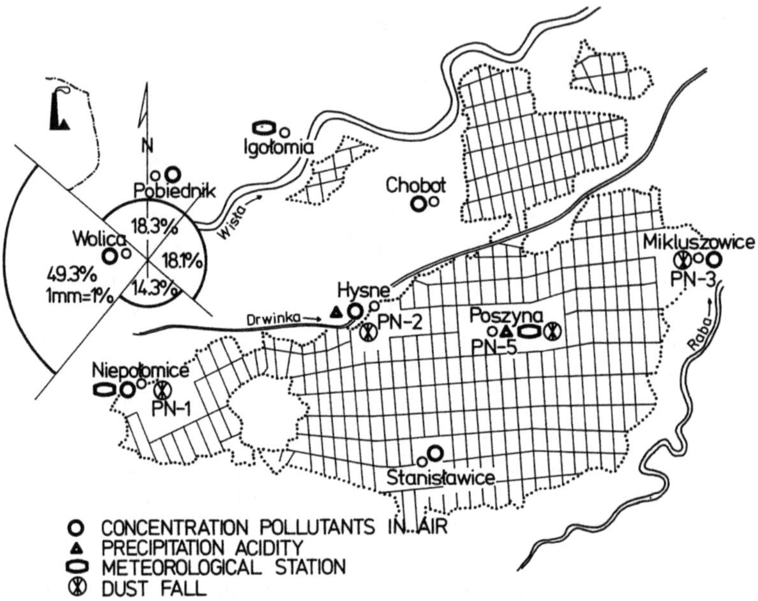

Fig. 2.10. Sampling network. *Heavy circle* air pollutant concentrations; *triangle* precipitation acidity; *oval* meteorological stations; *crossed circle* dust fall

both in the aerosol and in precipitation was measured by spectrophotometric titration with barium perchlorate and thorone used as the indicator (Yamamura and Fritz 1955). Samples were taken over a 24-h period. The aerosol was sampled with Whatman 40 paper, and precipitation with polyethylene funnels and bottles. Funnels with a diameter of 15.2 cm were positioned 1.5 m above the ground surface and were not closed during dry periods. Bottles and funnels were washed daily, regardless of the occurrence of precipitation, with double distilled water. The pH of precipitation was measured. In addition, the concentration of strong acids was determined (Liberti et al. 1972) applying Gran's function (Brosset 1978). Potential acidity accounting for the concentration of strong, weak, and Brønsted's acids was also measured (Galloway et al. 1976). Under routine conditions the measurement errors varied between 5 and 20%.

Results were separately evaluated for cool and warm half-years (October to March, April to September), with meteorological conditions taken into account. The latter included changes in wind pattern. The pollutant concentrations have been presented as cumulative frequency distributions. Where a highly significant correlation was noted, regression equations were computed. The significance was corrected with a Wilcoxon test.

The input of sulfur compounds to substrate was estimated using the sum of both dry and wet deposition. Dry deposition was computed as a product of the mean annual concentration of SO_2–S and SO_4^{2-}–S in the near-ground air layer and the average deposition velocity of these pollutants as recorded in the literature. Wet deposition was computed from the concentration of sulfur in the precipitation and precipitation amounts.

The sulfur budget in the atmosphere–ground system was computed using a "box model" (Mészárosz and Várhelyi 1977, Kasina 1978 a). The empirical data were obtained as presented in the sampling network shown in Fig. 2.10. Sampling stations were positioned to account for the location of emitters and in respect to the prevailing air circulation along the W–E axis. The data set includes mean daily concentrations of sulfur dioxide in the areas of the Cracow agglomeration and Niepołomice Forest, as measured over a 10-year period; all the other data are for shorter periods.

2.5.1 Concentration of Sulfur Dioxide and Sulfate Aerosol

Mean annual concentrations of sulfur dioxide calculated for the 10-year period range from 18.3 to 24.8 $\mu g\ m^{-3}$ (Table 2.3, Fig. 2.11) but single year means at particular stations may differ by a factor of 2 or 3. This confirms that the concentration of sulfur dioxide should be estimated only on the basis of multiannual measurements. The admissible concentration of 75 $\mu g\ m^{-3}$ for areas under special protection in Poland, is exceeded for 15 and 3% respectively of the time spans of cool and warm half-years (Fig. 2.12). Mean concentration in the cool half-year is approximately three times that in the warm. The inflow of sulfur dioxide to the Niepołomice area is predominantly from the western quadrant. This wind direction accounts for some 50% of total frequencies (Fig. 2.10) with calms included and amounting to 15–20%. The latter are caused by a marked decrease in wind velocity in contact with the front of the Niepołomice Forest. Despite increases in the total emission of sulfur dioxide in the last few years, no trend toward an increase in pollutant concentration has been observed in the same period (October 1967 to March 1978). It should be mentioned that increases in the emission of sulfur dioxide are confined to emitters higher than 100 m.

Variation in the concentration of sulfur dioxide was analyzed relative to meteorological conditions including air temperature, wind velocity, and precipita-

Table 2.3. Annual and seasonal concentrations of sulfur dioxide (in $\mu g\ m^{-3}$). Measurements from October 1967 to March 1978

Time	Concentration	Sampling station			
		Pobiednik	Wolica	Niepołomice	Hysne
Year	Mean	24.81	23.42	24.22	18.35
	Minimum	12.20	10.58	11.72	10.75
	Maximum	36.98	35.94	34.13	31.18
Warm half-year	Mean	11.77	10.25	11.98	10.03
	Minimum	5.41	3.36	3.76	5.74
	Maximum	19.04	17.79	14.81	18.22
Cool half-year	Mean	37.85	36.59	36.35	26.09
	Minimum	16.69	15.73	18.04	11.56
	Maximum	54.93	54.09	54.50	44.82

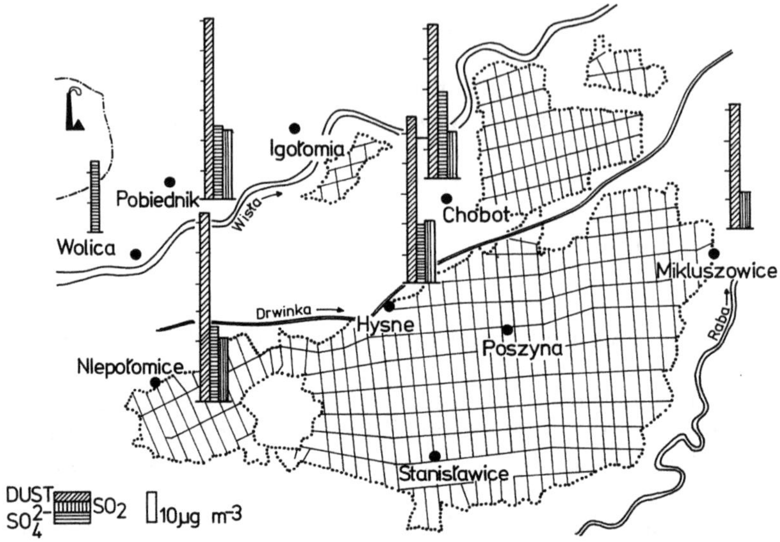

Fig. 2.11. Concentration of sulfur dioxide, particulate matter and sulfate aerosol. October 1967 to March 1978. *Vertical hatched* sulfur dioxide; *cross-hatched* dust; *horizontal hatched* sulfate aerosol

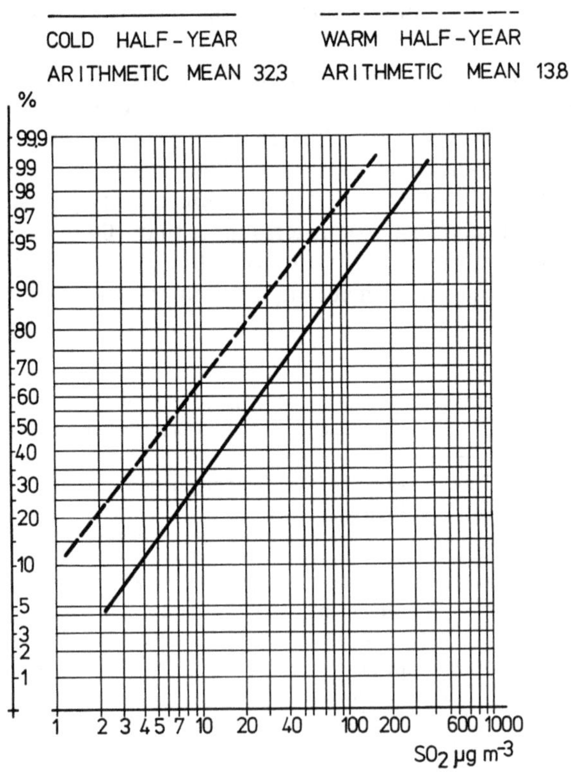

Fig. 2.12. Empirical distribution of sulfur dioxide at Hysne from 1975 to 1977. *Left* cool half-year; *right* warm half-year

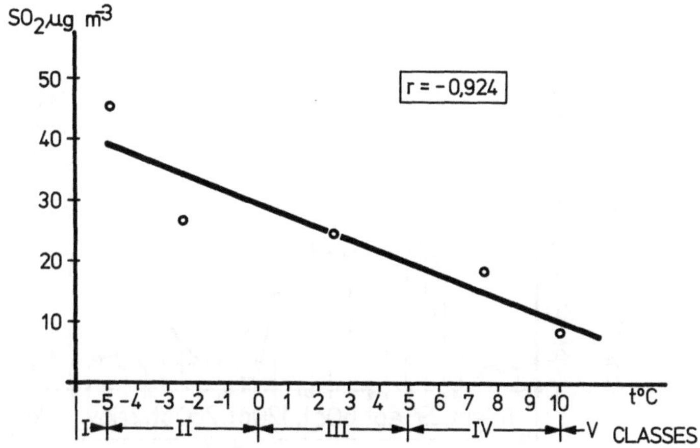

Fig. 2.13. Average concentrations of sulfur dioxide in specific air-temperature intervals at Hysne. October 1970 to March 1977

tion. Mean concentrations are generally dependent upon air temperature (Fig. 2.13), whereas they are only weakly correlated with wind velocity and amounts of precipitation. There is no decrease in concentration with increases in wind velocity, which indicates that sulfur dioxide comes to the forest from fairly distant emitters. In turn, a weak correlation of sulfur dioxide concentration in the near-ground air layer with amounts of precipitation would indicate that rainfall removes sulfur dioxide mostly from higher strata of the atmosphere.

Vertical gradients in the concentration of sulfur dioxide as measured by the lead peroxide and West-Gaeke methods permit recognition of the circulation of gaseous pollutants in the forest. Concentrations as measured by the lead peroxide method above the forest canopy exceed, on average, those measured in the forest interior by a factor of 1.5, however, the gradients measured by the West-Gaeke method do not indicate significant variation in concentration (Fig. 2.14).

With thermal inversion of the air in the forest area taken into account, one may claim that the circulation of pollutants depends first upon air dynamics. Where air currents meet a forest wall, a sliding movement occurs. Wind velocity above the forest exceeds that in the interior (Fig. 2.14, cf. also Niestierov 1960, Geiger 1961) so that the pollutants are also transported mostly above the forest. This is also the cause for the so-called wind coefficient of pollutant dispersion (Parczewski 1971) being higher above the forest than in the interior. Consequently, the absolute concentrations of sulfur dioxide per unit of air volume above and below the tops of trees, as measured by the West-Gaeke method, do not significantly differ from each other, whereas gradient concentrations obtained by the lead peroxide method reflect differences in air volume flowing above and below the tops of the trees per unit time.

Measurements of the sulfur dioxide concentration at the eastern margin of the forest for a year indicate that concentrations in the area between the Cracow agglomeration and the forest do not exceed those on the opposite side of the forest.

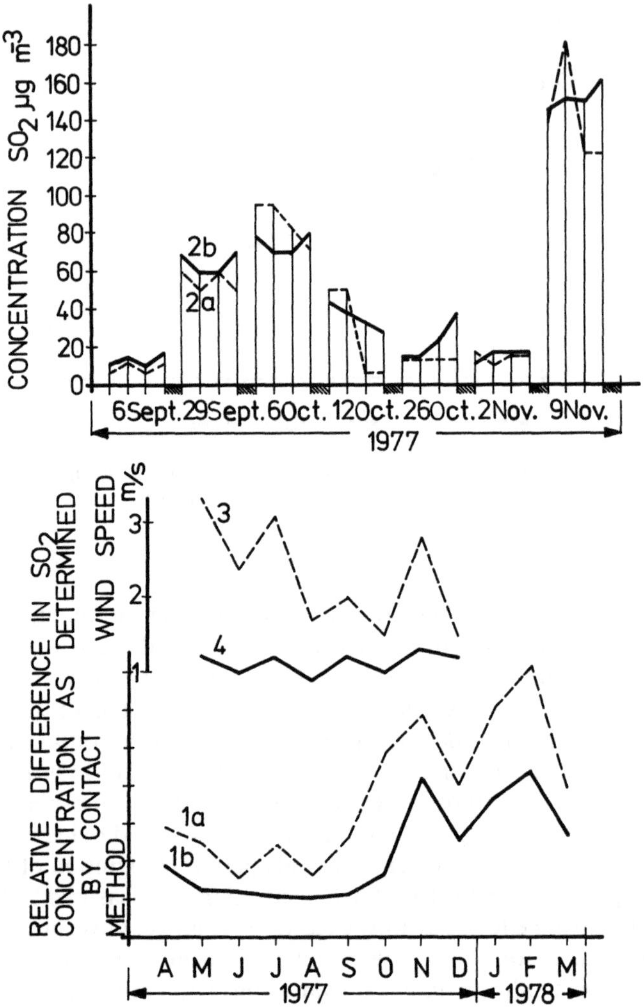

Fig. 2.14. Vertical pattern of sulfur dioxide in the forest at Stanisławice. April 1977 to March 1978. *1* mean monthly concentrations – lead peroxide method; *1a* 5 m above canopy; *1b* 7 m velow canopy; *2* concentrations – West-Gaeke method with air intake for 20-min period; *3* wind velocity above canopy; *4* wind velocity below canopy

However, as has been demonstrated above, one year of measurements of sulfur dioxide concentration is far from reliable.

A year of recording of sulfate aerosol concentration shows that mean concentrations at the western margin of the forest are approximately 45% higher than those on the eastern side. The annual variation in sulfate aerosol concentration is much smaller than that of sulfur dioxide. In general, the mean concentration of sulfate aerosol in the cool half-year is up to 40% higher than in the warm.

2.5.2 Estimation of the Dry Deposition of Sulfur

The crucial point in estimating the dry deposition of sulfur (i.e., sulfur dioxide and sulfate aerosol) in a forest area following measurement of concentration of the pollutants in the near-ground air layer, is in choosing the deposition velocity. When computing dry deposition of sulfur over a considerable area, a deposition velocity equal to 0.8 cm s^{-1} is commonly assumed for sulfur dioxide (Whelpdale and Shaw 1974, Garland et al. 1974), while the deposition velocity of sulfate aerosol is neglected because it is by an order of magnitude lower (Chamberlain 1975).

A velocity of 0.8 cm s^{-1} has been accepted in the present study for the deposition of sulfur dioxide in a forested area. It must be kept in mind that this value may differ widely from the actual value, because the deposition velocity of sulfur dioxide in forests ranges from 0.2 to 2.0 cm s^{-1} (ISSA 1978 b). One should keep in mind, however, the detailed research on sulfur migration in a North American forest ecosystem where the average deposition velocity of sulfur compounds was indirectly recognized as 0.73 cm s^{-1} (Eaton 1977).

There is no agreement among scientists as to the deposition velocity of sulfate aerosol. Wesely et al. (1977) gave a figure of 1 cm s^{-1} for aerosols ranging from 0.1 to 1 μm in size. Unsworth (1977) claimed that an increase in wind velocity above a forest may result in the deposition being greater at tree tops than on plants smaller in height. As recognized in the Cracow area, the sulfate aerosol occurs mostly in the form of calcium sulfate, which confirms the assumption that deposition velocity approximates 1 cm s^{-1}. An average deposition velocity of sulfate aerosol equal to that for sulfur dioxide (0.8 cm s^{-1}) has therefore been accepted for the computation of dry deposition of sulfur here.

The mean annual concentration of sulfur dioxide computed from multiannual measurements has been considered representative for the area. It is 20 μg m^{-3}. The annual concentration of sulfate aerosol has been computed as the mean of those measured at the eastern and western margins of the forest. It is 15 μg m^{-3}. The dry deposition of sulfur is thus 3.78 g S m^{-2} yr^{-1}, which approximates the lower limits recorded in highly industrial areas in Europe (ISSA 1978). Winds from the west account for 70% of this dry deposition of sulfur in the Niepołomice Forest area (Fig. 2.15).

2.5.3 Wet Deposition of Sulfur

On the basis of measurements over a 3-year period, the mean annual concentration in precipitation and the wet deposition of sulfur are 12.60 μg SO_4 cm^{-3} and 2.60 g S m^{-2} yr^{-1}, respectively, recorded at the sampling station at Hysne. The wet deposition is thus much lower than the dry.

The contribution of natural sulfates to precipitation rarely exceeds 10% in inland situations (Dovland et al. 1976). Thus, the wet deposition of sulfur at Hysne approximates the mean recorded in highly industrial areas in Europe (ISSA 1978). There is moderate variation in the wet deposition of sulfur in the Cracow agglomeration from year to year. The concentration of sulfate ions in precipitation is negatively correlated with precipitation (Fig. 2.16).

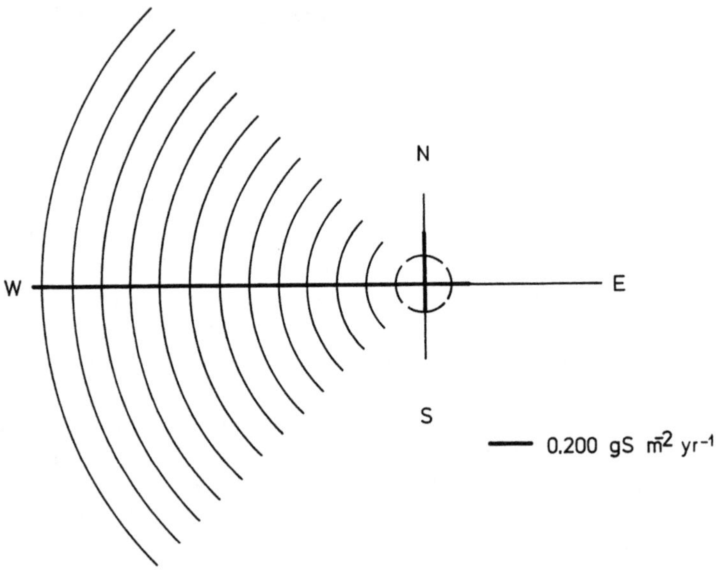

Fig. 2.15. Directional pattern of dry deposition of sulfur. Hysne, from January 1975 to December 1977

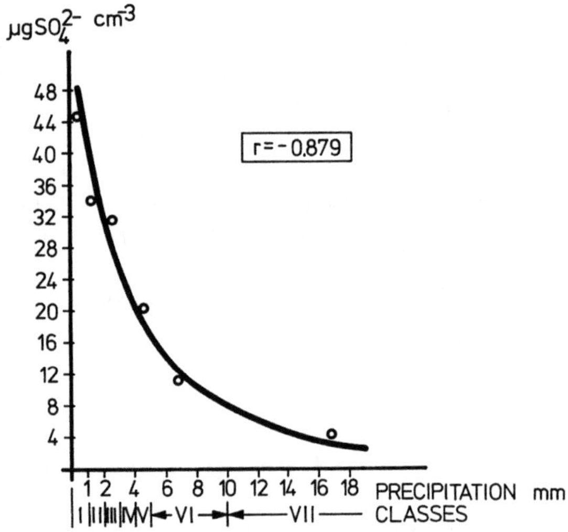

Fig. 2.16. Average concentration of sulfate ions in specific intervals of precipitation at Hysne. December 1974 to September 1978

The mean annual concentrations of sulfate ions in precipitation and the wet deposition of sulfur, measured April 1977 through March 1978, are higher by a factor of 1.3 and 1.4, respectively, than those in the center of the Niepołomice Forest (Fig. 2.17). Thus, wet deposition decreases with increasing distance from pollutant emitters. The negative correlation of sulfate with precipitation provides

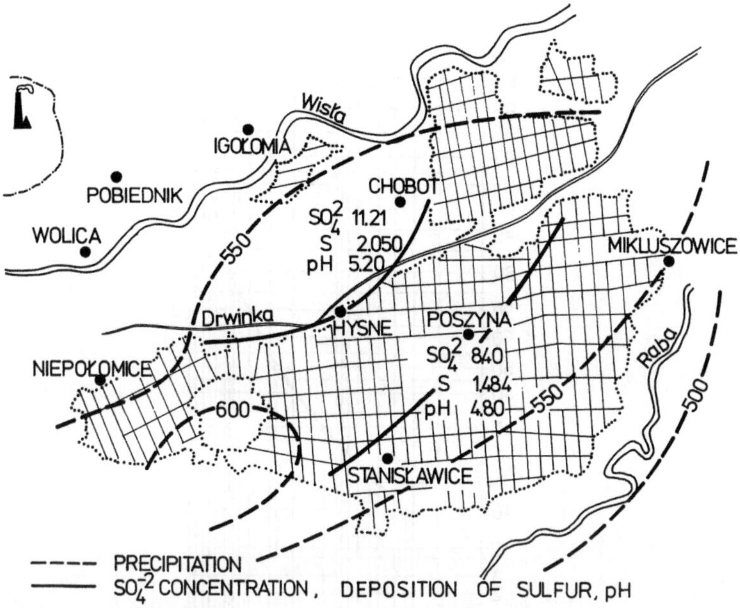

Fig. 2.17. Isolines of mean annual concentration of sulfate ions in precipitation (in μg cm^{-3}) and wet deposition of sulfur (in g S m^{-2}). Precipitation in mm from April 1977 to March 1978

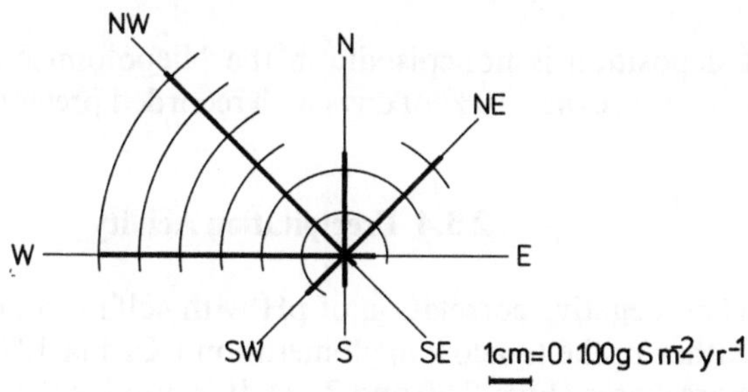

Fig. 2.18. Directional pattern of wet deposition of sulfur. Hysne, January 1975 to December 1977

an explanation for the concentration of sulfates being higher in the cool half-year than in the warm. Nevertheless, wet deposition is more or less constant throughout the year because the amounts of precipitation are greater in the warm half-year.

As indicated by the directional pattern in the distribution of the wet deposition, winds from the west are responsible for most of the input of sulfur compounds to the area (Fig. 2.18). Some 70% of the wet deposition originates from air masses from the western quadrant. As judged by Smith's criterion (OECD

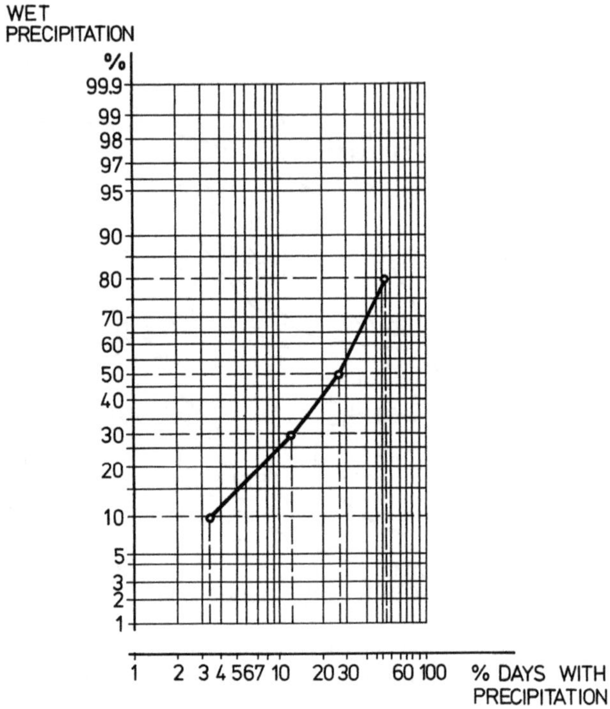

Fig. 2.19. Cumulative curve of wet deposition of sulfate at Hysne, January to December 1976

1977), wet deposition is nonepisodic in the Niepołomice Forest area, because 30% occurs on more than 10% of days with recorded precipitation (Fig. 2.19).

2.5.4 Precipitation Acidity

There is no negative correlation of pH with sulfate concentration in the precipitation either in the Cracow agglomeration (Kasina 1977a, b), or in the Niepołomice Forest area (Figs. 2.20 and 2.21). It is notable that higher pH values are associated with higher concentrations of sulfate ions in the proximity of pollutant emitters. As seen in Fig. 2.21, the contribution of precipitation with pH ≦ 5.6 in the center of the forest, is about 70% but only 30% at Hysne, and yet sulfate concentrations are higher at Hysne.

Mineralogic analysis of the dust fall sampled with precipitation demonstrated the presence of a high proportion of hydrated calcium sulfate $CaSO_4 \cdot 2H_2O$, which indicates that an efficient neutralization of sulfuric acid in the atmosphere occurs. Further research on the acidity of precipitation (Kasina 1978b) indicated that there are greater amounts of strong acids in precipitation with pH 4.0 to 5.6 than is indicates by the concentration of hydrogen ions computed from pH values (Table 2.4). The potential acidity of precipitation in the Niepołomice Forest area is very high (Table 2.4).

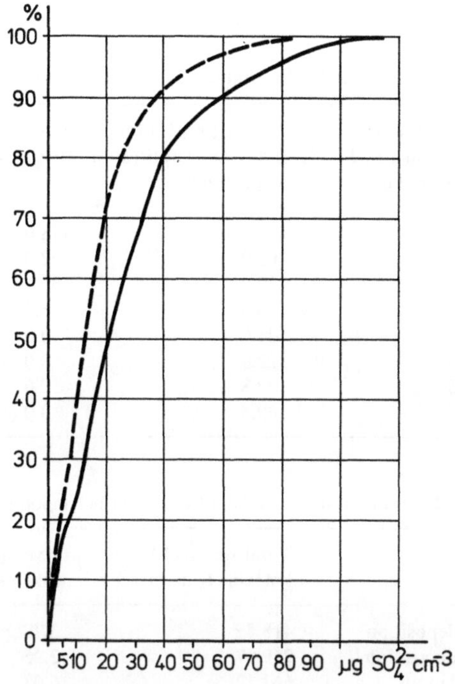

Fig. 2.20. Cumulative curves of sulfate concentration in precipitation, April 1977 to March 1978. *Full line* Hysne; *dashed line* Poszyna

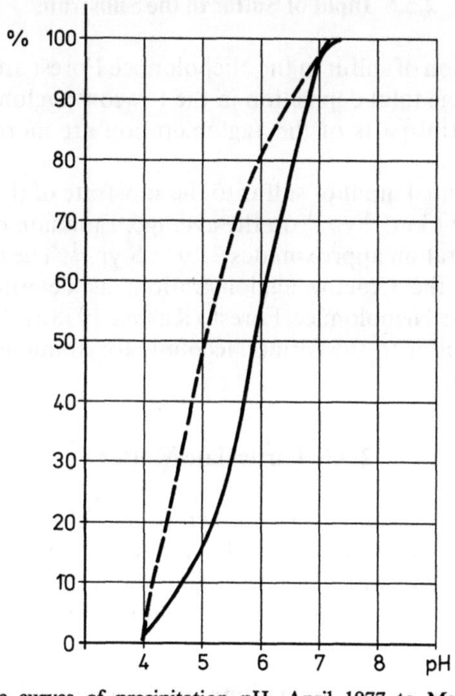

Fig. 2.21. Cumulative curves of precipitation pH, April 1977 to March 1978. *Full line* Hysne; *dashed line* Poszyna

Table 2.4. Precipitation acidity by titration

Sampling date	pH	Concentration of free H^+ ions (in μEq dm^{-3}) as measured by pH	Acidity (in μEq dm^{-3}) as measured by titration to pH equal 9.0	Strong acids (in μEq dm^{-3})	Sulfates (in μEq dm^{-3})	Conductivity (in μS cm^{-1})
Dec. 1, 1977	3.95	112.0	329.0	118.0	305.0	97.0
Nov. 2, 1977	4.50	31.6	202.0	52.0	302.0	97.0
Jan. 13, 1978	4.75	17.8	229.0	33.9	734.6	165.0
Sep. 18, 1977	5.40	3.98	126.0	8.4	50.0	39.0
Nov. 9, 1977	5.85	1.41	143.0	0.0	250.0	58.5
Nov. 18, 1977	6.40	0.40	122.0	0.0	343.7	98.5
Nov. 14, 1977	6.80	0.15	72.8	0.0	136.2	30.0
Nov. 3, 1977	7.25	0.05	49.3	0.0	322.0	94.0

Table 2.5. Input of sulfur to the substrate of the Niepołomice Forest

	Total area = 110.7 km^2 (in t S yr^{-1})	Average (in t S km^{-2} yr^{-1})
Dry deposition	418.45	3.78
Wet deposition	249.07	2.25
Total input	667.52	6.03

2.5.5 Input of Sulfur to the Substrate

The total deposition of sulfur in the Niepołomice Forest area is less by a factor of 1.8 than the average total deposition in the Cracow agglomeration. Note that the western and central parts of the agglomeration are more polluted than the eastern.

The estimated annual input of sulfur to the substrate of the Forest is 667.52 t, which is more than 6 t km^{-2} yr^{-1} on the average. Emission of sulfur dioxide by the Cracow agglomeration approximates 9.10^4 t S yr^{-1}. The total input of sulfur to the substrate of the Cracow agglomeration, as determined by the same methods used for the Niepołomice Forest (Kasina 1978a), is 3.58 10^4 t S yr^{-1}. Thus, the input of sulfur to the former accounts for about 40% of that emitted in the area (Table 2.5).

2.5.6 Particulate Matter

Samples of particulate matter were taken over a 24-h period with Whatman 40 filters. The sampling network included five stations (Fig. 2.10). The concentration of particulate matter was measured using the reflectometric method (Katz 1967). The concentrations of the following components of the samples were measured:

1. Iron, zinc, lead, manganese, potassium, and calcium, all with the atomic absorption spectrophotometer Pye Unicam;

Table 2.6. Mean annual and seasonal concentrations of particulate matter and selected components. Measurements April to September 1977 (warm half-year), and October 1977 to March 1978 (cool half-year)

Sampling station	Time interval	μg m^{-3}		ng m^{-3}						
		Dust	SO_4^{2-}	Fe	Zn	Pb	Mn	K	Ca	NH_4^{+}[1]
Pobiednik	Warm half-year	48.72	24.19	392	305	76	25	978	1366	4345
	Cool half-year	72.53	22.42	421	117	163	46	584	3344	7540
	Year	60.62	23.31	406	211	124	35	781	2355	5942
Hysne		35.57	19.02	418	263	110	38	867	2640	3426
		74.88	23.05	765	105	232	45	701	3474	2793
		55.25	21.04	591	184	171	41	784	3057	3109
Niepo-łomice		45.80	15.44	866	461	216	42	943	3107	4911
		81.16	25.59	636	160	277	40	614	4610	4245
		63.48	20.52	744	310	246	41	778	3858	4578
Chobot		38.45	12.35	585	214	87	82	800	1727	2033
		63.75	19.35	751	107	159	56	715	3583	6623
		51.10	15.85	668	160	123	69	757	2655	4328
Miklu-szowice		18.23	9.36	458	106	95	31	560	2406	642
		64.40	16.08	818	132	211	49	1260	5830	2176
		41.31	12.72	638	119	153	40	910	4118	1409

2. Ammonium ions, using the indophenolic method with T-chloramine and quaiacole (Hryniewicz 1973);
3. Sulfate ions (Yamamura and Fritz 1965).

The mean annual concentrations of particulate matter in the Cracow agglomeration and Niepołomice Forest areas remain more or less constant (Table 2.6 and Fig. 2.11), this is also the case with the concentration of sulfur dioxide. The maximum deviation of mean annual or seasonal concentrations of particulate matter for specific stations from means for the whole area investigated is less than 20%.

Their concentration decreases on the eastern side of the Forest. The difference between the eastern and western margins is 35% of the mean annual concentration, 15% of that in the cool half-year and 60% in the warm. The maximum mean daily concentration of particulate matter admissible for areas under special protection in Poland (75 μg m^{-3}) is exceeded in 40 and 10% of the time span of cool and warm half-years, respectively, in the Cracow and Niepołomice area. On the eastern side of the forest, the maximum admissible concentration is exceeded only in the cool half-year, by a frequency of 30%. The concentration of particulate matter is much less variable over time intervals than the concentration of sulfur dioxide so that one year of measurements may be sufficient to permit reliable determination of their spatial distribution.

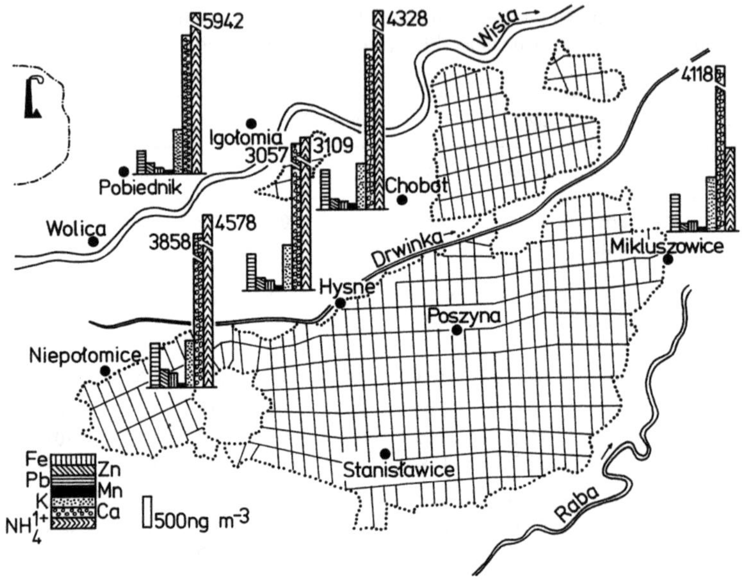

Fig. 2.22. Average concentration of metals and ammonium ions at different localities, April 1977 to March 1978

The aerial distribution of metals in the region indicates that the entire Niepołomice Forest area comes under the influence of the industrial emission of these pollutants (Fig. 2.22). Recorded concentrations of metals exceed considerably those observed in a control area located south of the Cracow agglomeration. Due to the prevalence of west–east air circulation in Poland, the control area is normally outside that influenced by industrial emissions of the Cracow agglomeration.

The mean annual concentrations of lead, manganese, and potassium in the Niepołomice Forest exceeded by a factor of 2.5 those recorded in the control area. This ratio ranges from 1.14 to almost 2 for the other metals sampled.

The concentration of alkaline ions (calcium, potassium, and ammonium) was compared in stechiometric terms to that of sulfate ions in the area. An excess of alkaline ions occurs in the cool half-year, equal to a mean of 2.42 $\mu g\ m^{-3}$ and a mean of 1.214 $\mu g\ m^{-3}$ sulfate ions in the warm half-year. This suggests that an acid aerosol is more likely to occur in the warm half-year than in the cool.

2.6 Input and Mineral Composition of Dust to the Substrate

A. Manecki and A. Skowroński

Geological processes (rock weathering, volcanic eruptions, cosmic processes, meteorite, and cosmic dust fall), and human activities both on earth and in space, contribute significantly to changes in the composition of the atmosphere through an input of gases and dusts. When analyzing dusts in the atmosphere, one must therefore distinguish between natural and artificial components. In this regard, mineralogical-geochemical methods are appropriate (Manecki 1976, 1978, Manecki and Skowroński 1970, 1976, Manecki et al. 1981).

Mineralogical investigation of dusts may provide an answer to the question of what is the mineral composition of the material and which phases contain the largest amounts of chemical elements or compounds detrimental to the natural environment.

In order to estimate the effect of a specific dust on changes in geochemistry in the natural environment, one should focus on those phases with considerable water solubility because they participate in the hydrologic cycle and hence are absorbed by the soil. The term aeromineralogy has been introduced to designate that branch of mineralogy which is concerned with natural and artificial dusts in the atmosphere (Manecki 1976).

Dusts in the urban industrial agglomeration of Cracow have been investigated (Bolewski et al. 1976, Wilczyńska-Michalik 1981).

Dusts were collected in the Niepołomice Forest area by free deposition in glass receptacles with a total area of 0.07 m^2 per sampling point, all situated 2 m above ground surface. Each individual sample represented a month of deposition. The location of sampling points is shown in Fig. 2.10.

Each sample was analyzed by various mineralogical and chemical techniques (Fig. 2.23). Field measurements and physical separation determined the amounts of fall and the proportions of fractions highly and slightly soluble in water. The latter were identified by optical features under a polarizing microscope in preparations fixed in Canada balsam. Grain size and shape and planimetric analyses of this fraction were also carried out under the microscope. Preparations necessary for microscopic identification of components highly soluble in water were obtained by slow crystallization of a drop of the solution on glass slides. X-ray methods, especially diffractometric DSH analysis, were used to identify the mineral composition of dust. Infrared spectroscopy also proved very useful, especially for dust components easily soluble in water. Shape of grains and grain aggregates were analyzed under scanning and transmitting electron microscopes. Chemical analyses in the microregion were carried out by X-ray fluoroscence. Of a wide variety of chemical-analytical methods, semiquantitative spectral analysis and quantitative analysis of selected elements by atomic absorption were those chosen.

2.6.1 Amounts of Dust Fall

Dust fall was studied over a period of 4 years, and continuous and systematic measurements were taken from 1977 to 1978. These conclusions are based mostly on data collected in the spring and summer of 1977 (warm half-year or vegetative season) and the fall of 1977 and winter of 1978 (cool half-year). The data were collected almost entirely at the following sampling points (Fig. 2.10); PN-1 – western margin of the Niepołomice Forest, PN-2 – northwestern margin, PN-5 – central area, and PN-3 – eastern margin of the forest. The stations are located along the W–E axis of the prevailing air circulation.

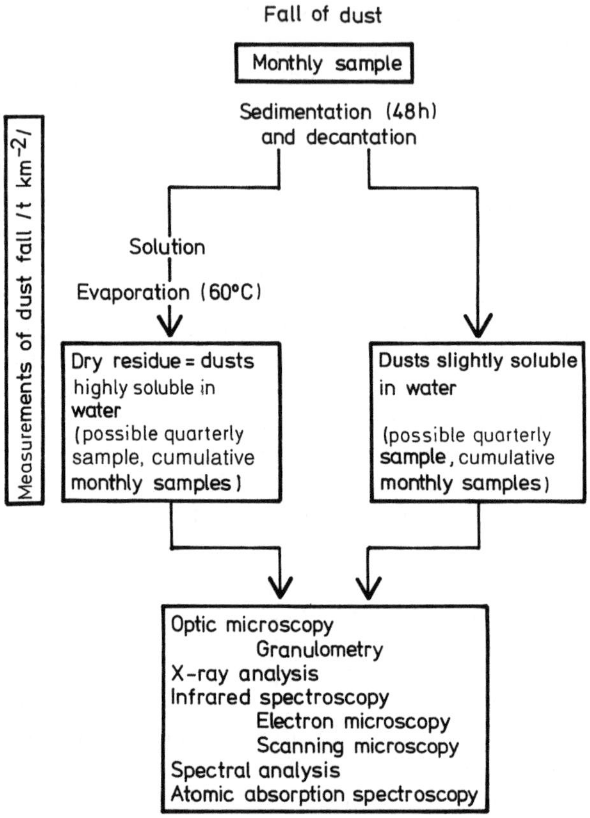

Fig. 2.23. Flow chart of mineralogical and chemical analysis of dust. (Manecki 1978, modified)

The mean monthly dust fall is as a rule greater in the warm half-year than in the cool. This may be due to resuspension of dust from the ground. The monthly fall ranges from 3.9 to 8.9 t km^{-2} $month^{-1}$ (6.4 t km^{-2} $month^{-1}$ average) in the warm half-year, while it ranges from 3.8 to 6.5 t km^{-2} $month^{-1}$ (5.2 t km^{-2} $month^{-1}$ average) in the cool. The mean annual fall approximates 59 t km^{-2}, or a total of 6480 t of dust yearly.

A considerable proportion of the dust components are quite soluble in water (Fig. 2.24). They vary from 24 to 45% by weight of total fall in the warm half-year, and from 24 to 33% in the cool. This proportion is equivalent to a total fall of 1.4–2.8 t km^{-2} $month^{-1}$, respectively. The lowest percentage of soluble components was measured on the north-western margin of the forest, or at the points nearest to the Cracow industrial area. This indicates that the soluble chemical components are separated from the solid components of dust pollutants and are transported as aerosols and deposited farther away from the emitters. Thus, considerable amounts of highly soluble compounds are introduced into the natural environment of the Niepołomice Forest. This input amounts to 1.4–2.6 t km^{-2} $month^{-1}$ in the warm half-year and to 1.5–1.8 t km^{-2} $month^{-1}$ in the cool, in the central part of the forest.

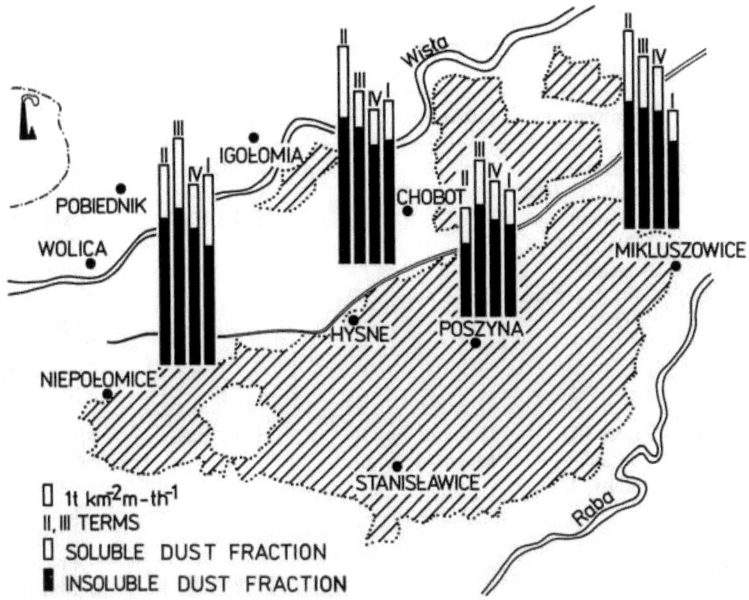

Fig. 2.24. Mean monthly dust fall in particular seasons (t km^{-2} month^{-1})

2.6.2 Highly Soluble Particulate Components

The solute from the evaporation of solutions of dust components soluble in water includes mostly calcium sulfate (gypsum) with minor amounts of hydrated magnesium sulfates (epsomite and hexahydrite), simple chlorides (halite, sylvite, and salmiac), and organic matter. Some colloidal, weakly crystallized, or amorphic silicates and aluminosilicates, as well as nitrates and carbonates, may also contribute to the soluble components of dust (Table 2.7). Thus the solutions include mostly the following ions: Ca^{2+}, Mg^{2+}, K^{+}, NH_4^{+}, Na^{+}, SO_4^{2-}, Cl^{-}, NO_3^{-}, and CO_3^{2-}. There are also heavy metals and the content of iron, manganese, nickel, lead, zinc, cadmium, and copper are given in Table 2.8.

2.6.3 Slightly Soluble Particulate Components

The deposition of dust particles slightly soluble in water varies from 2.5 to 5.8 km^{-2} month^{-1} and from 2.8 to 4.8 t km^{-2} month^{-1} in the warm and cool half-years, respectively. This comprises mostly fine dust. The size fraction below 10 μm accounts for 63–88% of the total. The maximum particle size is 120 μm. The coarse fraction is highest (23–25%) at the northwestern margin of the forest, that is at points nearest the Cracow agglomeration.

Natural and industrial dust components were distinguished in the samples. The industrial account for 52–71% of particles in the warm half-year and for 72–88% in the cool season. There is an increase in the proportion of industrial components from the warm to the cool half-year, due to the inhibition of natural dust

Table 2.7. Mineral composition of dust falling on the Niepołomice Forest area

	Components			
	Dominant	Subordinate		Accessory
Components highly soluble in water (24–45% by weight)	Gypsum $CaSO_4 \cdot 2H_2O$ Mascagnite[a] $(NH_4)_2SO_4$ Syngenite[a] $K_2SO_4 \cdot CaSO_4 \cdot H_2O$		Epsomite $MgSO_4 \cdot 7H_2O$ Sylvite KCl Halite NaCl	Salmiac NH_4Cl Hexahydrite $MgSO_4 \cdot 6H_2O$ Carbonates $CaCO_3$? Nitrates KNO_3? Organic matter Weakly crystallized or amorphic silicates and aluminosilicates
Components slightly soluble in water (55–76% by weight)	Quartz SiO_2 organic matter C	Hematite Fe_2O_3 Magnetite Fe_3O_4	Silica glass	Mullite $3Al_2O_3 \cdot 2SiO_2$ Feldspars (Na, K) $AlSi_3O_8$ Clay minerals (layer silicates and aluminosilicates of K, Mg, Al, and Fe) Micaceous minerals Gypsum $CaSO_4 \cdot 2H_2O$ Carbonates Heavy minerals

[a] Sporadic components (temporarily present at some sampling points)

Table 2.8. Input of selected metals to the Niepołomice Forest area

	Average input (in $t\,km^{-2}\,yr^{-1}$)		Total input (in $t\,yr^{-1}$ total area $= 110.7\,km^2$)	Input in form of compounds soluble in water in ($t\,km^{-2}\,yr^{-1}$)
	Range	Mean		
Fe	1.68 –4.32	3.03	334.0	0.024 –0.204
Zn	0.0624 –0.3120	0.1231	13.54	0.0216–0.2820
Mn	0.0480 –0.1200	0.0721	7.93	0.0072–0.0588
Pb	0.0084 –0.0540	0.0315	3.46	0.0096
Cu	0.0084 –0.0540	0.0187	2.06	0.0480
Ni	0.0012 –0.0396	0.0066	0.72	0.0372
Cd	0.00024–0.0156	0.0015	0.17	0.0011

resuspension from the ground in the latter period. The lowest content of the industrial fraction occurs in the central area of the forest, while the greatest proportion is at the northwestern margin (Fig. 2.25).

Transparent glassy and opaque spherules, irregular-shaped opaque particles (organic matter, hematite, magnetite), fine porous coke, and various aggregates

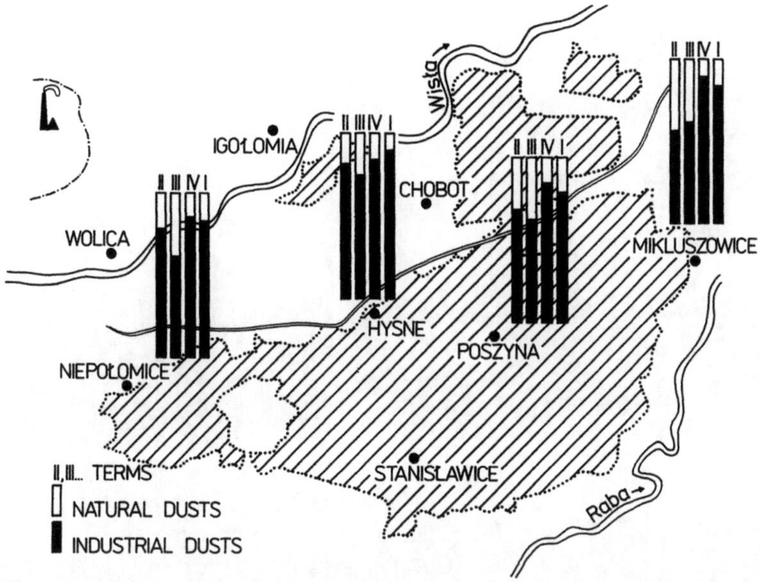

Fig. 2.25. Contribution of natural and industrial components to the dust fall in particular seasons (percent total volume)

of all these particle types cemented by a glassy material have been attributed to the industrial dust. Glassy spherules are less than 10 µm in diameter, either colorless, orange, brownish, or greenish in color, and sometimes with opaque inclusions. Opaque spherules and ovate particles are usually larger (average of 30–40 µm in diameter), often with a transparent glassy coating. Other opaque particles are in the form of irregular, generally sharp-edged plates or flakes. Porous, irregular particles of highly carbonized organic material have been recognized as fine coke.

The natural dust components are strongly represented by quartz associated with minor amounts of feldspars, micaceous minerals, heavy minerals (tourmaline, rutile, apatite, zircon), and aggregates of clay minerals (soil components).

The finest fraction has a strong representation of particles typical of industrial dust, mainly spherules and ovate particles with aggregates of spherules and small-sized plates. There are mostly glassy spherules, but opaque have also been recorded, the latter being visible even to a magnification of 10,000 under the electron microscope. The glassy substance of spherules includes mostly Si and Al, with small amounts of Ca and K, with Ti as a trace component, and a widely variable proportion of Fe as an addition. The metallic spherules consist entirely of Fe, rarely with an addition of Al, and show an uneven porous surface, making them different from the glassy types. Aggregates of spherules and plates include a sulfur-bearing glassy cementing substance. Natural dust includes mostly quartz and feldspar particles that are commonly coated with spherules of industrial origin. Microphotographs of the dust are included in Figs. 2.26–2.28.

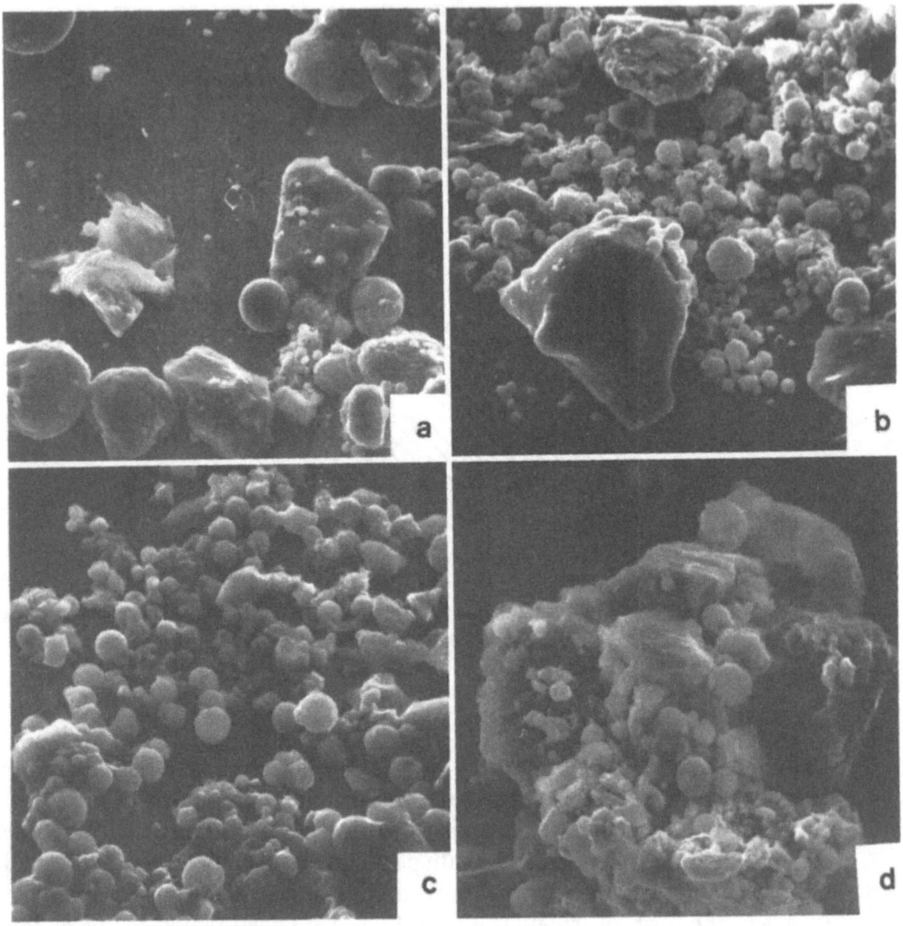

Fig. 2.26. a–d. Microphotographs of dusts. **a** Cluster of spherules and ovate particles. Plates and flakes probably consist of hematite. SEM micrograph, ×550; **b** Variations in particle-size of dust. Greatest variation shown by glassy spherules. Plates are usually larger-sized. SEM micrograph, ×550; **c** Spherules of variable sizes. SEM micrograph, ×1,000; **d** Typical aggregate of magnetic plates and silica glass, with quartz and micaceous minerals. SEM micrograph, ×1,000

2.6.4 Chemical Composition of the Dust Fall

Spectral analysis indicated the following proportions of chemical elements in the slightly soluble dust portion:

>10%: Si;
1–10%: Fe and Al;
0.1–1%: Mg, Na, K, and Ca;
0.01–0.1%: Ti, Zn, Cu, Mn, Cr, V, and Sr;
0.001–0.1%: Mo, Zr, B, Be, Sc, Ni, Cd, Sn, Ag, Ga, Ge, Ba, and Co.

The elements in the highly soluble dust components include:

>10%: Ca;
0.1–10%: Mg, K, Na, Si, Al, Fe, Zn, and Mn;

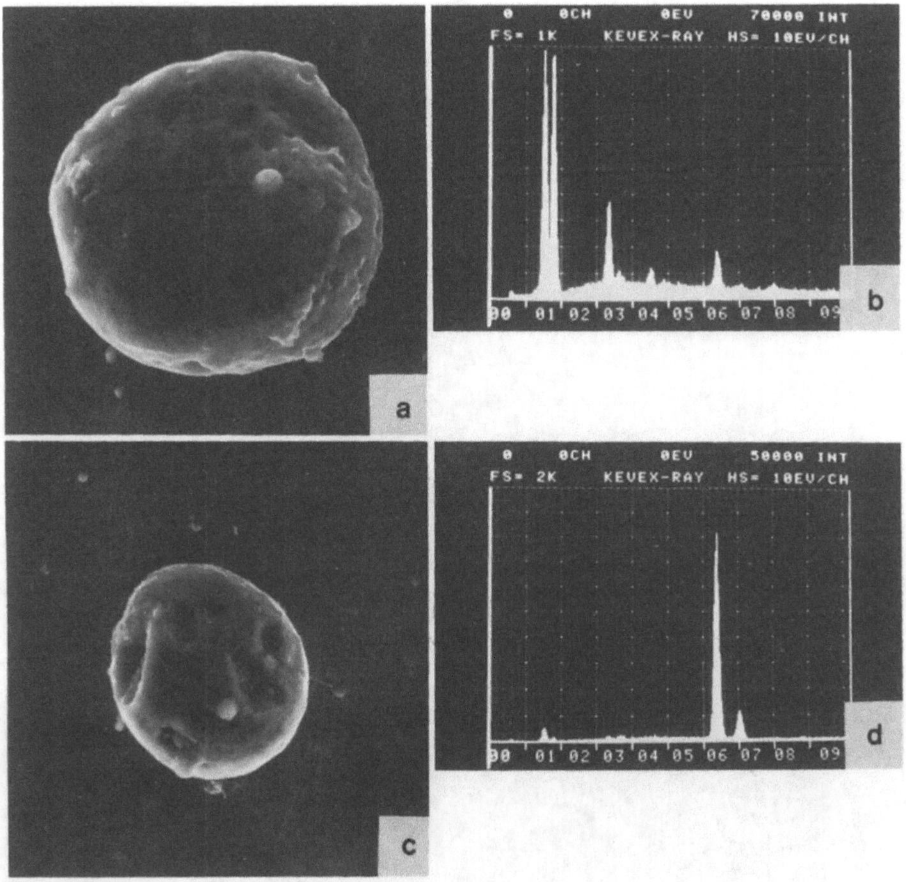

Fig. 2.27. a–d. Dust particle structure. **a** Glassy spherule with uneven surface. Note small-sized, irregular depressions with smooth margins. SEM micrograph, × 1,000; **b** X-ray spectrum of microarea of a glassy spherule shown in 2.27.a. Main peaks represent Al and Si; others represent Ca, K, Na, Fe, and Ti; **c** Almost pure iron spherule with very rough surface. Note elevations and depressions, the latter being larger and deeper than those on the surface of glassy spherules. SEM micrograph, × 1,000; **d** X-ray spectrum of a microarea of the iron spherule shown in 2.27.b. The two main peaks represent Fe, the smallest, Al.

<0.1%: Pb, Cu, Cd, Ni, Ag, Sn, Ti, B, Ba, Zr, and V;
sporadic: Be, Ga, Mo, and TL.

Quantitative analysis was carried out by the ASA method for selected heavy metals, including iron, manganese, zinc, lead, copper, cadmium, and nickel. For comparative purposes, proportions in dust have been transformed into deposition per unit area (Table 2.8).

The main heavy metal in the dust samples examined is iron. Deposition ranges from 1.7 to 4.3 t km^{-2} yr^{-1} and tends to be more or less uniform over the entire Niepołomice Forest area.

Zinc and manganese rank second and third, with annual depositions ranging from tens to hundreds of kilograms per square kilometer. Their deposition varies

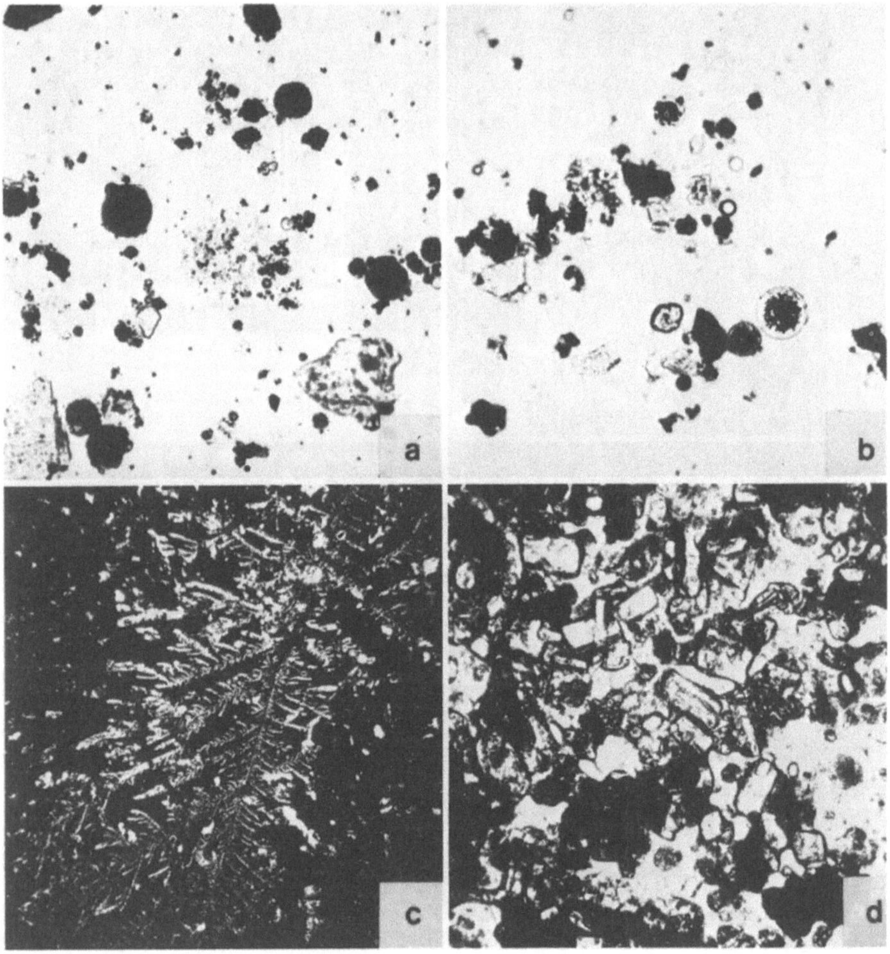

Fig. 2.28.a–d. a Industrial and natural dust. Industrial dust is represented by opaque spherules composed of iron oxides, and fine, transparent glassy spherules. Natural dust is composed of large sharp-edged quartz particles and very small carbonate particles. Polarizing microscope with parallel nicols, ×230; **b** Atmospheric dust. Note large-sized glassy spherule with fine inclusions of opaque plates, and small-sized spherules without inclusions. Polarizing microscope with parallel nicols, ×300; **c** Solute from rainwater (dust components very soluble in water). Note aggregates of gypsum crystals. Polarizing microscope with parallel nicols, ×80; **d** Rainwater solute (dust components very soluble). Polarizing microscope with parallel nicols, ×80

irregularly with time, resembling that of iron. It indicates a common emitter for iron, zinc, and manganese.

The annual depositions of lead and copper are comparable and range from 8 to 54 kg km^{-2}. Lead attains a minimum in summer but tends to increase over the time interval examined. Copper reaches minimum levels in autumn, whereas variations in time is irregular. Lead deposition is more or less uniform over the forest area. When a lack of correlation between lead and zinc deposition is also taken into account, the uniformity of spatial distribution of lead is indicative that emission occurs from a larger emitter and covers a considerable area (the urban area of Cracow or the Upper Silesian complex).

The annual nickel deposition increased from 1 to 40 kg km^{-2} over the period investigated. None of the other metals displays this trend, which may indicate the appearance of a new emitter. The lowest increase in nickel was recorded in the central part of the forest.

The level of cadmium is the least among the heavy metals studied, ranging from 0.2 to 16 kg km^{-2}, with a distinct maximum in the warm season.

The deposition of heavy metals is lowest in the central part of the forest, as a rule. This is especially true for copper, cadmium, and nickel.

It is important from the ecological viewpoint to analyze the deposition of chemical elements in the form of compounds very soluble in water, that is, those able to enter the geochemical cycle instantly. The highest is shown by iron, zinc, and manganese, with levels of 24 to 204 kg km^{-2} yr^{-1}, 22 to 282 kg km^{-2} yr^{-1}, and 7 to 59 kg km^{-1} yr^{-1}, respectively. Iron and manganese occur in significant amounts in soil so that their input may not result in significant changes in the natural environment. The input of lead, copper, and nickel in the form of soluble compounds amounts to some tens of kilograms per square kilometer per annum. These elements occur minimally in soil, so that their input may disturb the biological equilibrium. A similar effect may result from the deposition of cadmium. The input of this highly toxic element is approximately 1 kg km^{-2} yr^{-1}, but it is concentrated in the growing season.

2.6.5 Geochemical Characteristics of the Substrate

Heavy metals falling in dust in the Niepołomice Forest area migrate and are accumulated in the soil. Their soluble compounds are absorbed by forest litter (Chap. 5), as well as by organic matter and clay minerals in the soil. Of the clay minerals, the greatest sorption capacity is shown by the montmorillonite group. Amounts of the latter and organic matter in a soil determine therefore its sorption capacity.

As judged from analysis of soil minerals, the greatest sorption capacity of soil is to be expected in the deciduous forest complex because the proportion of the pelitic fraction amounts to some 20% of the soil with montmorillonite being the predominant clay mineral. In turn, the soil of the pine forest complex is dominated by sandy and aleuritic fractions. Clay minerals occur in minor amounts, with illite and kaolinite as the dominant phases and montmorillonite present only in the form of intergrowth with illite.

The accumulation of heavy metals can be estimated from the analysis of soil profiles in the Niepołomice Forest supplemented with hydrological observations (Chap. 6). One may conclude that iron and manganese in the form of soluble compounds are easily leached from the soil, whereas up to 80% of the zinc, lead, and cadmium and up to 98% of the copper and nickel may be accumulated.

The greatest sorption capacity is shown by the uppermost, organic layer of the soil which provides plants with nutrients. One may therefore expect that the most deleterious effects on the living part of the ecosystem will be caused by zinc, lead, cadmium, nickel, and copper. Even though these elements may not be directly detrimental to plants, they may seriously affect consumers because of their high rate of accumulation.

2.7 Conclusion

The Niepołomice Forest is influenced by air pollutants emitted by various industrial works. This area receives particulate and gaseous pollutants released in distant regions (Upper Silesian Industrial District) and especially in the urban industrial agglomeration of Cracow 15–30 km distant. This has been determined by both ground-based and remote-sensing methods of investigating the spread of air pollutants.

The deposition of sulfur in the area amounts to 667.52 t S yr^{-1}, or an average of 6.03 g S m^{-2} yr^{-1},with dry deposition equal to 3.78 g S m^{-2} yr^{-1}. The air circulation from the west accounts for 70% of the total deposition of sulfur.

There is no correlation between a decrease in pH and an increase in the sulfate concentration in precipitation. Furthermore, the precipitation shows a decreasing pH with an increase in distance from the pollutant emitters. Thus the sulfuric acid is more efficiently neutralized in the atmosphere in proximity to the emitters than farther away.

The annual dust fall in the Niepołomice Forest area approximates 6480 t, an average of 59 t km^{-2}. The proportion of dust components highly soluble in water ranges from 24 to 45% by weight. Solute remaining after the evaporation of rain water contains mostly hydrated calcium sulfate $CaSO_4 \cdot 2H_2O$, with minor amounts of $MgSO_4 \cdot 6\text{–}7H_2O$, NaCl, KCl, and NH_4Cl. There is also a sporadic amount of $(NH_4)_2SO_4$, as well as traces of organic matter, nitrates, and carbonates. The maximum monthly deposition of metals in the form of very soluble compounds is 0.017 t of iron, 0.024 of zinc, 0.0008 of lead, 0.005 of manganese, 0.004 of copper, 0.00009 of cadmium, and 0.003 t of nickel km^{-2}.

The fall of dust components slightly soluble in water ranges from 2.5 to 5.8 t km^{-2} $month^{-1}$, with the industrial dust proportion varying from 52 to 88% of particles. The industrial dust includes iron oxides (hematite, magnetite), glassy spherules, mullite, and carbonized matter. The natural dust is mainly quartz particles, with lesser amounts of micaceous minerals, feldspars, clay minerals, and heavy minerals.

The average monthly fall of dust is greater in the warm half-year than in the cool, as a rule 3.9 to 8.9 t km^{-2} $month^{-1}$ and 3.8 to 6.5 t km^{-2} $month^{-1}$, respectively. The maximum monthly fall of metals includes 0.36 t of iron, 0.026 of zinc, 0.005 of lead, 0.01 of manganese, 0.005 of copper, 0.0013 of cadmium, and 0.003 t of nickel km^{-2} of the forest.

The pelitic fraction amounts to 20% in soils of the deciduous forest complex, montmorillonite being the predominant clay mineral. This soil would therefore display the greatest sorption capacity for heavy metals. In turn, the soils of the pine forest complex are dominated by sandy and aleuritic fractions, and thus sorption capacity for toxic components should be inferior.

Chapter 3

Primary Production in the Niepołomice Forest Ecosystems

Chapter Editor: R. RIEGER

Contributors:

J. BANASIK, M. CZARNOWSKI, Z. DENISIUK, S. GRABCZYŃSKI, K. GRODZIŃSKA, W. MACZEK, S. ORZEŁ, J. RAIMER, R. RIEGER, and B. ZEMANEK

3.1 Introduction

R. RIEGER

The productivity of forest communities can be considered in economic as well as biological terms. The first relates to the volume of wood to be obtained, the second to the total biomass, or the mass of timber, irrespective of dimensions and usage, with leaves, needles, shrubs, and herbs. In the former case, volume units apply, whereas in the latter, weight and energetic value are measured.

This study was undertaken to estimate the biomass and productivity of the Niepołomice Forest ecosystems. This has been achieved by dendrometric and ecophysiologic methods. The basic dendrometric parameters were the volume of compact wood and increment, as estimated for the whole forest. These estimates were supplemented by detailed measurements of selected sample areas. Ecophysiologic studies consisted in measurements of the intensity of photosynthesis and respiration of selected tree species. Biomass of the herb layer and nonforest communities was estimated for selected communities and extrapolated for the entire forest. Additional data on the chemical composition of vegetation permit estimates of the balance of nutrients and pollutants as related to primary production.

3.2 Growing Stock and Increment of Tree Stands

R. RIEGER, S. GRABCZYŃSKI, S. ORZEŁ, and J. RAIMER

3.2.1 The Characteristics of Tree Stands of the Niepołomice Forest

The general characteristics of tree stands are provided from data presented in forest management plans (*Plan urządzenia gospodarstwa leśnego na okres 1968–1978*. Nadleśnictwo Niepołomice, msc. *Plan urządzenia gospodarstwa leśnego na okres 1969–1979*. Nadleśnictwo Damienice, msc.). To this end, all the available data on the relative abundance of specific tree species, as well as on age, quality, and volume of wood were used. The figures presented give the averages for the whole forest.

The forest covers ca. 110 km^2, of which 97 are productive tree stands. The main species is Scotch pine, and pine forest accounts for 70.2% of the wooded area and 71.3% of the tree volume. Oak stands cover 16.6% and represent 19.7% of the volume, whereas black alder stands account for 9.2% and 7.7% respectively. The remaining 3.8% of the wooded area and 1.3% of the tree volume are represented by young stands dominated by common birch, ash, hornbeam, poplar, linden, aspen, Norway spruce, and larch.

When the actual share of particular species is taken into account, these figures shift slightly in favor of deciduous species.

In general, three main categories of tree stands can be distinguished. These include:

Scotch pine stands with Norway spruce and larch stands, 71% of the wooded area (68.53 km^2) and 71% of the volume (1,253 × 10^3 m^3 in total volume over 7 cm);

Oak stands with hornbeam and ash stands, 18% of the area (17.72 km^2) and 20% of volume (353 × 10^3 m^3);

Black alder stands with the remaining deciduous species, 11% of the area (10.64 km^2) and 9% of volume (149 × 10^3 m^3).

The forest is dominated by stands of age classes II (21–40 years) and III (41–60 years), which jointly account for 44% of the wooded area. Tree stands of age classes I (1–20), IV (61–80), and V (81–100 years) occupy 45% of the area. The remaining 11% supports tree stands over 100 years old (classes VI and VII).

The age structure is more variable within particular species groups. Scotch pine stands are dominated by age class II. Black alder stands fall usually into classes I and III. Oak stands are more or less evenly distributed among age classes, except VI. The average age of black alder forest is 45 years, for Scotch pine 59 years and 72 for oak. The average age of tree stands of the entire Niepołomice Forest is 60 years.

In respect to wood volume, the term boles refers to trunks and tree branches larger than 7 cm diameter, the term branches to tree branches falling below this size. In the understory layer boles again refer to all stems larger than 7 cm in diameter and branches to those less than this size. Twigs refers to all smaller sapling tree members in the understorey.

Total volume of boles (7 cm + in diameter) of all tree stands amounts to 1,757 × 10^3 m^3, most represented by Scotch pine with Norway spruce and larch stands (Table 3.1). The total dry weight of this wood is estimated at 916,000 t (Table 3.1). Densities of dried wood are taken from Krzysik (1974) and are as follows: 0.49 t m^{-3} for Scotch pine, 0.65 t m^{-3} for oak, and 0.49 t m^{-3} for black alder.

Mean stand volume of all forests equals 181 m^3 (or 94.1 t) of bole wood ha^{-1} (Table 3.1).

Current annual increment has been calculated by the Zabielski method (1976). On the average, it is 6.0 m^3 (or 3.13 t) ha^{-1} yr^{-1}. This value is fairly constant for the basic groups of forests, but varies between particular age classes.

Current annual increment approximates 60 × 10^3 m^3, of which 43 × 10^3 m^3 is represented by Scotch pine, 8.7 × 10^3 m^3 by oak, and 6.9 × 10^3 m^3 by black alder.

3.2.2 Dendrometric Characteristics of Tree Stands in Selected Areas

Dendrometric characteristics have been estimated from detailed measurements of selected Scotch pine and oak forests typical of coniferous mixed forest

Table 3.1. Average stand volume and annual increment of boles in tree stands of the Niepołomice Forest

Species	Mean age (years)	Forested area		Stand volume					Annual increment			
		ha	%	$m^3 \times 10^3$	%	t (total area)	$m^3\ ha^{-1}$	$t\ ha^{-1}$	$m^3\ yr^{-1}$	$t\ yr^{-1}$	$m^3\ ha^{-1}\ yr^{-1}$	$t\ ha^{-1}\ yr^{-1}$
Pine (+spruce and larch)	59	6,853	71.0	1,252	71.0	613,536	182.7	89.5	43,069	21,102	6,3	3,1
Oak (+hornbeam and ash)	72	1,772	18.0	353	20.0	229,333	199.1	129.4	8,742	5,682	4.9	3.2
Alder (+other deciduous)	45	1,064	11.0	149	9.0	72,857	139.7	68.5	6,873	3,368	6.5	3.2
Total forested area (average)	(60)	9,689	100.0	1,754	100.0	915,726	181.0	94.1	58,684	30,152	6.0	3.1

and humid deciduous forest (humid hornbeam-oak forest) which predominate in the Niepołomice Forest. This accounts for deviations from averages for the forest as a unit.

Nine areas were investigated. Of these, six were in Scotch pine stands on coniferous mixed forest sites (three in stands of class age II; three in class V) and three were in oak stands of age class V on humid deciduous forest sites.

Total volume of boles in sample areas, both for the trees and the reproduction layer, have been computed from tree volume tables of Czuraj et al. (1966). Total volume of branches less than 7 cm in diameter have been computed from indices estimating the proportion in each 100 m^3 boles. In addition, the total volume of branches in the reproductive layer with diameter at breast height below 7 cm, which was beyond the scope of the former survey, was estimated on each sample area. This was achieved by computing the wood volume from weight (Krzysik 1974).

The volumes of specific parts of tree stands were multiplied by the density of dry wood (averages for particular species, Krzysik 1974) to obtain the dry wood weight of the tree layer and the understory (t ha^{-1}).

Finally, increment of boles of the tree layer has been computed (Grochowski 1973) and the other parameters (increment of branches of the tree, and under story) have been obtained by multiplying stand volume by increment percent of wood of boles as provided by Pressler's equation (Grochowski 1973). In the same way, increment of biomass, expressed as dry matter ha^{-1} yr^{-1}, has been computed with the density of dried wood taken into account.

The sample areas were chosen to permit analysis of forests of equal age but with different undergrowths. In pine stands of age class V, volume of boles is 85% of total volume, respectively (Table 3.2). Thus, total volume of standing trees ditionally, the dead volume of branches in these forests accounts for 7.5 and 6.7% of total volume, respectively (Table 3.2). Thus, total volume of the standing trees averages 378.5 m^3 (195 t) ha^{-1} in pine and 512.8 m^3 (342 t) ha^{-1} in oak stands, or it accounts for 92% and approximately 96% of total volume, respectively (Table 3.2).

Volume of boles in pine stands of age class II are almost half those of age class V; approximating 83% of total volume (Table 3.2). Volume of branches in the standing crop approximates 15% of total volume, or twice that in older forests. Thus, total volume of the tree layer averages 229 m^3 (112 t) ha^{-1}, which is 98.5% of total volume.

In the sample areas, the understory is dominated by oak, hornbeam, common birch, Norway spruce, linden, European beech; and the shrub layer by rowan, bird cherry, and alder buckthorn.

The volume of this layer is much greater in older pine stands (31.2 m^3 ha^{-1}) than in oak (22.3 m^3 ha^{-1}), accounting for 7.6 and 4.1% of total volume, respectively (Table 3.2). In pine stands of age class II, however, the volume of the reproduction layer is one tenth that of older forests amounting to 3.4 m^3 ha^{-1} (1.5% of total volume).

Annual increment of the reproductive layer varies from 0.2 to 0.46 m^3 ha^{-1} yr^{-1}, which accounts for a minor proportion of the total increment of wood in the sample areas (Table 3.2).

Table 3.2. Mean volume and volume increment of timber on study plots

Species and age	Tree layer				Reproduction layer						Total stand	
	Boles Trunks and branches >7 cm diameter		Branches Branches <7 cm diameter		Breast-height diameter ≧7 cm				Breast-height diam. <7 cm		Volume	Increment
					Boles – trunks and branches >7 cm diameter		Branches <7 cm diameter		Twigs			
	Volume	Increment	Volume	Increment	Volume	Increment	Volume	Increment	Volume	Increment		
Volume units	$m^3\ ha^{-1}$	$m^3\ ha^{-1}\ yr^{-1}$	$m^3\ ha^{-1}$	$m^3\ ha^{-1}\ yr^{-1}$	$m^3\ ha^{-1}$	$m^3\ ha^{-1}\ yr^{-1}$	$m^3\ ha^{-1}$	$m^3\ ha^{-1}\ yr^{-1}$	$m^3\ ha^{-1}$	$m^3\ ha^{-1}\ yr^{-1}$	$m^3\ ha^{-1}$	$m^3\ ha^{-1}\ yr^{-1}$
Pine age class V	347.9	4.78	30.6	0.44	12.3	0.18	13.0	0.19	5.9	0.08	409.7	5.67
Oak age class V	477.0	7.85	35.9	0.63	5.1	0.10	5.1	0.10	12.0	0.23	535.1	8.91
Pine age class II	192.4	9.50	35.5	2.00	–	–	–	–	3.4	0.20	232.4	11.70
Mass units	$t\ ha^{-1}$	$t\ ha^{-1}\ yr^{-1}$	$t\ ha^{-1}$	$t\ ha^{-1}\ yr^{-1}$	$t\ ha^{-1}$	$t\ ha^{-1}\ yr^{-1}$	$t\ ha^{-1}$	$t\ ha^{-1}\ yr^{-1}$	$t\ ha^{-1}$	$t\ ha^{-1}\ yr^{-1}$	$t\ ha^{-1}$	$t\ ha^{-1}\ yr^{-1}$
Pine age class V	179.0	2.53	16.0	0.23	9.0	0.13	9.3	0.14	3.6	0.05	217.0	3.05
Oak age class V	318.0	5.51	24.3	0.43	3.3	6.06	3.3	0.06	10.0	0.19	358.7	6.25
Pine age class II	94.3	4.66	18.0	1.02	–	–	–	–	1.9	0.08	114.2	5.76

The biomass of leaves and needles was not investigated specifically, and the available data from the literature, mostly from studies in the Niepołomice Forest, have been utilized.

The extent of photosynthetic material in pine forests has been estimated after Lemke (1973, 1978a, b), expressed in terms of the weight of green-needled twigs (tons) or of the number of needles (millions) per unit area. The weight of green-needled twigs ranges from 28 to 18 t ha^{-1} in stands of age classes I and V respectively. The number of needles ranges from 600 to 350 million ha^{-1} respectively. Needle dry matter averages 950 mg/100 units (Maczek 1977).

Consequently, total dry needle biomass varies from 3.84 to 3.32 t ha^{-1} in pine forests of age classes II to V, respectively. The biomass of leaves approximates 4.0 t ha^{-1} in oak stands on humid oak-hornbeam sites (Bandoła-Ciołczyk 1974).

The annual production of photosynthetic materials has been considered to be represented by the dry leaf biomass in oak forests and half of the dry needle biomass in pine stands.

3.2.3 Productivity of Tree Stands

This study presents the biomass and productivity values for pine and oak forests excluding roots. The biomass has been assumed to represent the total volume of boles and branches of the tree layer and the understory plus that of leaves and/or needles (Table 3.3). The annual productivity is represented by annual increment of wood of the overstory and the understory plus the biomass of leaves or half that of needles (Table 3.4).

Total biomass varies from 118, 220, and 363 t ha^{-1} in pine stands of age classes II and V and in oak stands respectively; bole wood of the tree layer accounts for approximately 80% of these values (Fig. 3.1). The share of dried biomass of the understory layer is 1.6, 10.0, and 4.6% respectively; and that of the dry biomass of leaves (needles) is 3.2, 1,5, and 1.1% respectively.

Annual productivity in pine forests declines from 7.7 to 4.7 t ha^{-1} yr^{-1} in age classes II and V respectively. In contrast to the standing biomass, only 60.5 and

Table 3.3. Aboveground standing biomass of tree stands in study area

Species and age	Dried mass	Tree story		Understory		Leaves	Total	Index
		Boles	Branches	Boles	Branches			
1	2	3	4	5	6	7	8	8:3
Pine	t ha^{-1}	179.0	16.0	9.0	12.9	3.2	220.1	1.23
age	%	81.3	7.3	4.1	5.9	1.5	100.0	
class V	%	88.6		10.0		1.5		
Pine	t ha^{-1}	94.3	18.0	–	1.9	3.8	118.0	1.25
age	%	79.9	15.3	–	1.6	3.2	100.0	
class II	%	95.2		1.6		3.2		
Oak	t ha^{-1}	318.0	24.3	3.3	13.3	4.0	362.9	1.14
age	%	87.6	6.7	0.9	3.7	1.1	100.0	
class V	%	94.3		4.6		1.1		

Table 3.4. Aboveground productivity of tree stands in study areas

Species and age	Dry mass	Tree layer		Undergrowth		Leaves	Total	Index
		Boles	Branches	Boles	Branches			
1	2	3	4	5	6	7	8	8:3
Pine	$t\ ha^{-1}$	2.5	0.2	0.1	0.2	1.7	4.7	1.88
age	%	53.3	4.3	2.1	4.2	36.2	100.0	
class V	%	57.5		6.3		36.2		
Pine	$t\ ha^{-1}$	4.7	1.0	–	0.1	1.9	7.7	1.64
age	%	61.0	13.0	–	1.3	24.7	100.0	
class II	%	74.0		1.3		24.7		
Oak	$t\ ha^{-1}$	5.3	0.4	0.1	0.2	4.0	10.0	1.89
age	%	52.8	4.3	0.6	2.5	39.8	100.0	
class V	%	57.1		3.1		39.8		

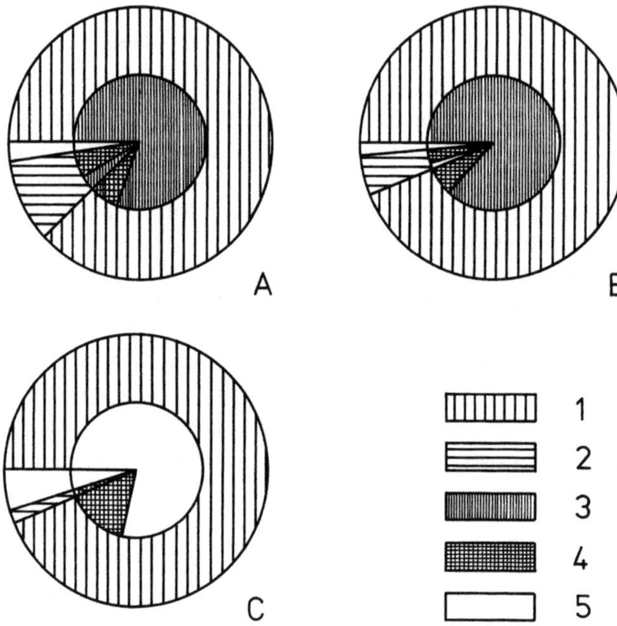

Fig. 3.1. Standing aboveground biomass of forests studied. A Pine forests, age class V. B Oak forests, age class V. C Pine forests, age class II. *1* tree layer; *2* understory; *3* boles; *4* branches, twigs; *5* leaves

53% of this increment is accounted for by wood of the tree layer, while 33.2 and 54.6% respectively is accounted for by needles (Fig. 3.2).

Annual productivity is twice as high in oak forests of age class V than in pine stands of the same age. It approximates 10 $t\ ha^{-1}yr^{-1}$, of which 52.8% is accounted for by wood of the tree storey and almost 40% by leaves.

The total biomass of boles, which accounts for more than 80% of the entire forest, was measured directly during the inventory, as well as dendrometrically

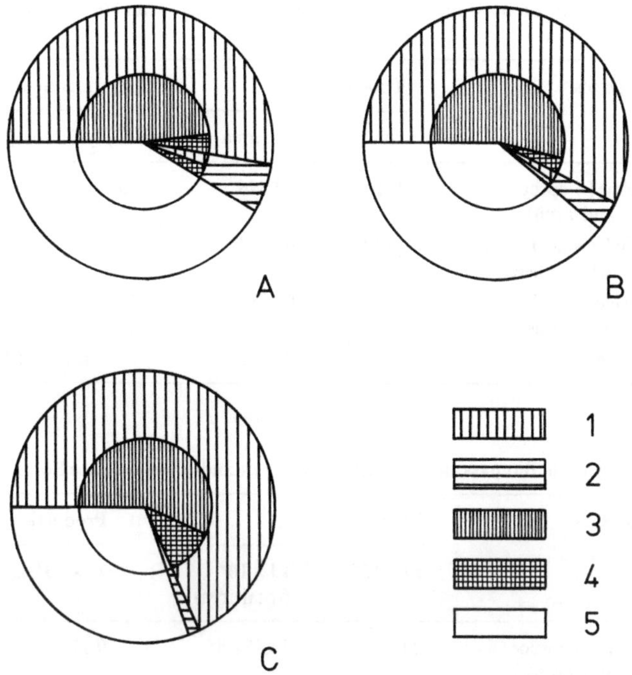

Fig. 3.2. Aboveground productivity of forests studied. A Pine forests, age class V. B Oak forests, age class V. C Pine forests, age class II. *1* tree layer; *2* understory; *3* boles; *4* branches and twigs; *5* leaves

in the sample plots. Similarly, the bole increment is the main component (over 50%) of total productivity, and it was estimated for the whole Niepołomice Forest, as well as measured on the sample plots.

With these data it has been possible to compute total productivity of forests in the complex and, to this end, some computational indices were employed (Tables 3.3 and 3.4). Total biomass of pine forests was estimated by multiplying by an index of 1.24; oak forests by 1.14 (Table 3.3). To estimate total productivity, the indices are 1.76 and 1.89, respectively (Table 3.4). For black alder stands the oak forest index was applied because no original data for such forests is available. The values for total productivity are given by weight (Table 3.5) and in terms of energy (Table 3.6).

Average standing biomass of forests is 115.5 t (2.15 yr 10^9 kJ) ha^{-1}, and average productivity 5.6 t ha^{-1} yr^{-1}. The greatest biomass occurs in oak forests, and the lowest in black alder (Table 3.5). The same pattern holds for productivity (Table 3.5).

Total biomass of the Niepołomice Forest exceeds 1,000,000 t, of which more than 760,000 are accounted for by pine, 260,000 by oak, and over 83,000 t by black alder forests. Total annual productivity exceeds 54,000 t, of which 37,000 are accounted for by pine, almost 11,000 t by oak, and the remaining 6,000 t by black alder forests.

Table 3.5. Biomass and productivity of tree stands of the Niepołomice Forest (weight units)

Species	Total biomass		Net production	
	$t\ ha^{-1}$	t (total area)	$t\ ha^{-1}\ yr^{-1}$	$t\ yr^{-1}$(total area)
Pine (+spruce and larch)	111.0	760,546	5.46	37,417
Oak (+hornbeam, ash and beech)	147.5	261,398	6.0	10,632
Alder (+other deciduous)	78.1	83,088	6.0	6,384
Total		1,105,032		54,433

Table 3.6. Biomass and productivity of tree stands, Niepołomice Forest (energy units)

Species	Biomass		Production	
	$kJ \times 10^9\ ha^{-1}$	$kJ \times 10^9$ (total area)	$kJ \times 10^9\ ha^{-1}$	$kJ \times 10^9$ (total area)
Pine (+spruce and larch)	2.01	14,982.55	0.11	715.59
Oak (+hornbeam and ash)	2.90	5,150.36	0.11	211.58
Alder (+other deciduous)	1.54	1,636.22	0.11	125.51
Total (average)	(2.15)	21,769.13	(0.11)	1,052.68

These figures do not include the roots, which represent 20 to 30% of the aboveground parts of the tree volume (Borowski 1974). Roots account for approximately 16% of the total biomass of a forest, and for about 10% of its productivity (Hanley 1976).

3.3 Photosynthetic Productivity

M. Czarnowski and W. Maczek

Deciduous and coniferous forests of the Niepołomice Forest grow under the increasing pressure of industrial pollution of the Cracow industrial complex. Gas and dust pollutants disturb the physiological activities of plants, primarily photosynthesis. The ecophysiological study conducted was aimed at recognizing changes in the photosynthetic activities and productivities of selected deciduous and coniferous trees existing under conditions of variable air pollution.

3.3.1 Seasonal Changes in the Photosynthetic Activity of Leaves in Areas with Variable Air Pollution

The oak-hornbeam forests (Tilio-Carpinetum typicum) are dominated by *Quercus robur, Carpinus betulus,* and *Tilia cordata* (Dziewolski 1974, Ferchmin and Medwecka-Kornaś 1976). Their rate of photosynthesis ranges from 6 to 10 mg CO_2 dm^{-2} h^{-1}. The rate of photosynthesis may reach 12 mg CO_2 dm^{-2} h^{-1}, but only in the leaves of *Carpinus betulus* and only under strong insolation. Seasonal changes in the photosynthetic activities of these three species have been investigated and compared to a control area at Czartak (near Wadowice), beyond the reach of industrial emissions, and in Branicki Park, under conditions of very high air pollution due to the close proximity of the Nowa Huta steel works. The results indicate a significant difference between the fall-off in net photosynthesis due to natural aging of the assimilation apparatus during the vegetative period, and that affected by air pollution (Czarnowski 1977 a).

If the net photosynthetic rate (FN) of leaves from Czartak is assumed to be 100, then that of leaves from Niepołomice is maintained almost at the same levels during the first half of the vegetative period, but drops at the rate of 10% in *Tilia* and 20% in *Carpinus* during the second half. In Branicki Park, this decline in net

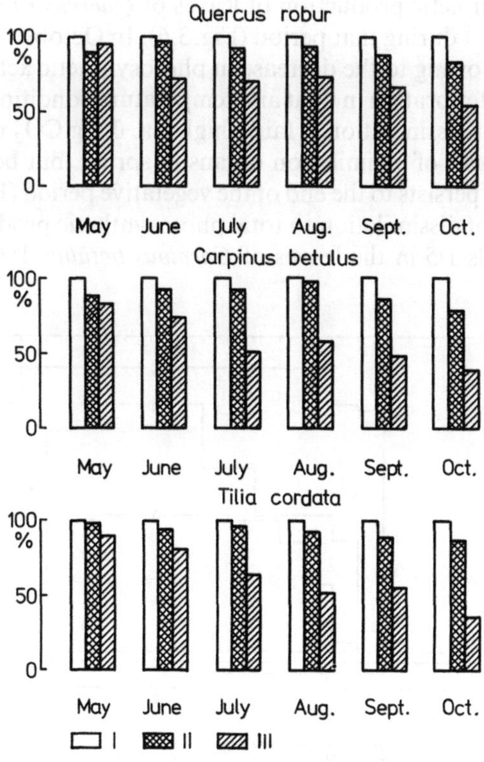

Fig. 3.3. Net photosynthesis of deciduous trees under varying levels of air pollution. *I* Czartak; *II* Niepołomice Forest; *III* Branicki Park

photosynthesis ranges from 43% in *Quercus* to 64% in *Tilia* (Fig. 3.3). These data indicate that industrial emissions of low concentration (average daily range 0.01 to 0.03 mg SO_2 m^{-3}) cause only a slight decrease in the rate of gas exchange in leaves of trees at Niepołomice. However, in Branicki Park, photosynthesis is considerably reduced by high concentrations of phototoxic dusts and gases. The leaves of *Tilia cordata* and *Carpinus betulus* are most sensitive to air pollution, and those of *Quercus robur* are less so.

3.3.2 Seasonal Photosynthetic Production and Respiration in Leaves of Selected Tree Species

In order to estimate the photosynthetic production of trees in oak-hornbeam forests, the ecophysiological estimative method was used (Czarnowski 1972, 1973, 1974, 1975a, b, 1977a, b, 1978, Starzecki 1975). The effective daily photosynthetic production depends on weather conditions. For *Quercus robur*, it varies from 23.3 mg CO_2 dm^{-2} day^{-1} under high cloudiness to 61.4 with high insolation (Fig. 3.4). As a comparison total photosynthetic production varies from 29 to 85 mg CO_2 dm^{-2} day^{-1} and total dissimilation from 6 to 24 mg CO_2 dm^{-2} day^{-1} (Fig. 3.5).

The investigated trees assimilate most intensely from June to September. The total photosynthetic production of leaves of *Quercus robur* achieved 1.98 g CO_2 dm^{-2} $month^{-1}$ during that period (Fig. 3.6). In October, the production declines considerably, owing to the decrease in photosynthetic activity of aging leaves, as well as to a deterioration in light and temperature conditions.

The rate of dissimilation is fairly high (ca. 0.6 g CO_2 dm^{-2} $month^{-1}$) during the rapid growth of assimilation organs in spring, but begins to decline in June and this trend persists to the end of the vegetative period (Fig. 3.6).

The ratio of dissimilation to total photosynthetic production under strong insolation equals 1:5 in the leaves of *Carpinus betulus*, 1:3 in *Quercus robur*, and

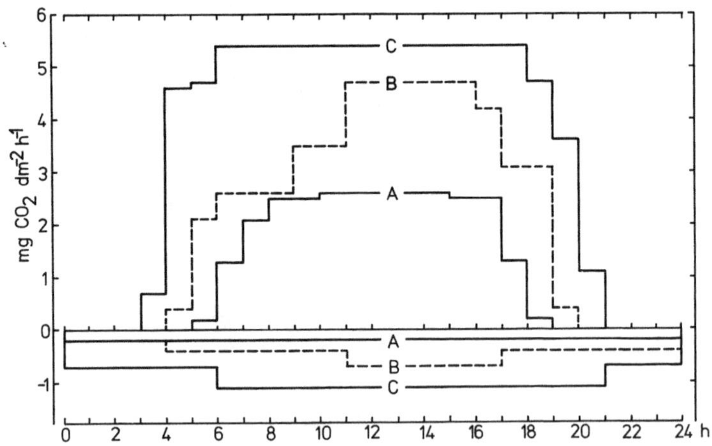

Fig. 3.4. Daily photosynthetic production and respiration in leaves of *Quercus robur* under varying weather conditions. *A* cloudy; *B* moderately cloudy; *C* sunny

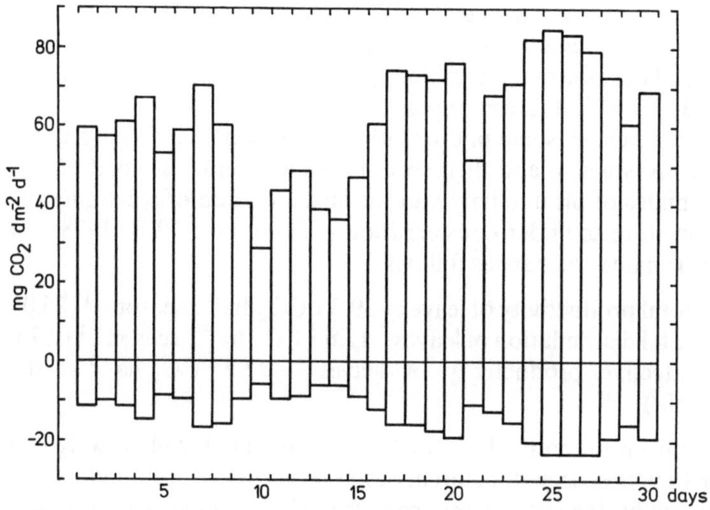

Fig. 3.5. Daily photosynthetic production and respiration in leaves of *Quercus robur* in June

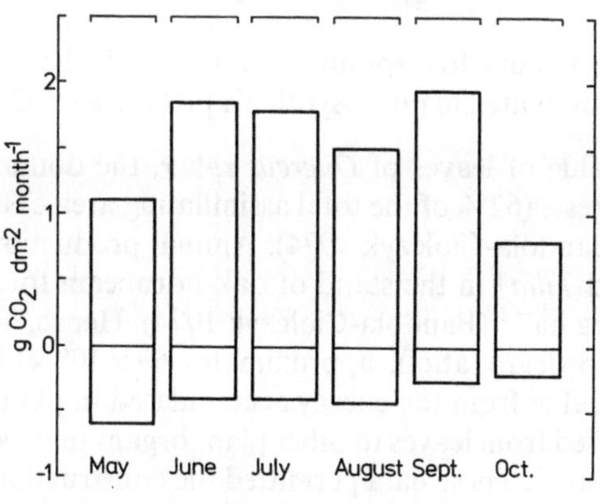

Fig. 3.6. Total photosynthetic production and respiration in leaves of *Quercus robur* in specific months of the vegetative season

1:2.5 in those of *Tilia cordata*. This ratio is lower in shade, where it equals 1:4 in *Carpinus betulus*, and 1:2.6 in seedlings of *Quercus robur*.

The average total photosynthetic production of leaves in the oak-hornbeam forest investigated achieved 8.14 g CO_2 dm^{-2} season^{-1}, while dissimilation reached 2.40, and effective photosynthetic production 5.74.

3.3.3 Productivity and Accumulation of Solar Energy in Oak-Hornbeam Forests

At the period of maximum leaf density, the leaf area index achieved the following values for selected tree species of oak-hornbeam forests: 2.99 for *Quercus robur*, 1.28 for *Carpinus betulus*, and 0.56 for *Tilia cordata* (Bandoła-Ciołczyk 1974). *Quercus* and *Carpinus* jointly contributed more than 88% of the total leaf area. Thus, the maximum foliage density of *Quercus, Carpinus,* and *Tilia* reached 4.83 ha ha^{-1} of forested area.

Average seasonal photosynthetic production and dissimilation of the tree species investigated and their specific leaf area indices was the basis for preliminary estimates of productivity and the accumulation of solar energy in the oak-hornbeam forests under consideration (Czarnowski 1973, 1978). The results in approximate values are as follows:

1. total productivity of leaves – 39.3 t CO_2 ha^{-1} $season^{-1}$ (26.0 t dry weight);
2. total dissimilation of leaves – 11.6 t CO_2 ha^{-1} $season^{-1}$ (7.7 t dry weight);
3. effective productivity of leaves – 27.7 t CO_2 ha^{-1} $season^{-1}$ (18.3 t dry weight).

Given the equivalence of 1 g CO_2 to 0.66 dry plant weight (Květ et al. 1971) or 10.9 kJ energy (for starch; Sěstak et al. 1971), accumulation and losses of solar energy by the specific tree species could be computed. The approximate values are:

1. amounts of total energy used for photosynthesis – 428×10^6 kJ ha^{-1} $season^{-1}$;
2. losses in energy due to respiration – 126×10^6 kJ ha^{-1} $season^{-1}$;
3. energy accumulated in photosynthetic products – 302×10^6 kJ ha^{-1} $season^{-1}$.

Caloric value of leaves of *Quercus robur*, the dominant tree species in oak-hornbeam forests (62% of the total assimilating area of trees), equals 21.3 kJ g^{-1} dry weight (Bandoła-Ciołczyk 1974). Annual production of leaf litter *(Quercus, Carpinus,* and *Tilia)* in the stand of oak-hornbeam forest investigated approximates 3,255 kg ha^{-1} (Bandoła-Ciołczyk 1974). Hence, total caloric value of the litter, prior to degradation, approximates 69×10^6 kJ ha^{-1} $season^{-1}$. By subtracting this value from the energy accumulated in photosynthetic products, energy transported from leaves to other plant organs may be calculated as 233×10^6 kJ ha^{-1} $season^{-1}$. These data permitted the construction of an energy flow model through the oak-hornbeam forest ecosystem (Medwecka-Kornaś et al. 1974). As demonstrated by Woodwell (1970), respiration of trees accounts for 55% of the total production of pine-oak mixed forests. Given this value and the effective photosynthetic productivity of leaves of oak-hornbeam forests (18.3 t dry weight ha^{-1} $season^{-1}$), total annual production is 9.15 t dry weight and 10.1 t pure oxygen ha^{-1}, because the effective photosynthetic productivity should be halved to account for dissimilation of shoots and roots. Precise data on the respiration of these plant parts are not available.

Photosynthetically active radiation energy falling upon the upper surface of the tree canopy of oak-hornbeam forests of Niepołomice is approximately 134×10^8 kJ ha^{-1} $season^{-1}$ (Czarnowski 1973, 1978). Under conditions of maximum leaf density, only 9.2% of this energy reaches the ground surface in oak-hornbeam forests (Klein 1977, 1978). Total amounts of energy used in photosynthesis is approximately 428×10^6 kJ ha^{-1} $season^{-1}$. Thus, the average photosynthetic efficiency is 1.7% energy PhAR in this community.

3.3.4 Seasonal Changes in the Rate of Photosynthesis and Respiration of Pine Shoots

In order to determine the effects of air pollution on coniferous forests here, four sites (A, B, C, D) were chosen at variable distances from the emission source, in the direction of prevailing winds (west-east). A control area was chosen at Nowe Rybie (NR), south of Bochnia, beyond the area affected directly by industrial emissions of the Cracow agglomeration.

There are seasonal changes in the rate of gas exchange in needles of *Pinus silvestris* (Maczek 1977). Production of organic matter is low in winter, with the maximum rate of photosynthesis achieved at the end of April. Beginning in May, new shoots are produced so that a shift in needle generations takes place. At the beginning of July the youngest generation attains a rate of photosynthesis of 4.7 mg CO_2 g^{-1} h^{-1}, while respiration falls to about 1.2 mg CO_2 g^{-1} h^{-1} (Fig. 3.7).

No significant variation in the rate of photosynthesis of pine needles has been found in July in the sites investigated at Niepołomice or in the control area at Nowe Rybie. This holds true for shoots of the youngest as well as the next youngest generation. These results are consistent with those found by Żelawski and Łotocki (1974) in other regions. A drop in the rate of net photosynthesis of two year old shoots relative to those of the youngest generation is entirely consistent (Fig. 3.8) between the Niepołomice Forest (59% on the average) and Nowe Rybie (60%).

The data cited above do not suggest any directional change in the rate of photosynthesis depending upon the geographical location. There are, additionally, considerable differences between generations. One may conclude that needles remaining on long shoots maintain their characteristic rate of photosynthesis.

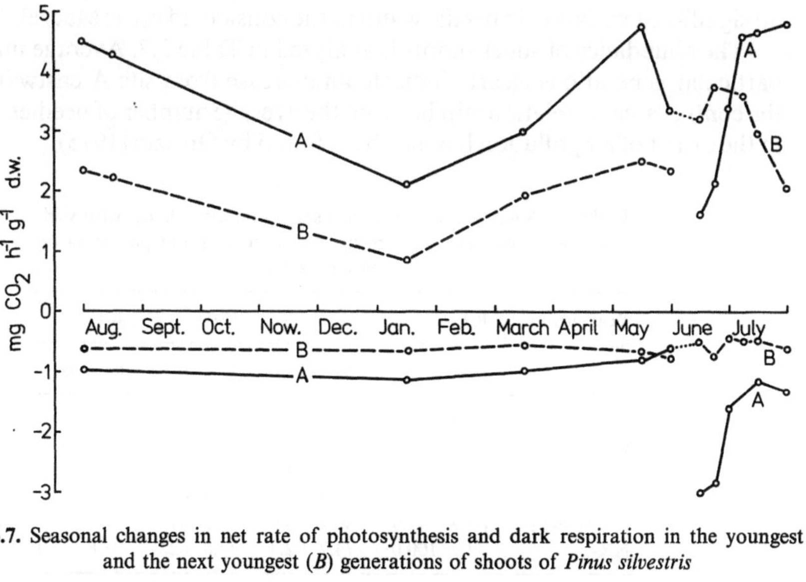

Fig. 3.7. Seasonal changes in net rate of photosynthesis and dark respiration in the youngest (*A*) and the next youngest (*B*) generations of shoots of *Pinus silvestris*

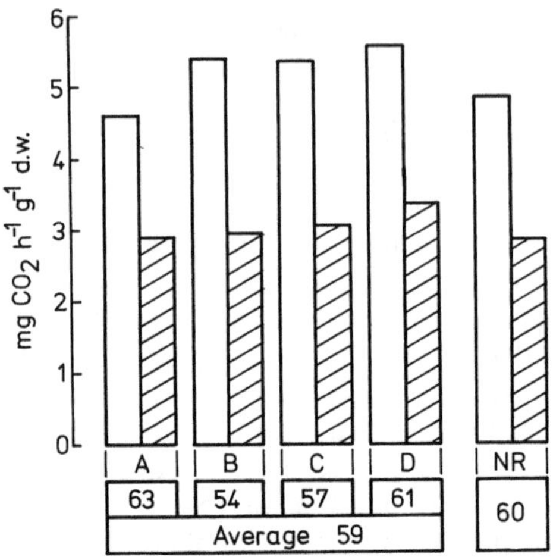

Fig. 3.8. Net rate of photosynthesis under constant photosynthetic activity of the youngest (*light*) and next youngest (*striped*) generations of shoots of *Pinus silvestris*. *A, B, C, D* locations in the Niepołomice Forest; *NR* Nowe Rybie

3.3.5 Effects of Industrial Air Pollution on the Morphology and Abundance of Pine Needles

Pine needles are only slightly variable in length for the sites investigated. Average length of needles ranges from 64 to 71 mm in the youngest generation. However, needles from the control area at Nowe Rybie reach 80 mm in length and are thus 16% longer than the average pine needles in the Niepołomice Forest. A similar pattern occurs in two year old needles. Needle thickness ranges from 570 to 620 µm at Niepołomice, whereas it is 620 µm at the control site. There is no significant variation in needle width in the considered area (Maczek 1977).

The abundance of short shoots is analyzed in Table 3.7. Average numbers for particular generations clearly indicate an increase from site A eastwards and to the control area. A relationship between the average number of needles on a shoot to the extent of air pollution has also been found by Greszta (1975).

Table 3.7. Average number of new shoots on long shoots with various generations (a), and percent of new short shoot persisting on long shoots (b)

Year-group Site	R I		R II		R III		R IV	
	a	b	a	b	a	b	a	b
A	60.2	91.9	66.9	81.3	54.0	32.5	—	0
B	80.8	92.1	76.2	81.8	65.0	46.0	—	0
C	96.0	97.0	86.2	84.0	78.6	44.8	—	0
D	86.4	98.4	92.2	84.7	71.2	43.6	—	0
A–D means	80.9	94.9	80.4	83.0	67.2	41.7	—	0
NR	79.7	100.0	74.2	92.7	74.8	62.2	73.8	11.0

As indicated by the numerical ratio of newly produced needles to old on long shoots, the short shoots of the youngest generation have been lost by summer at Niepołomice, just after their production. This is a phenomenon which has not been observed in the control area at Nowe Rybie. Short shoots are more numerous on sites more distant from the industrial emission source and in the control area. Older short shoots show this pattern more distinctly (Table 3.7). Phytotoxic effects of air pollution increase the aging rate of assimilating organs and thus promote the fall of needles. Consequently, the number of needles that have fallen from long shoots, and, more precisely, the ratio of needles fallen to those initially folded, reflects the degree of air pollution.

The dry weight of needles and average weight of short shoots of the two youngest generations is given in Table 3.8.

3.3.6 Potential Photosynthetic Production of Long Pine Shoots

The relationship of photosynthetic activity to observed morphologic changes in needles of the two youngest generations has been employed to compute an index of potential photosynthesis. Starting with net photosynthesis (Fig. 3.8) and needle weight (Table 3.8), an index of rate of photosynthesis has been obtained for all the sites investigated (Table 3.9). By multiplying the index by the average number of short shoots remaining on long shoots, potential photosynthesis of long shoots can be estimated (Fig. 3.9). This parameter rapidly increases eastward, from 4.9 to 9.4 mg CO_2 h^{-1} g^{-1} dry weight for needles of the youngest generation, and from 3.1 to 5.1 for needles of the next youngest generation. Average values for the Niepołomice Forest are between 13 and 18% lower than in the control area, for the youngest and the next youngest generations, respectively (Maczek 1977).

The observed relationship permits one to identify the causes of decreased production of pines in areas affected by industrial air pollution.

Table 3.8. Dry weight of needles and average weight of short shoots in the Niepołomice Forest study areas

Year-group Site	R I		R II	
	a	b	a	b
A	1061.5	19.3	1042.2	19.3
B	1524.2	20.6	1190.4	19.2
C	1581.0	17.0	1274.4	17.7
D	1708.5	20.1	1521.5	19.5
A–D means	1482.2	19.3	1263.5	19.0
NR	1821.6	23.0	1642.2	23.8

[a] Total dry weight of needles on long shoots (in mg)
[b] Average dry weight of a short shoot (in mg)

Table 3.9. Rate of photosynthesis and number of short shoots in the study areas

Year-group Site	R I		R II	
	a	b	a	b
A	89.35	55	56.54	54
B	110.82	74	56.06	62
C	91.12	93	53.99	72
D	111.35	85	65.33	78
A–D means	100.66	76.8	57.98	66.5
NR	109.94	79	67.83	69

[a] Rate of photosynthesis of a short shoot (in mg $CO_2 g^{-1} h^{-1}$)
[b] Average number of short shoots per long shoot

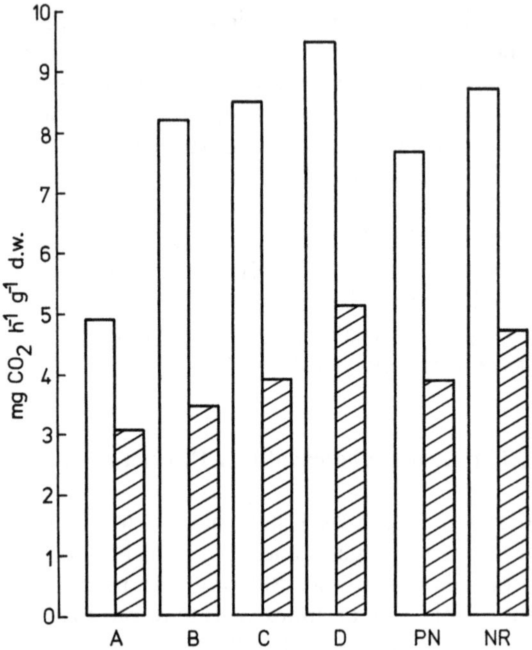

Fig. 3.9. Potential rate of photosynthesis of long shoots of *Pinus silvestris*. *A, B, C, D* locations in the Niepołomice Forest; *PN* average for Niepołomice; *NR* Nowe Rybie

3.3.7 Estimation of Photosynthetic Production of Pine Forests

When phytoclimatic characteristics of the Niepołomice Forest are taken into account, the photosynthetic productivity of the pine forests can be estimated. The basic parameters are as follows: total number of hours for the season, April to November, with thermal-light conditions adequate for photosynthesis (3,492 h) and respiration (5,856 h) of pine (Czarnowski 1978); average net photosynthesis during the vegetative season (2.5 mg CO_2 h^{-1} g^{-1} dry matter); and average dark respiration of needles during the vegetative season (0.7 mg CO_2 h^{-1} g^{-1} dry mat.). On this basis, total photosynthetic production of pine needles is 11.1 g CO_2 g^{-1} dry mat. season^{-1}; global dissimilation is 4.1 g CO_2 g^{-1} dry mat. season^{-1}; and effective photosynthetic production is 7.0 g CO_2 g^{-1} dry mat. season^{-1}.

The average number of pine needles in pine stands of age class V is 350×10^6 per ha (Lemke 1978a, b), and each needle averages 9.6 mg dry mat. (Maczek 1977). Thus the size of the assimilation apparatus approximates 3.4 t dry mat. ha^{-1} in the pine forests.

From the data given above, other characteristics can be approximated for pine needles. These include:

1. total photosynthetic production – 37.7 t CO_2, or 24.9 t dry mat., or 411×10^6 kJ hy^{-1} season^{-1};

2. global dissimilation – 13.9 t CO_2, or 9.2 t dry mat., or 152×10^6 kJ ha^{-1} season^{-1};
3. effective photosynthetic production – 23.8 t CO_2, or 15.7 t dry mat. or 259×10^6 kJ ha^{-1} season^{-1}.

According to Woodwell (1970), dissimilation of shoots and roots of the pine accounts for 50% of the effective photosynthetic production. Consequently, annual production of the pine forest at Niepołomice is approximately 7.85 dry mat. and 8.7 t pure oxygen ha^{-1}.

3.4 Productivity of the Herb Layer

J. Banasik and B. Zemanek

Data on species composition and net productivity of the herb layer in various forest communities of the Niepołomice Forest, which have been published (Tumidajowicz 1973, Banasik 1973, 1978 a, Jankowska 1978, Banasik and Jankowska 1978), relate only to the deciduous forests of the northern sections.

The present study, carried out in 1977–1978, was organized to estimate the productivity of herbs in four stands in mixed oak-pine forest (Pino-Quercetum), and also to estimate total production of herbs as shown on the map of existing vegetation (cf. Chap. 1).

3.4.1 Study Areas

Area I. This is a stand of the fresh variant of Pino-Quercetum (0.12 ha). The tree stand consists of *Pinus silvestris* with lesser amounts of *Quercus robur*. The cover is 30%. The undergrowth is very sparse, consisting mostly of *Frangula alnus*. The herb layer is poor in species, with low coverage (60%), dominated by *Vaccinium myrtillus* accompanied by *Trientalis europaea, Calamagrostis epigeios, Agrostis tenuis,* and other species. Moss layer consists of *Pleurozium schreberi* with coverage approximately 20%.

Area II. This is a moist stand of Pino-Quercetum with *Molinia coerulea* (0.12 ha). The tree stand includes *Pinus silvestris, Quercus robur,* and *Betula verrucosa* and cover is 40%. The undergrowth with 10% cover consists of *Quercetus robur* and *Frangula alnus*. The herb layer (70% cover) is dominated by *Vaccinium myrtillus, Trientalis europaea,* and *Molinia coerulea*. Moss layer is very scanty (5% cover), dominated by *Pleurozium schreberi*.

Area III. This is a stand of Pino-Quercetum dominated by *Carex brizoides* (0.08 ha). The tree stand consists of *Pinus silvestris* with minor amounts of *Quercus robur* and cover approximately 50%. The undergrowth is very scanty, including *Frangula alnus* and *Betula verrucosa*. The herb layer (95% cover) consists almost exclusively of *Carex brizoides,* while other species *(Pteridium aquilinum, Calamagrostis arundinacea, Molinia coerulea)* occur only as single individuals. No moss layer is present, but *Carex brizoides* grows in the litter of previous generations.

Area IV. This is a fresh, rich variant of Pino-Quercetum, located on the southern slope of a dune (0.20 ha). The tree stand consists of *Pinus silvestris* (50% cover). The undergrowth, rich in places, consists mostly of *Frangula alnus* associated with *Quercus robur* and *Sorbus aucuparia*. A lush, mossy herb layer contains large amounts of grass, produces 100% cover. The dominant herbs and moss are *Pleurozium schreberi, Trientalis europaea, Moehringia trinerva, Vaccinium myrtillus, Festuca ovina, Agrostis vulgaris,* and *Vaccinium vitis-idaea.*

3.4.2 Methods

The methods of sampling and estimating productivity were somewhat variable in the different study areas. All plant materials, however, were treated in the same way. Plants were cleaned, sorted by species, counted, oven-dried at 85 °C, and then weighed.

In study areas I and II, the biomass was sampled over 2 years (1977–1978), three times per year (spring, summer, and autumn). In each sampling run, 60 samples were taken at random from each study area. Each sample was 0.17 m^2 in area and included the soil to a depth of 25 cm. Production of aboveground and belowground parts of herbs has been assumed to be represented by the difference between the maximum and the spring biomass. Mosses were not taken into account.

In study area III, 15 samples of aboveground herbs were taken at random in August 1977. Each sample was 0.1 m^2 in extent. This biomass has been assumed to represent the production of *Carex brizoides* and associated species.

Finally, 20 samples of the aboveground parts of herbs were taken at random in June and August 1978 from study area IV, each sample was 0.1 m^2. Herb production has been assumed to be represented by biomass at the maximum point of generative development for particular species. For the bilberry, however, current shoots were distinguished directly on the plants. Current increment in the thickness of multiannual shoots was estimated from the literature (Traczyk et al. 1973). The same data were used to estimate production of the few mosses and seedlings of trees and shrubs present. Of the moss biomass present in summer 35% has been assumed to represent annual moss productivity.

Total production of the herb layer in forest communities has been estimated from the present data and the results of other detailed studies and extrapolated for the whole Niepołomice Forest, in respect to the proportions occupied by specific forest communities as recorded by Lesiński (cf. Chap. 1). For those forest communities in which production of herbs has not been studied in detail, approximate figures have been consulted.

3.4.3 Production of the Herb Layer in Forest Communities

The results are provided in Table 3.10. Production of aboveground parts of herbs in mixed oak-pine forests of the southern sections of the Forest varies from 141.4 kg ha^{-1} yr^{-1} in a poor Pino-Quercetum to 1,738.7 kg ha^{-1} yr^{-1} in

Table 3.10. Productivity of herb layer in mixed oak-pine forests of the Niepołomice Forest

Sample plot No.	Association type	Aboveground production	Belowground production	Year
I	Pino-Quercetum fresh, poor variant	141.4	179.1	1977
	(without mosses)	151.3	300.3	1978
II	Pino-Quercetum, moist variant	376.4	397.1	1977
	(without mosses)	273.3	162.2	1978
III	Pino-Quercetum with *Carex brizoides*	1738.7	–	1977
IV	Pino-Quercetum fresh, rich variant	1293.8	–	1978

Pino-Quercetum dominated by *Carex brizoides*. These values are consistent with those obtained in other Polish forests (Rajchel 1965, Traczyk 1967, Plewczyńska 1970, Traczyk et al. 1973, Izdebski and Popiołek 1976).

Productivity of underground parts of herbs in mixed oak-pine forests in the same area averages 259.7 kg ha^{-1} yr^{-1}.

On this basis, an attempt has been made to estimate the total aboveground production of herbs (Table 3.11). Belowground production of herbs has been estimated from the means of the available data (cf. Tumidajowicz 1973, Banasik 1973, and Table 3.10 in the present paper).

Total annual production of aboveground parts of herbs in the forest communities of the Niepołomice Forest is 6,529.7 t, while total annual production of belowground parts is 3,344.9 t. These figures are equivalent to 117.2×10^9 kJ and 67.0×10^9 kJ, respectively (caloric values of plant materials from Banasik 1978 a).

Thus, the total annual productivity of the herb layer is 10,274.6 t, i.e., 1.05 t (or 184.2×10^9 kJ ha^{-1}).

These data are supplemented with additional information presented in Table 3.12. One can find there the average aboveground production of the herb layer in the most important forest communities – oak, black alder, and pine forests.

3.4.4 Discussion

The map of the existing forest vegetation of the Niepołomice Forest (Chap. 1), which was the basis for this study, is considerably simplified. This is well exemplified by comparison with the map of the Grobla site which had been completed previously (Denisiuk and Medwecka-Kornaś 1976, Denisiuk et al. 1977). One may nevertheless assume that this simplification does not significantly affect an estimate of the total herb production.

This generalization (Table 3.11) of detailed studies is biased also by the sizeable proportion of disturbed forest communities in the forest. This disturbance

Table 3.11. Area and above-ground productivity of the herb layer in various forest communities of the Niepołomice Forest

Association	Area covered ha	Mean production kg ha^{-1}	Total production t
Vaccinio uliginosi-Pinetum	68.4	700[a]	47.9
Pino-Quercetum:			
Moist	4361.9		
Fresh	694.4	662.5[b]	3356.3
Dry	9.8		
Tilio-Carpinetum/ Pino-Quercetum	2405.9	625.7[c]	1505.4
Tilio-Carpinetum typicum	19.6	589.1[d]	11.5
Tilio-Carpinetum stachyetosum	1242.1	485.9[e]	603.5
Circaeo-Alnetum Tilio-Carpinetum Ficario-Ulmetum stachyetosum	420.5	789.9[f]	332.2
Circaeo-Alnetum	107.6	3070.1[g]	330.3
Ficario-Ulmetum	29.3	1500.0[h]	43.9
Carici elongatae-Alnetum	39.1	750.0[i]	29.3
Carici elongatae-Alnetum/ Pino-Quercetum	381.4	706.2[j]	269.4
Total	9780.0		6529.7

[a] Approximate average production for Pino-Quercetum
[b] Mean of 6 analyses of Pino-Quercetum and P.-Q./Vaccinio myrtilli-Pinetum (Table 3.4); production has not been estimated for particular variants of Pino-Quercetum because the areas dominated by *Carex brizoides* are unknown; Pino-Quercetum in its dry variants occupies minor areas and thus, it has been considered jointly with the other variants
[c] Mean of 2 analyses of average production in Tilio-Carpinetum and Pino-Quercetum
[d] Mean of 4 analyses of production in Tilio-Carpinetum typicum (Banasik 1973, 1978a, Jankowska 1978, Jankowska and Banasik unpubl.)
[e] Mean of 5 analyses of production in Tilio-Carpinetum stachyetosum and T.-C. stachyetosum/T.-C. typicum (Tumidajowicz 1973, Banasik 1978a, Banasik and Jankowska 1978, Jankowska and Banasik unpubl.)
[f] Production of Fraxino-Ulmetum/Tilio-Carpinetum stachyetosum after Jankowska and Banasik (unpubl.)
[g] Production of Circaeo-alnetum after Jankowska and Banasik (unpubl.)
[h] Estimated as half the production of Circaeo-Alnetum
[i] Estimated as a quarter of the production of Circaeo-Alnetum
[j] Mean of 2 analyses of average production in Carici elongatae-Alnetum and Pino-Quercetum

of the biological balance is commonly expressed in the herb layer being largely or even almost totally dominated by *Carex brizoides* (Ferchmin 1976, Ferchmin and Medwecka-Kornaś 1976). Under such conditions, the herb layer of the oak-hornbeam forest may produce twice the amount of aboveground parts as normally (Tumidajowicz 1973). In mixed oak-pine forests, the aboveground productivity of the herbs may increase several times relative to productivity in undisturbed conditions (Table 3.10). Unfortunately, the actual area of disturbed communities is unknown.

The results are biased also by a lack of precise data on the herbaceous productivity of some phytosociologic units distinguishable on the map of existing vegetation. In such cases, values had to be approximated.

Table 3.12. Average annual production of the herb layer of the main forest types of the Niepołomice Forest

Forest type	Average production kg ha^{-1}	
	Aboveground	Belowground
Oak-hornbeam forests	487.5	629.4
transitional	625.8	?
Pine forests	663.0	259.7
transitional	706.2	?
Alder forests	1233.4	

The existing productivity of forest communities of the Niepołomice Forest is certainly considerably lower than it was several hundred years ago, prior to human acitivities. At that time, the contribution of humid and fertile communities (carrs) was much higher than it is now (Chap. 1). As a point in fact, the aboveground production of herbs is very high in alder forests and carrs, compared to other forest communities (Tables 3.11 and 3.12).

3.5 Productivity of Nonforest Vegetation

Z. Denisiuk

3.5.1 Nonforest Communities

The nonforest communities of the Niepołomice Forest have not yet been fully monographed and thus their productivity cannot be reliably evaluated. It is only in the northern section that the ecology and production of some nonforest communities have been studied (Baradziej 1974, Kotańska 1973, 1975). Therefore, this attempt to estimate the productivity of nonforest communities in the whole area utilizes first the modern studies carried out in the northern part of the forest complex (Denisiuk 1976a, Denisiuk et al. 1977, Dubiel 1973), as well as comparative data from other regions in Poland and abroad.

The point of departure for recognizing the areal relationship of various nonforest communities in the Forest has been the extent of the two major complexes, separated from each other by the Drwinka River valley. In both complexes, as well as in their vicinity, all nonforest communities were investigated. These include aquatic, swamp, meadow, and open field communities. The areas occupied by these communities in the northern forest complex has been established from published reports (Denisiuk 1976a, Denisiuk and Medwecka-Kornaś 1976, Denisiuk et al. 1977). In the southern forest complex, the communities have been mapped (scale 1:25,000), and the areas covered by each has also been measured planimetrically.

These data indicate that nonforest communities account for 45% of the area in the northern complex, and 17% in the southern. In all, somewhat more than 22% of the area of the Forest is occupied by nonforest communities.

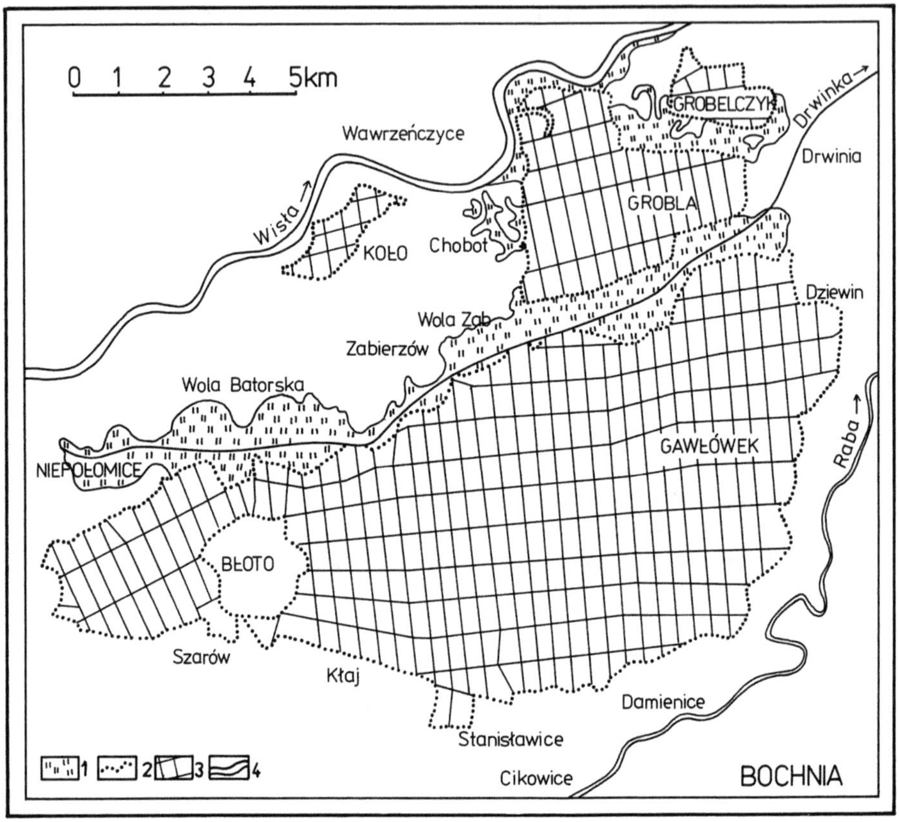

Fig. 3.10. Distribution of the main nonforest communities in the Niepołomice Forest. *1* grasslands; *2* forest boundary; *3* forest divisions; *4* rivers and ditches

The investigated nonforest communities (Table 3.13) can be separated into six main habitat-floristic categories equivalent to six phytosociologic orders (Table 3.14). Twenty-five plant communities and two groups of field associations have been distinguished within these orders, and their areal relationships are given in Table 3.14. These data demonstrate that, according to area, meadow, reed-sedge and field communities are dominant, even though the latter play only a minor role in the total energy balance of the Niepołomice Forest.

3.5.2 Productivity of Nonforest Communities

Only a few nonforest communities here have been studied in terms of their productivity (Baradziej 1974, Kotańska 1973, 1975) and thus, data from other regions had to be applied in the present research.

The productivity of aquatic communities (order Potametalia) has been established by Krzywański and Samosiej (1976), who studied plant associations in the old bed of the Pilica River which closely resemble those observed in the Nie-

Table 3.13. Nonforest communities of the Niepołomice Forest

No.	Nonforest communities	Area (ha)	% of nonforest area
1.	Aquatic communities, mainly in old beds of the Vistula and Drwinka	5	0.2
2.	Flooded meadows along Vistula River in the northern sectors	85	2.7
3.	Nonforest meadows in overgrown old beds in the northern sectors	105	3.4
4.	Meadows between the Grobla and Grobelczyk forest areas	255	8.2
5.	Meadows along the Drwinka River between Niepołomice and Drwinka Village	1385	44.7
6.	Nonforest meadows on peat-bogs of the Błoto complex	360	11.6
7.	Nonforest meadows in the southern sectors	170	5.5
8.	Field communities within the northern forest complex	352	11.3
9.	Field communities within the southern forest complex	383	12.4
Total		3100	100.0

połomice Forest. These data, expressed in terms of fresh matter, are assumed to be 15% dry weight. In addition, the work of Pieczyńska (1972) on plant associations of the lacustrine littoral had also been taken into account.

The productivity of reed-sedge communities (order Phragmitetalia) has been assessed partly from reports from the Niepołomice Forest (Baradziej 1974, Kotańska 1973, 1975) and partly from other studies. Thus, the productivity of the Scirpo-Phragmitetum community has been established from data from Kowalczewski and Wasilewski (1966) and Pieczyńska (1972); the Glycerietum maximae community according to Westlake (1966) and Hajkova and Květ (1970); the Caricetum gracilis community according to Baradziej (1974) and Baszyński et al. (1978); and the Phalaridetum community according to Hajkova and Kvě (1970). The productivity of other investigated communities has been estimated as analogous to similar communities investigated in Wielkopolska (Denisiuk 1976b, 1980).

The productivity of humid meadows (order Molinietalia) has been estimated after Traczyk (1971) and Nowiński (1967), while the productivity of mesic meadows has been considered according to data provided by Jankowska (1967, 1971, 1975), Kotańska (1967, 1970), and Bray (1963). The productivity of pastures has been computed from values given by Leipnitz (1970), Sauerbrey (1975), and Voisin (1964). In order to estimate the productivity of field weed communities (order Secali-Violetalia), the papers of Herbichowa (1969a, b), Pasternak (1974), and Wójcik (1973) were consulted.

Table 3.14. Average production (in dry weight) of nonforest communities of the Niepołomice Forest

Production of association groups	Area (ha)	% of non-forested area	Above-ground production (t ha^{-1} yr^{-1})	Below-ground production (t ha^{-1} yr^{-1})	Total production (t ha^{-1} yr^{-1})	Below-ground biomass (%)	Production of biomass (kJ × 10^7 ha^{-1} yr^{-1})	Total production of associ-ations (t × 10^2 yr^{-1})	Total production of associ-ations (kJ × 10^7 yr^{-1})	% of total pro-duction
I. Aquatic communities (order Potametalia)	5	0.2	1.35	1.01	2.36	42.8	4.82	0.12	24.1	0.05
II. Reed-sedge communities (order Phragmitetalia)	430	13.9	8.35	7.49	15.84	47.3	27.81	68.11	11,959,6	26.36
III. Swamp communities (order Caricetalia fuscae)	15	0.5	4.00	1.20	5.20	23.1	9.09	0.78	136.4	0.30
IV. Humid meadows (order Molinietalia)	953	30.7	5.58	2.59	8.17	31.7	13.24	77.86	12,615.8	27.81
V. Mesie meadows (order Arrhenateretalia)	962	31.0	6.48	4.65	11.13	41.8	20.09	107.07	19,326.4	42.61
VI. Field communities (order Secali-Violetalia)	735	23.7	0.83	0.18	1.01	17.8	1.77	7.42	1,301.1	2.87
Total	3,110	100.0	–	–	–	–	–	261.36	45,363.4	100.00

In some cases, production could be assessed only by comparison with communities similar in geomorphologic position and phytosociologic characteristics; thus, the productivity of the Scirpetum silvatici community has been compared to the Caricetum gracilis, the Cirsio-Polygonetum community to the Cirsietum rivularis, and the Carici-Agrostietum, and to the community with *Agrostis stolonifera*.

The production in joules has been computed from the energetic value of dried plant materials in meadows, which averages 17.84 kJ g^{-1} for aboveground (Jankowska 1975) and 16.24 kJ g^{-1} for belowground parts (Kotańska 1970). As demonstrated by Table 3.14, the greatest share of total production of the nonforest communities of the Niepołomice Forest is held by mesic meadows (43%). This is followed by humid meadows (28%) and reed-sedge communities (over 26%). These communities contribute strongly to the area occupied by nonforest vegetation. Additionally, field communities, which contribute only slightly less to the total area than the meadows, have productivity more than an order of magnitude smaller than the latter communities.

Total annual net primary production of nonforest communities of the Niepołomice Forest amounts to 26,135 t (or 453.6×10^9 kJ), of which 6,810 t are contributed by reed-sedge communities and 18,490 t are accounted for by meadow and pasture communities. Productivity per unit area is highly variable for each particular community, with respect to both aboveground and belowground biomass. The widest variability occurs in reed-sedge communities. Their total annual primary production ranges from 7 to 31 t ha^{-1}, with aboveground from 5.0 to 12.8 t and belowground from 2.0 to 21.5 t ha^{-1}. In humid and mesic meadows, total annual production ranges from 2.6 to 13.5 t ha^{-1}, with aboveground from 2.0 to 8.0 t and belowground from 0.7 to 5.5 t ha^{-1}. The lowest variation in production occurs in field and aquatic associations with the exception of the Hydrocharo-Stratiotetum community, especially with respect to aboveground parts.

3.6 The Concentration of Nutrients and Pollutants in Plant Materials in the Niepołomice Forest

K. Grodzińska

In this report primary productivity is considered in terms of dry matter and energy. In order to achieve a general model of matter and element flow through the ecosystem (Chap. 7), the concentrations of macroelements and pollutants in plant materials must be known.

Here data on the concentrations of 12 elements (six macroelements: N, P, S, Mg, Ca, K; and six heavy metals: Cd, Ni, Cu, Pb, Zn, Fe) in various parts of trees, shrubs, herbs, and meadow vegetation are presented (Grodzińska 1980, Rudnik 1980, Grodzińska et al. 1983, Bobek unpubl.).

Two dominant tree species, *Pinus silvestris* and *Quercus robur*, were analyzed. Samples were taken from trees 60 to 80 years old, and included bark, wood (heartwood and sapwood), and needles or leaves. Bark was sampled at a high of 1.5 m

Table 3.15. Concentration of heavy metals (ppm dry wt) and nutrients (%) in selected plant materials, Niepołomice Forest

Species	Material	Element concentrations											
		ppm						%					
		Cd	Ni	Cu	Pb	Zn	Fe	Mg	Ca	K	N	P	S
Quercus robur	Bark	2.07	8.53	19.47	82.0	69.10	1,923	0.034	1.342	0.039	0.677	0.019	0.342
	Wood	0.83	1.83	1.86	4.42	1.38	15	0.017	0.030	0.010	0.125	0.007	0.026
	Leaves	1.19	5.17	11.30	14.35	45.88	974	0.118	0.542	0.322	2.043	0.174	0.203
Pinus silvestris	Bark	1.26	5.53	11.50	54.74	35.40	1,740	0.010	0.274	0.002	0.001	0.000	0.132
	Wood	1.19	2.86	1.37	2.60	9.82	22	0.023	0.038	0.017	0.017	0.003	0.021
	Needles	1.76	9.35	5.08	23.11	57.36	607	0.093	0.219	0.242	1.52	0.121	0.177
Salix caprea	Browse	10.5	7.2	11.08	14.4	367	655	0.104	0.480	0.146	–	–	0.112
Frangula alnus	(twigs buds)	0.6	4.4	11.10	18.3	20.7	605	0.118	0.656	0.175	–	–	0.087
Quercus robur		0.85	3.6	13.11	11.6	60	289	0.150	0.245	0.156	–	–	0.073
Carex brizoides	Above-ground	2.44	8.1	7.7	16.7	118.0	806	0.121	0.088	0.359	1.64	0.17	0.315
Calamagrostis arundinacea	parts	1.28	11.3	6.9	9.4	57.4	501	0.112	0.051	0.421	1.24	0.13	0.248
Vaccinium myrtillus		1.7	9.2	7.1	22.6	79.8	613	0.098	0.159	0.263	1.30	0.13	0.262

Table 3.16. Concentration of heavy metals (ppm dry wt.) in selected plant meterials of the Niepołomice (NF) and Białowieża (BF) Forest

Species	Location	Elements concentrations					
		Cd	Ni	Cu	Pb	Zn	Fe
Pinus silvestris (needles)	N.F.	1.76	9.35	5.08	23.11	57.36	607
	B.F.	1.0	1.7	1.85	3.3	40.0	80.5
Salix caprea (browse)	N.F.	10.5	7.2	110.0	14.4	367	655
	B.F.	5.8	0.3	8.8	7.7	115	75
Vaccinium myrtillus (aboveground parts)	N.F.	1.7	9.2	7.1	22.6	79.8	613

above the ground, in the form of thin slips up to 3 mm; needles and leaves were taken from the tops of canopies; wood was discs every 2 m along the whole length of the tree.

Two shrub species, *Frangula alnus* and *Salix caprea*, and young sapling oaks, *Quercus robur*, were sampled for twigs up to 10 cm in length. Such twigs are intensively browsed by game.

Of the herb layer, the dominant species (*Vaccinium myrtillus, Carex brizoides, Calamagrostis arundinacea*) were sampled. In addition, all herbs found on selected circular sample plots were collected for analysis. The results are actually the averages for both types of samples.

Eighteen plant species in fresh meadows were sampled and the result presented are the averages for these species.

The plant materials were dried at 80 °C and treated with concentrated HNO_3 and $HClO_4$. The following elements were measured in the resulting solutions: total sulfur by the Butters-Chenery nephelometric method using the spectrometer Specol (Zeiss, Jena); Mg, Ca, K, Cd, Ni, Cu, Pb, Zn, Fe by atomic absorption spectrometry using the Varian-techtron 1000 A spectrophotometer; phosphorus by the molybdate method using the Specromom 204 colorimeter; and nitrogen by potentiometric tritration with sodium-hypobromite. All samples were collected in 1977–1978, from early spring to autumn, from ten areas located in mixed oak-pine forest.

Considerable variation exists in the concentrations of specific elements in plant species and their parts (Table 3.15). Thus, oak contains more sulfur, nitrogen and copper than pine, and its bark displays high concentrations of calcium. Oak leaves and pine needles contain many more nutrients than the bark and wood. In turn, the concentration of heavy metals reaches a maximum in the bark, which shows also the highest concentrations of calcium and sulfur. The latter two elements, as well as the heavy metals, are emitted by the Cracow industrial complex. The browse of *Salix caprea* is unusually rich in iron and zinc, whereas *Calamagrostis arundinacea* and *Carex brizoides* are enriched with nickel.

The concentrations of pollutants (heavy metals) in the investigated plants from the Niepołomice Forest are much higher than in those of the Białowieża Forest control area. This pattern is apparent also in the concentrations of sulfur and calcium

The concentrations of ten elements in *Vaccinium myrtillus, Salix caprea,* and pine needles from Niepołomice and Białowieża are given in Table 3.16. Plants from the former area are 5 to 24 times richer in nickel, 1.8 to 7 richer in lead, 1.4 to 3 times richer in zinc, and up to 8.7 richer in iron than those from Białowieża. Certainly, the high concentrations of heavy metals, sulfur, and calcium in plants in the Niepołomice Forest are effected by the influence of industrial air pollution.

Only the concentrations of nutrients have been analyzed in meadow plants. The average figures are: 6.68% for nitrogen, 0.25% for phosphorus, 1.76% for potassium, and 0.57% for calcium. Meadow plants are considerably richer than those of the forest herb layer in potassium and nitrogen.

3.7 Conclusions

Wooded areas account for 88% of the entire Niepołomice Forest. The dominant tree species is Scotch pine. Pine stands represent 71% of the wooded area, oak 18%, and black alder 11%. The remaining 12% is occupied by nonforest communities, of which 62% are meadow, 24% field, and 14% reed-sedge communities. The highest standing biomass is observed in oak forests (149 t ha^{-1} average). It is lower in pine (112 t ha^{-1} average), and lowest in black alder forest (79 t ha^{-1} average). The mean standing biomass approximates 115 t ha^{-1} for the whole area.

Total standing biomass of aboveground parts of trees and shrubs (wood of the tree and understory layer with needles and leaves) amounts to 1,115,000 t, of which 767,000 are contributed by pine forests, 264,000 by oak, and 84,000 t by black alder forests. Annual production of pure oxygen in the Niepołomice Forest averages 8.7 t ha^{-1} in pine and 10.1 t in deciduous forests. The whole forest complex produces 91,200 t (or 78.4 10^6 m^2) of oxygen per year.

The potential energy of a forest phytocoenosis can be measured by primary production. Annual productivity of oak forests has been estimated as 10.0 t dry matter ha^{-1} by dendrometric methods, and as 9.15 t dry matter ha^{-1} by ecophysiologic methods. The median value (9.58 t dry matter ha^{-1} yr^{-1}) was obtained for a similar forest by Dziewolski (1974).

The annual production of pine forests has been estimated as 4.7 t dry matter ha^{-1} dendrometrically, and as 7.8 t dry matter ha^{-1} by ecophysiological methods. This wide discrepancy may be due to the application of ecophysiological methods different from those used for oak forest, as well as to the application of Woodwell's (1970) method of estimating shoot and root respiration. The total annual productivity of tree stands of the forest (wood and leaves and/or needles of the tree layer and undergrowth) is 65,400 t dry matter, i.e., 5.8 t dry matter ha^{-1} on the average.

Net primary production is lower in the Niepołomice Forest than in most other forest ecosystems of the temperate zone. For instance, annual net productivity of oak-pine forests in the United States is 12 t ha^{-1}. This value attains 13.6 t ha^{-1} in beech forests in Denmark, 14.5 t in spruce forests in West Germany, and 16.0 t ha^{-1} in pine forests of Great Britain (Woodwell 1970).

Hanley (1976) investigated the primary production of *Abies grandis, Thuja plicata,* and *Tsuga heterophylla* forests in northern Idaho. There annual production varies from 7.7 to 17.5 t dry matter ha^{-1} (12.0 t ha^{-1} average), of which 55% is accounted for by wood, 40% by bark, and 5% by needles and leaves. Root production averages 10% of aboveground productivity in those forests.

Total annual productivity of the aboveground parts of herbs in the Niepołomice Forest approximates 6,530 t dry matter, of which 4,200 t are represented by coniferous communities and the remaining 2,330 t by deciduous forest communities. Average annual production of the herb layer varies from 0.49 t ha^{-1} in oak forests to 0.66 in pine forests and to 1.23 t ha^{-1} in black alder forests. Annual productivity of the herb layer averages 0.67 t dry matter ha^{-1} for the entire Niepołomice Forest, which represents 12% of the tree productivity.

Average annual productivity of aboveground plant organs in nonforest communities varies from 0.8 t dry matter ha^{-1} in fields to 6.0 t in meadows and over 8.0 t in reed-sedge communities. Total annual aboveground production of nonforest communities approximates 15,280 t dry matter. Total annual production of aboveground plant parts approximates 87,700 t dry matter for the whole Niepołomice Forest, of which 74% are represented by trees, 8% by herbs, and 18% by nonforest communities.

Chapter 4

Energy and Matter Flow Through Consumers in the Niepołomice Forest Ecosystem

Chapter Editor: A. GÓRECKI

Contributors:
B. BOBEK, K. BORUSIEWICZ, Z. GŁOWACIŃSKI, A. GÓRECKI, W. GRODZIŃSKI, J. KOZŁOWSKI, K. PERZANOWSKI, K. SAWICKA-KAPUSTA, J. WASILEWSKI, J. WEINER, and Z. WITKOWSKI

4.1 Introduction

W. GRODZIŃSKI

All the transfers presented in this chapter concern only selected groups of homeo- and poikilothermic animals, phytophagous as well as higher order consumers. Poikilotherms are represented by phyto- and cambiophagous insects. Homeotherms include "small birds", raptors, rodents, ungulates, and carnivores. Herbivore consumers are thus represented by insects, rodents, and ungulates, while higher order consumers include carnivorous mammals, raptorial and insectivorous birds.

A similar scheme of data presentation has been used for all the animal groups investigated. First, the size and biomass of populations were estimated directly in the Niepołomice Forest. The methods of estimation, however, have of necessity been variable. A single tree was considered the basic sampling plot for insects. Abundance of small mammals and birds was estimated on plots ranging from 1 to 6 and from 10 to 100 ha, respectively. Driving census of cervids and wild boar was performed every year over an area of 400 to 1,500 ha. Raptorial birds were censused throughout the whole area, as was also the case with inhabited burrows of badgers and foxes. Consequently, the methods used and the reliability of estimating population size for the entire Niepołomice Forest are widely variable among the investigated groups. Birds were investigated in such detail that total population sizes could be computed by simple addition, taking variation in type and age of forest ecosystem into account. Rodents were investigated in 12 sample plots in coniferous and deciduous forests, and their abundance extrapolated over the whole area, taking variation in type and age of tree stands into account. In turn, the estimates of raptorial birds and carnivorous mammals apply directly to the whole Niepołomice Forest. Additional difficulties were experienced at the ecotones at the borders between the forest complex and arable fields and meadows. In most cases, the meadows were also taken into account.

Having estimated the sizes and biomass of populations (or communities), energy and matter flow through consumers could be analyzed. This was also done for certain pollutants. Bioenergetic parameters and concentrations of macroelements and pollutants in consumers were derived from the author's studies, except for minor data from the literature. The paratrophic effects of consumers were also analyzed because they may considerably affect the functioning of forest ecosystems (Sect. 4.5).

The estimates presented in this chapter are not only incomplete because of the absence of data for several consumer groups, but are also variable in precision. Nevertheless, their extrapolation for the whole Niepołomice Forest is highly instructive. Many parameters describing the role of consumers in energy and matter flow through the ecosystem show wide variation, and means deviate considerably from data obtained during the IBP from single "typical" sampling plots (Medwecka-Kornaś et al. 1974).

4.2 Ecology, Energetics and the Significance of Phytophagous Insects in Deciduous and Coniferous Forests

Z. WITKOWSKI and K. BORUSIEWICZ

Species composition of the insect fauna, as well as ecology and significance in the ecosystem are variably understood in the Niepołomice Forest. The best known are the phytophagous insects, especially those injurious to wood (Starzyk 1973, 1976, 1977a, b, Starzyk and Witkowski 1981). Insects feeding on the assimilatory apparatus of trees were studied by several authors (Bandoła-Ciołczyk and Witkowski 1976, Lipa and Madziara-Borusiewicz 1976, Nunberg 1946, Witkowski 1975b, Witkowski and Kosior 1974, Witkowski and Langer 1978). Furthermore, Witkowski (1980) and Witkowski and Mazur (in press) investigated the species composition and ecology of weevils (Curculionidae). Other insect groups were studied only fragmentarily (Medwecka-Kornaś et al. 1974, Kazmierczak 1978, Starzyk 1975, Strojny 1961).

This study was aimed at recognizing species composition and dominance, energy and matter flow, nutrient resources, and trophic relationships in three phytophagous insect communities. Of these one included weevil species, which have rhizophagous larvae. Another community included beetles of the families Buprestidae, Cerambycidae, Curculionidae, and Scolytidae, the larvae of which are cambio- and xylophagous. The third included mostly butterflies with phyllophagous larvae.

4.2.1 Sampling Plots

The study was carried out on two plots representative of the dominant forest ecosystems. Plot 1 (3.2 ha) was located in the northern part of the forest complex and represented 90-year-old oak-hornbeam forest (Tilio-Carpinetum). The tree layer was dominated by oak (*Quercus robur* – 77% of standing crop), with minor components including lime (*Tilia cordata* – 10%), pine (*Pinus sylvestris* – 9.6%), hornbeam (*Carpinus betulus* – 2.5%), and alder (*Alnus glutinosa* – 0.8%). The herb layer was dominated by *Aegopodium podagraria, Urtica dioica, Impatiens nolitangere,* and *Anemone nemorosa.*

Plot 2 (1 ha) was located in the southwestern section of the Forest and represents mixed pine forest (Pino-Quercetum). The tree stand was dominated by pine (*Pinus sylvestris* – 67.4% of standing crop) and oak (*Quercus robur* – 32.6%). The undergrowth included hornbeam (*Carpinus betulus*), oak (*Quercus robur*), and beech (*Fagus silvatica*). The herb layer was dominated by sedges, mostly *Carex brizoides,* and grasses. Of the dicotyledons, the greatest share in the herb biomass was contributed by bilberry (*Vaccinium myrtillus*).

For comparative purposes other plots in both sections of the forest were surveyed.

4.2.2 Species Composition, Bioenergetics, and Biomass Structure

Rhizophagous Insects (Weevils)

Weevils were collected on both plots using a dip net (Witkowski 1969). Abundance varied seasonally, with two maxima, one in spring (May/June) and the other in autumn. The latter was less than the former. This pattern is typical of weevils (Witkowski 1975a).

Maximum abundances of weevil imagines were almost constant between plots (Table 4.1). Most of the species have larvae that feed on herb roots. The portion of larvae which feed on flowers, seeds, and/or leaves was negligible (see Witkowski and Kosior 1974).

After several assumptions and simplifications, the energetic budgets of weevils were estimated. To this end, the average individual biomasses and maximum abundances of imagines were used (Table 4.1).

According to Banasik (1978a), the standing crop of belowground parts of herbs in an oak-hornbeam wood adjacent to the Niepołomice Forest approximates 11.3×10^6 kJ ha^{-1}. Of this amount, weevils consume 596.7×10^3 ha^{-1} annually. Over 12.6×10^3 kJ ha^{-1} are transferred annually to the imago stage.

Table 4.1. Characteristics of the weevil communities in the oak-hornbeam (site 1) and the pine-oak (site 2) forests in the Niepołomice Forest, southern Poland

Characteristic of community	Site 1	Site 2
Dominant species	1. *Cidnorhinchus quadrimaculatus* L. 2. *Ceutorhynchus scrobicollis* Meresh. et Wagn. 3. *Sciaphilus asperatus* Bonsd. 4. *Liophloeus lentus* Germ.	1. *Phyllobius arborator* (Herbst) 2. *Apion flavipes* (Payk.) 3. *Stroptohosoma capitatum* (Deg.)
Number of species	23	21
Number of individuals	129	96
Average dry mass of an individual (mg)	27.53	29.55
Maximum number of individuals (indiv. ha^{-1})	18.200	15.300
Weevils feeding on underground organs of plants (percent of the total biomass)	99.4	98.8

In the mixed pine forests, the standing crop of belowground parts of herbs is 4.7×10^6 kJ ha^{-1} (Chap. 3.4). The weevils consume 545.0×10^3 kJ ha^{-1} annually, of which 109.0×10^3 kJ ha^{-1} is used to build body tissues, whereas 11.3×10^3 kJ ha^{-1} is transferred annually to the imago stage. Thus the weevil larvae use 5.3% of consumed plant biomass in oak-hornbeam forest, and 11.6% in mixed pine forest.

When expressed in terms of matter flow, these values indicate that the standing crop of belowground parts of herbs in oak-hornbeam forests is 629.4 kg ha^{-1} (Chap. 3.4), of which 33.3 kg ha^{-1} are consumed annually by weevils. Annually, 4.2 kg ha^{-1} of the biomass produced by weevil larvae is eliminated; 0.5 kg ha^{-1} (or 10.6%) is transferred to the stage of imago; 12.8 kg ha^{-1} are returned to the ecosystem in the form of larval excrement.

In mixed pine forests the standing crop of belowground parts of the herb layer is 259.7 kg ha^{-1} (Chap. 3.4), of which 30.4 kg ha^{-1} is consumed annually by weevil larvae. Annually, 11.7 kg ha^{-1} is returned to the ecosystem in the form of excrement, whereas 10.3% of the biomass produced by weevil larvae is transferred to the stage of imago.

Cambio- and Xylophagous Beetles

To estimate the abundance of cambio- and xylophagous beetles, all trees on the sample plots were surveyed intensively and the number of buprestid, cerambycid, and curculionid larvae was determined. The abundance of scolytids was estimated from the number of inhabited larval tunnels, or maternal tunnels and festive chambers. In addition, imagines feeding on herb flowers were collected with a dip net on plot 1.

Consumption by cerambycid, buprestid, and curculionid larvae was estimated from the volume of wood, bast, and bark eaten by particular species (Starzyk and Witkowski 1981).

Total consumption by larvae of the five species considered was computed according to the formula:

$$C = \sum_{i=1}^{s} (N_i \times 0.8C_i),$$

where N_i denotes the average larval density of i-th species (or density of feeding patches for scolytids), and C_i denotes the average annual individual larval consumption of i-th species (or consumption per 1 feeding patch for scolytids).

In three scolytid species, viz. *Scolytus intricatus* in oak-hornbeam forests and *Blastophagus piniperda* and *B. minor* in pine forests, supplemental feeding by juvenile imagines has been observed. This concerned the feeding on top shoot buds of pines, and small chambers excavated by *S. intricatus* in the base of the second youngest shoots of oak. Taking into account the average volume of wood consumed in this manner and the density of imagines, the biomass and caloric value of this supplemental food could be estimated. On the average, an individual of *S. intricatus* attacks and consequently destroys, because the attacked twigs fall, 1 shoot with 6 leaves annually, while an individual of *Blastophagus* sp. attacks 2.5

Table 4.2. Cambio- and xylophagous species in the oak-hornbeam (site 1) and pine-oak (site 2) forests of the Niepołomice Forest

Site	Species	Number of larvae ha^{-1}
	Cerambycidae	
	Leiopus nebulosus L.	65
	Oplosia fennica Payk.	3
	Plagionotus arcuatus L.	108
	Rhagium mordax Deg.	85
	Rh. sycophanta Schrk.	14
	Saperda scalaris L.	37
1.	*Stenostola dubia* Laich.	14
	Buprestidae	
	Agrilus angustulus Ill.	26
	Agrilus biguttatus F.	1
	Scolytidae	
	Scolytus intricatus Ratz.	90
	Xyleborus dispar F.	6
	X. saxeseni Ratz.	2
	Cerambycidae	
	Criocephalus rusticus L.	118
	Rhagium inquisitir L.	240
	Curculionidae	
	Pissodes piniphilus (Herbst)	2,540
2.	Scolytidae	
	Blastophagus piniperda L.	737
	B. minor Htg.	356
	Pityophthorus pityographus Ratz.	2
	Pityogenes bidentatus (Herbst)	11

shoots with 89 needles per shoot annually. It has also been assumed that 16.5 scolytid imagines hatch annually at a single feeding ground.

The biomass produced by larvae of the investigated cerambycid, buprestid, and curculionid species has been calculated assuming a cylindrical shape, according to the formula:

$$p = \sum_{i=1}^{s} \frac{N_i P_i}{T_i},$$

where N_i is the average larval density of i-th species, P_i the biomass produced by the average larva of i-th species, and T_i denotes time (in years) necessary to produce a standing crop of i-th species.

The oak forest of plot 1 was attacked by 12 species of cambio- and xylophagous beetles; on plot 2, only 7 species were observed (Table 4.2).

The caloric value of trees attacked by cambio- and xylophagous insects amounts to 5.6×10^6 kJ ha^{-1} in the investigated oak-hornbeam forests. The annual consumption by cerambycid and buprestid larvae is 6.7×10^3 kJ ha^{-1}, of which 0.18×10^3 kJ ha^{-1} is used to build body tissues. The maximum standing

Table 4.3. Parameters of the energy budget of the cambio- and xylophagous insects feeding on oaks (site 1) and pine trees (site 2) in the Niepołomice Forest (kJ ha^{-1} yr^{-1})

Biomass of infected trees (B)	Oak-hornbeam (site 1) 5.59×10^6			Mixed pine forest (site 2) 15.02×10^6		
	Scolytidae	Cerambycidae and Buprescidae	Total	Scolytidae	Other species	Total
Damaged tree biomass (D)	7.50×10^3	8.42×10^3	15.92×10^3	116.31×10^3	69.09×10^3	185.4×10^3
Consumption of larvae (C)	6.56×10^3	6.75×10^3	13.31×10^3	87.11×10^3	55.27×10^3	142.4×10^3
Supplementary feeding by imagines	0.86×10^3	–	0.86×10^3	75.06×10^3	–	
Production of larvae (P)	0.17×10^3	0.18×10^3	0.35×10^3	6.04×10^3 [a]	3.83×10^3	9.87×10^3
Maximal standing crop of imagines (B_1)	–	5.70×10^3		–	–	
Elimination	–	0.17×10^3	–	–	–	

[a] Estimated from the consumption to production ratio in the other cambio- and xylophagous groups

crop of imagines is 5.7 kJ ha^{-1}, indicating that 0.17×10^3 kJ ha^{-1} of the biomass produced by larvae is eliminated annually (Table 4.3).

The scolytids feeding on plot 1 consume 7.50×10^3 kJ ha^{-1} annually, with larval consumption amounting to 6.56×10^3 kJ ha^{-1}. The caloric value of the supplemental food of *S. intricatus* approximates 0.86×10^3 kJ ha^{-1} annually (Table 4.3). Together, these insect larvae consume less than 0.001% of standing crop of the tree stand (Starzyk and Witkowski 1981).

In the mixed pine forests investigated the cerambycid and curculionid larvae consume 55.25×10^3 kJ ha^{-1} annually, while the annual production is 3.83×10^3 kJ ha^{-1} (or 6.9% of the consumed biomass). The scolytids consume 116.3×10^3 kJ ha^{-1} annually, of which the larvae consume 87.1×10^3 kJ ha^{-1} (Table 4.3). In total, the insect larvae considered consume 0.005% of the standing crop of the tree layer.

Phyllophagous insects

This group of insects has been studied continuously since 1968, partly within the framework of the IBP. The major aim was to recognize the ecology and bioenergetics of the dominant phyllophagous species, the oak leaf roller moth, *Tortix viridiana* L., and to study its role in the oak-hornbeam forest ecosystem (Witkowski 1975b). Associated species include other leaf roller moths, *Archips podana* Scop., *A. crataegena* Hb., *Pandemis ribeana* Hbn., moths and butterflies *Amphipyra pyramidea* L., *Cosmia trapezina* L., *Thecla quercus* L. (Lipa and Madziara-Borusiewicz 1976).

Table 4.4. Biomass flow ($kg\ ha^{-1}\ yr^{-1}$) and parameters of the energy budget ($kJ\ ha^{-1}\ yr^{-1}$) of caterpillars feeding on oak leaves in the oak-hornbeam forest in the Niepołomice Forest

	High number of larvae		Low number of larvae	
	$kg\ ha^{-1}\ yr^{-1}$	$kJ\ ha^{-1}\ yr^{-1}$	$kg\ ha^{-1}\ yr^{-1}$	$kJ\ ha^{-1}\ yr^{-1}$
Leaf production (P)	–	71.31×10^6		48.14×10^6
Consumption (C)	1,328	27.82×10^6	312	6.54×10^6
Production (P)	410	10.31×10^6	95	2.39×10^6
Respiration (R)	–	6.37×10^6	–	1.47×10^6
Excreta (FU)	531	11.15×10^6	124	2.60×10^6
Elimination	325	8.17×10^6	88	2.22×10^6

The abundance of phyllophagous larvae ranges from 2.02×10^6 to 8.7×10^6 individuals ha^{-1}, the upper limit being achieved during mass outbreaks. The gradations are periodic, but fairly irregular and strongly correlated with weather conditions, especially those prevalent during the hatching of larvae (Witkowski 1975b). There is also a correlation between the average weight of pupae and their vulnerability to parasites on the one hand and the phase of gradation on the other. The pupae are heavier and less affected by parasites in the initial phase of the outbreak than in the decline (Witkowski 1975b). Such changes in vulnerability to parasites were probably responsible for the recent infestation of the population being investigated, which has been considered to indicate that larval abundance is controlled by pathogens at least in the final stages of gradation (Lipa and Madziara-Borusiewicz 1976).

A recent study of energy flow through phyllophagous insects feeding on oaks in the Niepołomice Forest (Bandoła-Ciołczyk and Witkowski 1976) demonstrates that the larvae consume 6.5×10^6 to 27.8×10^6 kJ ha^{-1} annually, depending on the phase of gradation (Table 4.4). The larvae thus consume 13.6 to 37.0% of the total production of oak leaves, and 36.5% of this is used for the building of body tissues (Table 4.4).

The larvae consume 312 to 1,328 kg of leaves annually, of which 124 to 531 kg ha^{-1} are returned to the ecosystem in the form of excrement. Carnivores and parasites consume 88.3 to 325 kg ha^{-1} of the larval biomass ha^{-1} annually (Table 4.4).

A few years ago, the mixed pine forests of Niepołomice also contained other abundant phyllophagous insect species (Nunberg 1946), but these have been eliminated by pesticides.

4.2.3 Nutrients (Nitrogen, Phosphorus, Potassium, and Calcium) in Insect Bodies and Excrement

The amounts of nitrogen, phosphorus, potassium, and calcium contained in the groups of phytophagous insects investigated have been estimated (Table 4.5) from data on nutrient contents in insect bodies and excrement, as taken from Spector (1956), Zlotin and Hodašova (1974).

Table 4.5. Amounts of major nutrients in the production and excreta of insect groups investigated (g ha^{-1} yr^{-1})

		Body mass increment			Excreta	
		Phyllophagous insects	Cambio- and xylophagous beetles	Rhizophagous weevils	Phyllophagous insects	Rhizophagous weevils
Oak-hornbeam forest (site 1)	nitrogen	46,864.0	1.30	881.5	6,877.5	269.2
	phosphorus	6,262.0	0.17	117.8	2,084.7	81.6
	potassium	2,020.0	0.06	38.0	5,436.5	212.8
	calcium	134.7	0.01	2.5	50,559.5	1,957.8
Pine-oak forest (site 2)	nitrogen	–	28.2	803.6	–	245.9
	phosphorus	–	3.8	107.0	–	74.6
	potassium	–	1.2	34.6	–	194.4
	calcium	–	0.08	2.3	–	1,807.9

In the oak-hornbeam forest studied, the rhizophagous larvae assimilate 882 g nitrogen, 118 g phosphorus, 38 g potassium, and 3 g calcium ha^{-1} annually. The cambio- and xylophagous larvae assimilate 1 g nitrogen and less than 1 g phosphorus, potassium, and calcium ha^{-1}. Finally, the phyllophagous larvae, which predominate among phytophagous insects in oak-hornbeam forests, assimilate approximately 47 kg nitrogen, over 6 kg phosphorus, 2 kg potassium, and 135 g calcium ha^{-1} (Table 4.5).

The amounts of nutrients in excrement have been estimated (Table 4.5) only for rhizophagous and phyllophagous insects, because the energy and biomass produced by cambio- and xylophagous insect larvae could not be determined (Table 4.3). However, as judged from other energy and matter budget parameters in the latter group, amounts of nutrients in the excrement are negligible.

The relationship of calcium to other nutrients in insect bodies and excrement is notable (Table 4.5). Contrary to the vertebrates (Sects. 4.3–4.5), insects do not assimilate calcium. Hence, the amounts of calcium incorporated in body tissues are one-fifteenth those of potassium, but the amount of calcium in excrement exceeds that of potassium by a factor of 10.

Together, in the oak-hornbeam forests the phytophagous insect groups investigated assimilate 71.6×10^3 kg nitrogen, 9.6×10^3 kg phosphorus, 3.1×10^3 kg potassium, and 0.21×10^3 kg calcium annually, and 98% of this accounted for by phyllophagous insects. In the mixed pine forests, in turn, phytophagous insects assimilate 7.4×10^3 kg nitrogen, 0.98×10^3 kg phosphorus, 0.3×10^3 kg potassium, and 0.02×10^3 kg calcium annually.

The excrement of phytophagous insects in the oak-hornbeam forests contains 10.7×10^3 kg nitrogen, 3.2×10^3 kg phosphorus, 8.5×10^3 kg potassium, and 78.8×10^3 kg calcium annually. In the mixed pine forests, only the excrement of rhizophagous weevils has been studied in this respect. It contains 2.2×10^3 kg nitrogen, 0.7×10^3 kg phosphorus, 1.7×10^3 kg potassium, and 16.0×10^3 kg calcium annually.

4.2.4 Phytophagous Insects in the Ecosystems of the Niepołomice Forest

The significance of phytophagous insects in the forest ecosystems may be considered from various perspectives. The affects of cambio- and xylophagous insects on tree species are well known in forestry (Escherich 1923). Generally, they destabilize artificially managed forest ecosystems.

The caloric value of their damage to tree stands considerably exceeds their consumption. Insect tunnels in the wood account for only 0.29% of the standing crop in deciduous forests and for 1.23% in coniferous forests. Greater damage is effected by the scolytids whose imagines feed supplementarily on young shoots. What is really eaten by imagines of *Scolytus intricatus* represents only 1.9% of the caloric value of the leaves destroyed by their feeding. For *Blastophagus piniperda* and *B. minor* feeding on pines, this proportion amounts to 3.7% (Table 4.6). Cambio- and xylophagous larvae do not attack healthy trees, but only those infected or weakened by phyllophagous insects or by competition for light and nutrients. Scolytid imagines, however, mostly attack the canopies of healthy trees adjacent to those infected previously by larvae. In this way, new trees are being prepared as candidates for subsequent attack by scolytid larvae.

Outbreaks of phyllophagous insects also contribute to the destabilization of forest ecosystems, because they weaken the trees and thus facilitate their infection by cambio- and xylophagous species. This is especially important in coniferous forests, because recovery of the assimilation apparatus is seldom, if at all possible (Koehler 1978).

Another important aspect of the role of phytophagous insects in forest ecosystems is that they fertilize the soil with nutrient-rich excrement. On the average, over 7 kg nitrogen, almost 6 kg potassium, over 2 kg phosphorus, and almost 52 kg calcium ha^{-1} is deposited annually in the oak-hornbeam forests studied (Table 4.5). These values increase considerably during mass outbreaks of phyllo-

Table 4.6. Damage to tree trunks infected by cambio- and xylophagous insects, and to tree canopies attacked by scolytid imagines (supplementary feeding) in the oak-hornbeam (site 1) and pine-oak (site 2) forests of the Niepołomice Forest (kJ ha^{-1} yr^{-1}), southern Poland

	Site 1	Site 2
Biomass of infected trees (B_1)	5.59×10^6	15.02×10^6
Damage by larval and imaginal feeding (C_1)	15.92×10^3	185×10^3
Percent of tree biomass damaged (C_1/B_1)	0.28%	1.23%
Leaves (needles) destroyed by imaginal feeding (Z)	44.43×10^3	2011.40×10^3
Supplementary feeding by scotylid imagines (C_2)	0.86×10^3	75.06×10^3
Percent of leaf biomass damaged (C_c/Z)	1.94%	3.73%

phagous insects (Bandoła-Ciołczyk and Witkowski 1976). The increased concentration of easily soluble excrement accelerates litter decomposition by a factor of 2–3 (Bandoła-Ciołczyk 1974, Zlotin 1975). These data confirm the role of phytophagous insects in the regulation of forest ecosystems (O'Neil and Reichle 1980). Zlotin (1975) suggests that phyllophagous outbreaks in oak forests only occur during warm dry summers when the decomposition rate of litter is decreased and hence the nutrient supply for trees declines. A phyllophagous outbreak causes short-term weakening of the trees, but contributes many easily available nutrients to the soil and accelerates decomposition of the litter, which increases the rate of matter flow through the ecosystem.

In addition to that of phyllophagous insects, a regulatory function in forest ecosystems is served also by rhizophagous weevils. As demonstrated by Schauermann (1977), they increase the rate of remineralization in the soil in beech forests.

Phytophagous insects are a keystone link in the trophic chain of forest ecosystems; a method of energy transfer from plant producers to secondary consumers.

The role of phytophagous insects in energy flow through the forest ecosystem investigated represents only a part of the role played by the whole insect fauna. For the purpose of constructing a model of energy flow through the oak-hornbeam ecosystem (Medwecka-Kornaś et al. 1974), it was estimated that saprophagous insects use 34.4×10^6 kJ ha^{-1} in the form of litter annually. Given that they process 80% of this matter, their consumption amounts to 27.52×10^6 kJ ha^{-1} annually. This value closely approximates the consumption by the phytophagous insects investigated during mass outbreaks of phyllophagous forms (Table 4.6).

The significance of such browsing on the energy flow between producers to consumers and reducers has thus far not been observed in forest ecosystems (Wiegert and Owen 1971, Odum 1971). It has been generally assumed that consumers process only 10% of the plant-produced energy in these ecosystems (Odum 1971).

Notable observations also follow from a comparison of the browsing rate and the leaf productivity in the ecosystem. Total annual production of leaves averages 81.71×10^6 kJ ha^{-1} (Medwecka-Kornaś et al. 1974). Under low populations of phyllophagous insects, only 8% of the energy contained in the leaves is consumed, but during mass outbreaks, however, this proportion may reach 34%.

The phytophagous groups considered in this study are only a part of the species-rich insect fauna at the primary consumer level. Insects sucking tree and herb sap, as well as rhizophagous midges and insects feeding on galinuts and aboveground parts of herbs, have been omitted from this study. These groups would not significantly change the estimates of the role of phytophagous insects in the ecosystem because they are subordinate in abundance and biomass to phyllophagous types. Nevertheless, they may be of some importance for forest ecosystems, as for example the destabilization of tree stands.

Unknown is also the role of carnivorous and parasitic insects in this ecosystem. As shown by Witkowski (1975a), 10% of the larvae of the oak leaf roller moth are consumed by insectivorous birds, carnivorous insects, spiders, and amphibians. Concerning infection by parasites, 0.9 to 6.3% of the pupae of this species are infected, depending on the phase of outbreaks. The pupae outside oak forest canopies (on oak trunks, herbs, etc.) are twice as liable to infection as those in the canopy. Kaźmierczak (1978) recorded 135 parasitic species of the family

Ichneumonidae in the Lipówka Reserve in the Niepołomice Forest. Of these, 58% feed on phyllophagous insects, primarily on the oak leaf roller moth (24 ichneumonid species), while 18% feed on cambio- and xylophagous insects, mostly on *Leiopus nebulosus* (10 ichneumonid species).

4.2.5 Effects of Industrial Pollution on the Insect Fauna

K. Borusiewicz

The most injurious insects of the pine forests at Niepołomice are the cambiophagous insects, which are assigned as secondary pests in forestry. Their incidence is increased in areas most severely affected by air pollution. The most abundant species are *Blastophagus piniperda* L. and *Pissodes piniphilus* Hbst. The former especially causes great damage to tree stands. The associated assemblage of cambiophagous species includes *Blastophagus minor* Htg., *Pityogenes bidentatus* Hbst., *Pityophthorus pityographus* Rotz., *Pissodes pini* L., *Monochamus galloprovincialis* Ol., *Acanthocinus aedilis* L., *Criocephalus rusticus* L., *Rhagium inquisitor* L., and *Callidium aenum* Deg.

The occurrence of such a diversified and abundant assemblage of pests results in considerable injury to the trees. The weakened trees attract cambiophagous insects, especially *B. piniperda*. This may be due either to decreased water content of the tree (Madziara-Borusiewicz and Strzelecka 1977) and hence, decreased osmotic pressure, or to increased content or modified composition of etheric oils (Madziara-Borusiewicz and Strzelecka 1977).

In the areas affected by industrial air pollution, the concentration of pollutants in plant organs (bast, needles) consumed by phytophagous insects is high. Consequently, the pollutant elements are incorporated into the plant-insect-excrement cycle.

This cycle has been examined in respect to Zn, Cd, Ni, Fe, Mn, Cu, and Pb in the imagines of *B. piniperda* (Table 4.7; Madziara-Borusiewicz, in press). The elements contained in the bast are to various extents assimilated and excreted by the insects. Their concentrations measured in the imagines reflect cumulative concentrations achieved throughout the life cycle: at the larval stage, at the imagine stage during supplementary feeding, and during the development of maternal

Table 4.7. Concentration of elements in food, body, and excrement of *Blastophagus piniperda* L.

	Elements (ppm)						
	Zn	Cd	Ni	Fe	Mn	Cu	Pb
Phloem	69	3.6	5.3	137	152	3.4	4.8
Insect body	375	9.3	62.5	281	46	87.5	[a]
Excreta	312	2.6	[a]	2589	72	89.3	[a]

[a] Indiscernible

tunnels. The concentration in excrement, however, concerns only the latter stage (Madziara-Borusiewicz, in press).

Nickel, cadmium, and zinc are accumulated in insect bodies. Thus, these toxic elements are returned to the ecosystem only after the death of the insects, as a rule. They may be partially transferred along the trophic chain, firstly to woodpeckers.

The amounts of iron passed in feces by the insects exceed by one order of magnitude those assimilated and incorporated into body tissues. This may be due to the antagonistic relationship of iron to zinc, the latter being accumulated by the insects.

Manganese and copper are also evacuated rather than assimilated by the insects. Lead seems to be not assimilated at all (Madziara-Borusiewicz, in press).

Industrial dusts and gases also effect changes in the action of phytophagous insects on tree stands. Some pests intensify their injurious affects, for example, *Exoteleia dodecella* and *Rhyacionia buoliana* in young pine plantations; others which previously did not cause injury initiate it. This concerns pine as well as oak stands. An increase in the abundance of cynipids, mostly *Cynips lignicola* Htg., has been recently observed in young oak plantations at Niepołomice. These insects cause damage to buds (Madziara-Borusiewicz, in press). Owing to their mass occurrence, the height growth of trees is inhibited, or they are reduced to shrubs.

4.3 Energy and Matter Flow Through Homeothermic Consumers

4.3.1 Energy and Matter Flow Through Rodents

A. Górecki

Field work on small rodents of the Niepołomice Forest was conducted from 1966 to 1971 in oak-hornbeam and alder forests, and between 1975 and 1978 in coniferous forests. Two rodent species are major dominants in all sampled plots. These are the bank vole (*Clethrionomys glareoulus*) and the yellow-necked field mouse (*Apodemus flavicollis*), which jointly account for more than 94% of the total (Table 4.12). The associated species include *Microtus arvalis, M. agrestis, Apodemus agrarius, Micromys minutus, Pitymys subterraneus,* and *Muscardinus avellanarius.* Other Gliridae and squirrels were beyond the scope of this study (Gębczyński et al. 1972). The sampling plots were chosen to include old as well as young tree stands in deciduous and coniferous forests. The variation in density and species composition of rodent assemblages related to forest age and type has been taken into account in computing the mean densities of rodents in the Niepołomice Forest. Only the two dominant species have been used for the purpose of estimating energy and matter flow in the forest ecosystems.

Variation in the abundance of rodents in the coniferous forests has been determined from spring (April/May) and autumn (October) sampling using the "Standard Minimum" (Grodziński et al. 1966). The data have been interpolated with the curve obtained in 1976 from three additional sampling periods (July,

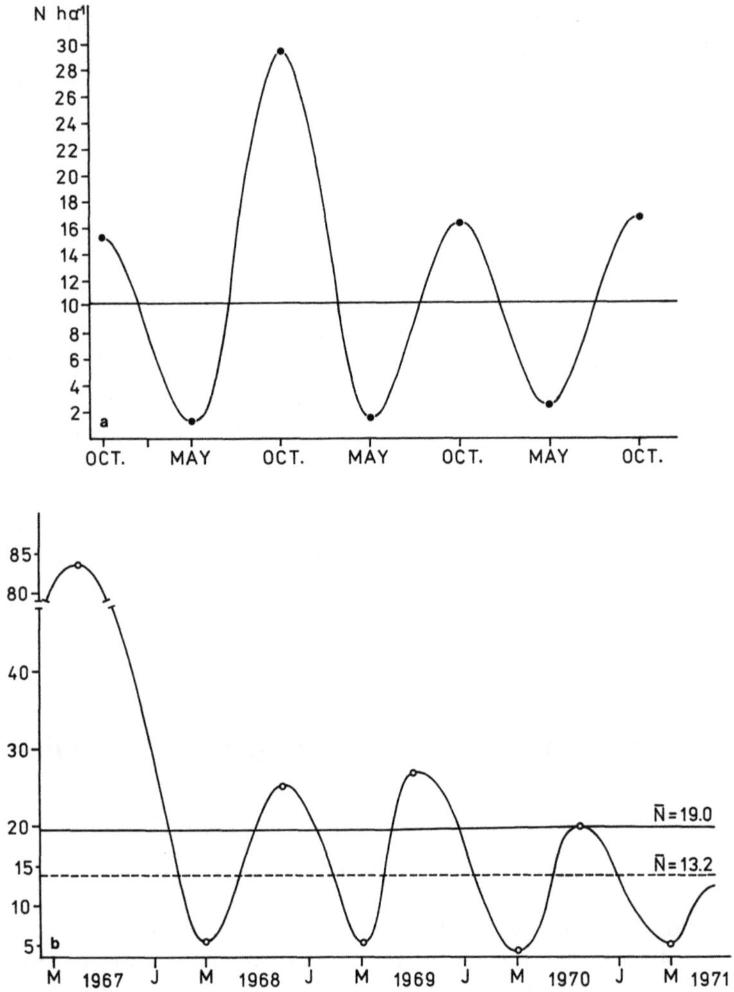

Fig. 4.1. a, b. **a** Rodent abundance in coniferous forests in 1975–1978. **b** Rodent abundance in deciduous forests in 1967–1971 (Bobek 1973)

September, and December). The pattern of variation is presented in Fig. 4.1 a. Maximum density was observed in autumn 1976 (29.5 individuals ha^{-1}), the minimum in spring 1976 (fewer than 1.5 individuals ha^{-1}). The average density throughout the time interval investigated was 10.3 individuals ha^{-1} (Fig. 4.4). No massive increases of rodents have been observed.

A similar study was undertaken between 1966 and 1971 in the deciduous forests at Niepołomice (Bobek 1973). The patterns obtained are given in Fig. 4.1 b. The maximum density was recorded in May and June 1967 (87 individuals ha^{-1}), and the minimum in the spring of 1970. The average density for the period investigated was 19.0 individuals ha^{-1}, of which 66% were accounted for by the bank vole (Table 4.8). The peak of rodents in the summer of 1967 was due to a massive

Table 4.8. Percent occurrence of particular species in the rodent communities of different forest types in the Niepołomice Forest (Górecki unpubl.)

	Coniferous forests 1975–1978	Deciduous forests 1966–1971	Average
Clethrionomys glareolus	81.7	65.9	73.8
Apodemus flavicollis	14.3	26.7	20.5
Apodemus agrarius	1.3	6.8	4.05
Micromys minutus	1.6	–	0.8
Microtus agrestis	0.9	0.2	0.55
Microtus arvalis	0.2	–	0.1
Pitymys subterraneus	–	0.3	0.15
Muscardinus avellanarius	–	0.1	0.05

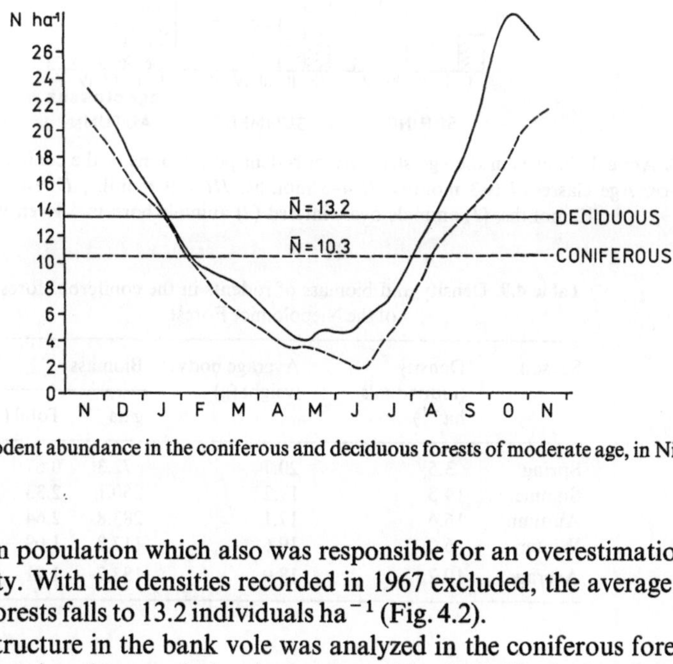

Fig. 4.2. Rodent abundance in the coniferous and deciduous forests of moderate age, in Niepołomice

increase in population which also was responsible for an overestimation of average density. With the densities recorded in 1967 excluded, the average in the deciduous forests falls to 13.2 individuals ha^{-1} (Fig. 4.2).

Age structure in the bank vole was analyzed in the coniferous forests (Zejda 1961). In spring the population is dominated by individuals 6 to 9 months old, while juveniles account for only 7.5% of the total number. In summer, juveniles (2.3 months old on average) predominate and account for 44% of the population. In autumn, 57% are accounted for by animals ranging from 4 to 6 months, which indicates that the maximum birth rate is achieved in May (Fig. 4.3).

The share of specific age classes in the rodent population served as a basis for estimating the average standing crop in the coniferous forests. This is 180 g biomass ha^{-1}, which represents 1.7 t over the whole coniferous complex (Table 4.9).

Because of the predominance of the bank vole, the annual turnover rate of biomass (3.2 after Bobek 1973) has been assumed to estimate net productivity of

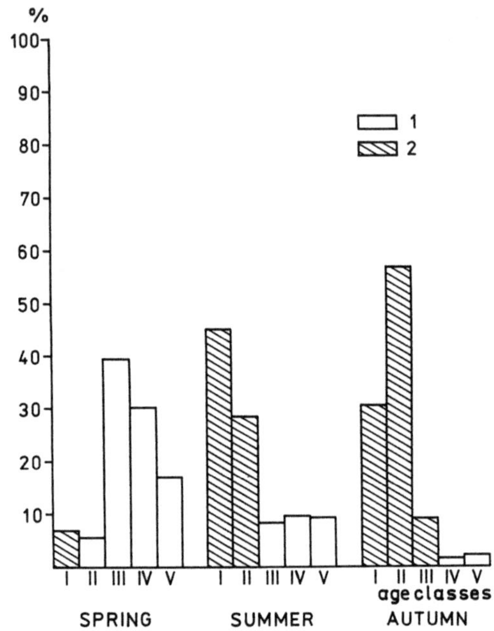

Fig. 4.3. Annual changes in the age structure of rodent populations in the coniferous forests of Niepołomice. Age classes: *I* 0–3 months; *II* 4–6 months; *III* 7–9 months; *IV* 10–12 months; *V* over 12 months. (*1*) animals overwintered (*2*) animals born in a given year

Table 4.9. Density and biomass of rodents in the coniferous forests of the Niepołomice Forest

Season	Density (individuals ha^{-1})	Average body weight (g)	Biomass	
			$g\ ha^{-1}$	Total (t)
Spring	3.5	20.7	72.3	0.67
Summer	14.5	17.2	250.1	2.33
Autumn	16.6	17.1	283.8	2.64
Winter	6.0	19.6	117.8	1.09
Average	10.2	18.0	183.7	1.71

rodents (Petrusewicz et al. 1983). The total annual net production then is 587.8 $g\ ha^{-1}$ or 4,073.7 $kJ\ ha^{-1}$ (caloric value is 6.3 $kJ\ g^{-1}$; Górecki 1965), which represents 5.2 t for the coniferous complex.

Bobek (1973) also estimated rodent productivity in the oak-hornbeam forests. Standing crop averages 283 $g\ ha^{-1}$, and total annual net productivity 5,940 kJ^{-1}, which represents 1.4 t for rodent biomass for the deciduous complex. Thus, the total net annual production of rodents exceeds 6.6 t for the whole Niepołomice Forest.

Seasonal variation in the diet of rodents has been determined from stomach contents (Zemanek 1972, Sierek 1977, Siemieńska 1977). According to Zemanek,

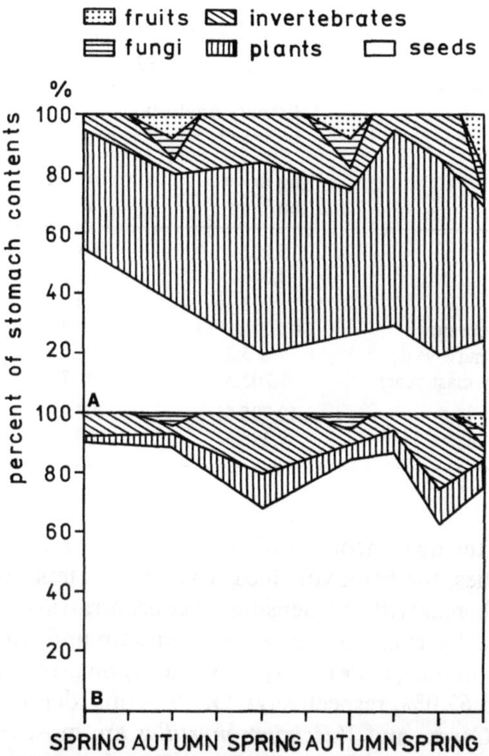

Fig. 4.4. A, B. Seasonal changes in the stomach contents in *C. glareolus* (**A**) and *A. flavicollis* (**B**) in a deciduous forest at Niepołomice (Zemanek 1972)

the mice in deciduous forests eat mostly seeds (62–90% of stomach contents), and invertebrates and green plant tissues play only a subordinate role (4.5–26% and up to 12%, respectively). There is no distinct predominance of any single kind of food in the diet of the bank vole. Seeds (19–55% of stomach contents) as well as green plant tissue (38–65%) and invertebrates (up to 16%) are eaten (Fig. 4.4). Of the herbs, *Anemone nemorosa, Mycelis muralis, Oxalis acetosella,* and *Ranunculus lanuginosus* are most commonly eaten by the bank vole, while the mouse feeds preferentially on *M. muralis* and *R. lanuginosus*.

The annual primary production of the oak-hornbeam forests of the Niepołomice Forest has been estimated by Zemanek (1972) as approximately 180×10^6 kJ ha^{-1} in a poor mast year and over 201 kJ ha^{-1} with heavy mast (Table 4.10). The amounts of food available for rodents depend on the crop of acorns and range from 6.8 to 16.9% of primary production for the bank vole, and from 0.9 to 11.7% for the mouse.

In the coniferous forests, the amounts of seeds in rodent stomachs are largely independent of availability (Górecki and Gębczyńska 1962, Sierek 1977, Siemieńska 1977). The diet of the vole includes the green parts of 45 herb species, 2 shrubs, and 3 trees. The plants that are the most commonly fed upon are: *Moehringia, Carex,* and *Genista.* Mosses and ferns have also been commonly found in vole

Table 4.10. Primary production (kg dry weight ha^{-1} yr^{-1}) and food supply of rodents investigated ($10^6 \times kJ$ ha^{-1} yr^{-1}) in a deciduous forest (Tilio-Carpinetum) in the Niepołomice Forest (Zemanek 1972)

Food	Primary production		Available food	
			Voles	Mice
	kg dry wt ha^{-1} yr^{-1}	$10^6 \times kJ$ ha^{-1} yr^{-1}	$10^6 \times kJ$ ha^{-1} yr^{-1}	
Wood	5,962	108.0	0.4	–
Leaves	3,026	62.0	3.1	–
Herbs	602.36	10.7	8.4	1.2
Fungi and invertebrates	–	0.2	0.2	0.2
Tree seeds (light mast year)	5.3	0.1	0.1	0.1
Tree seeds (heavy mast year)	1,205.3	22.2	22.2	22.2
Total heavy mast year	10,795.66	202.9	34.3	23.7
Total light mast year	9,595.66	180.8	12.2	1.6

stomachs. Of the mast, acorns and pine are most commonly used. With high population densities, the bank vole feeds to a large extent on materials with low caloric value, whereas with low densities, it concentrates on various seeds (Fig. 4.5).

The mouse feeds mostly on seeds in autumn and winter (86.5 and 78.8% of the stomach contents, respectively), but mainly on green plants in spring and summer (77.7 and 67.0%, respectively). Of 19 plants identified in stomach contents, *Moehringia, Carex,* and *Vaccinium myrtillus* are most commonly, while acorns and pine seeds are most frequent (Fig. 4.5).

The diets of the two rodents are very similar but the mouse eats half the green materials of the vole (Sierek 1977, Siemieńska 1977, Górecki unpubl.). The preference for seeds may significantly affect the process of forest regeneration, especially during mass outbreaks of rodents. The diets of both species in coniferous and deciduous forests are presented in Table 4.11.

The size of the rodent population in the Niepołomice Forest is highly variable over time, consequently, the parameters of energy flow through the population (production, respiration, consumption, and food utilization) are variable as well (Górecki 1968, Grodziński and Górecki 1967, Drożdż 1968). A simulation model has been developed to attempt to estimate this pattern of variation (Kozłowski et al. 1981).

The parameters of energy flow through rodent populations in the coniferous and deciduous forests are given in Table 4.12. Rodent production is much higher in the deciduous forests. This is due to the higher density of rodents, as well as to greater average individual weight caused by a higher proportion of mice (Grodziński et al. 1970). The total annual consumption of rodents in the forest is $2{,}465 \times 10^6$ kJ, of which only 46.9×10^6 kJ (1.9%) are used for production.

The level of nutrients, and their flow through the rodents investigated have been estimated from the data on body and excrement composition presented by Górecki and Szafranek (1981). The average concentrations of N, P, K, and Ca are given in Table 4.13. With species and age structure of the population taken into account, the average level of nutrients in rodent bodies has been estimated. It ap-

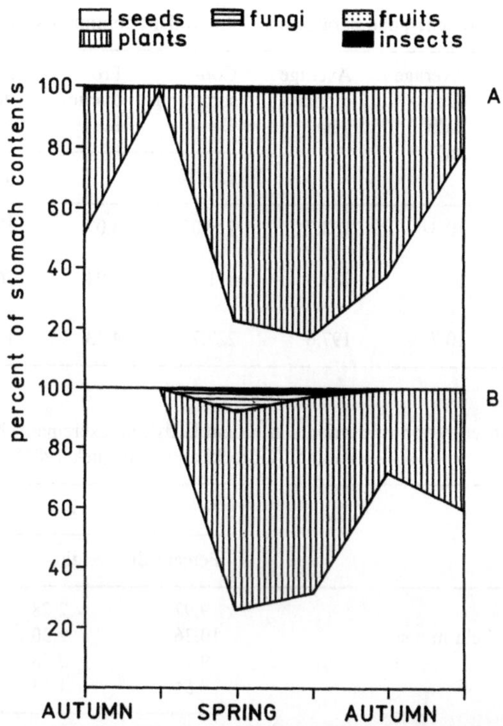

Fig. 4.5. A, B. Seasonal changes in the stomach contents in *C. glareolus* (**A**) and *A. flavicollis* (**B**) in coniferous forest at Niepołomice (Siemieńska 1977, Sierek 1977)

Table 4.11. Average annual food consumption for the bank vole and the yellow-necked field mouse in the deciduous and coniferous forests of the Niepołomice Forest (Siemieńska 1977, Sierek 1977, Górecki unpubl.)

	Clethrionomys glareolus	*Apodemus flavicollis*
	Percent share of stomach contents	
Coniferous forests		
Plants	54.7	40.3
Seeds	43.5	56.9
Insects	0.8	0.6
Fungi	0.5	2.2
Fruits	0.5	–
Deciduous forests		
Plants	53.2	7.5
Seeds	29.8	79.0
Insects	13.2	11.9
Fungi	2.0	1.3
Fruits	1.8	0.3

Table 4.12. Bioenergetic parameters of the rodent community in the Niepołomice Forest

	Average density indiv. ha^{-1}	Average biomass $g\ ha^{-1}$	Consumption C	Production P	Assimilation A	Respiration R	P/R %
			$kJ \times 10^3$		$ha^{-1}\ yr^{-1}$		
Coniferous forests	10.3	183.7	214.02	4.07	177.61	173.64	2.3
Deciduous forests	13.1	283.3	311.43	5.94	264.90	259.05	2.3
Average	10.7	197.6	227.70	4.33	189.86	185.63	

Table 4.13. Concentration of nutrients in rodent body and excrement, Niepołomice Forest, southern Poland. (Górecki and Szafranek 1981)

		N	P	K	Ca
		Percent dry weight			
Bank vole		9.92	2.28	0.99	2.92
Yellow-necked field mouse		10.36	2.30	1.09	2.88
Average		9.96	2.28	1.00	2.92
Feces		7.54	1.13	1.43	0.60
		$g\ ha^{-1}\ yr^{-1}$			
Total amounts of nutrients in excrement	Coniferous forest	580	21.0	26.6	11.1
	Deciduous forest	740	27.0	34.1	14.3

proximates 9 g ha^{-1} in the coniferous forests of which almost half is accounted for as nitrogen. Taking into account the turnover rate, the annual contribution of nutrients to production is 32.4 g ha^{-1} (19.8 g nitrogen, 4.5 g phosphorus, 2.1 g potassium, and almost 6 g calcium). This contribution is higher in the deciduous forests where it amounts to 41.2 g ha^{-1} (25.2 g nitrogen, 5.7 g phosphorus, 2.7 g potassium, 7.6 g calcium) (Górecki and Szafranek 1981).

The total annual amount of nutrients in rodent feces has been computed from average composition (Górecki and Szafranek 1981). The daily amounts of feces are after Drożdż (1968). The total annual amount of nutrients evacuated in rodent feces reaches 200 g ha^{-1} in the coniferous forests, of which 140 g are accounted for as nitrogen. In the deciduous forests, these levels increase to 250 g ha^{-1} and 180 g ha^{-1}, respectively.

The concentration of nitrogen is much higher in urine than in feces. Annually, the rodents excrete in urine 440 and 560 g nitrogen ha^{-1} in the coniferous and deciduous forests respectively. The total annual amounts of nutrients in rodent excrement approximates 580 g ha^{-1} and 740 g ha^{-1} in the coniferous and deciduous forests respectively.

The paratrophic significance of rodents for various ecosystems has been intensively studied (see Golley et al. 1975 for review). It is due first of all to the large amounts of organic matter which are destroyed but not eaten. To investigate this problem in forest ecosystems is very difficult, and hence the available data are from arable fields and grasslands. In the latter ecosystems, the rodent consumption may represent as little as 10% of destroyed organic matter. The latter, however, is transferred immediately to the reducers which considerably accelerates its cycling (Abaturov and Zubkova 1969, Tertil 1977, Abaturov et al. 1975, Zlotin 1975). This effect seems to be much weaker in forest ecosystems.

Rodent activities are highly disadvantageous to forestry, because large amounts of seeds and seedlings are destroyed (Golley et al. 1975). In this way, they counteract renewal of the forest. Their affects on forest ecosystems resemble those of fire, since they accelerate the decomposition of organic matter (Golley et al. 1975).

On the other hand, small rodents are highly mobile and may transport seeds, organic compounds, and chemical elements. Over 12% of tree seeds may be sown by rodents (Golley et al. 1975). Invasion of some habitats by new tree species may actually be due to zoochory.

The transport of organic chemicals is much more difficult to observe. However, it has been reported in several ecosystems, especially grasslands, that plants show a larger biomass and higher concentration of various nutrients in the vicinity of rodent colonies (Chew 1974, Batzli 1978, Weiner et al. 1982). This is caused in part by an increased concentration of rodent excrement, and in part by a considerable increase in the abundance of plant matter destroyed and decaying nearby.

Burrowing activities of rodents contribute considerably to the enrichment of soil nutrients, as the nutrient-rich matter is dug from deeper horizons. The amounts of this depend on animal size, and the weight of it exceeds the animal weight by a factor of 50–150. For instance, *Apodemus flavicollis* may dig 3 kg soil within 2 h, while microtines may transport 10 m^3 soil ha^{-1} annually under a density of 400 individuals ha^{-1} (Golley et al. 1975). Burrowing has also a scarifying effect on the soil and, what may be of even greater importance, it may influence the amounts of water retained in the soil.

4.3.2 Energy and Matter Flow Through Ungulates

B. Bobek and K. Perzanowski

The ungulate fauna of the Niepołomice Forest includes three species, the roe deer (*Capreolus capreolus*), the red deer (*Cervus elaphus*), and the wild boar *(Sus scrofa)*. The extent and location make it well suited for studies on cervids and investigations have been carried on there from 1970 (Bobek et al. 1972).

The population densities of roe deer and wild boar were determined by two techniques. In winter, just after the hunting season, tracking was carried out in the deciduous forests (1,500 ha) as well as in part of the coniferous forest complex (500 ha). In early spring or late autumn, a supplementary check was made by run-

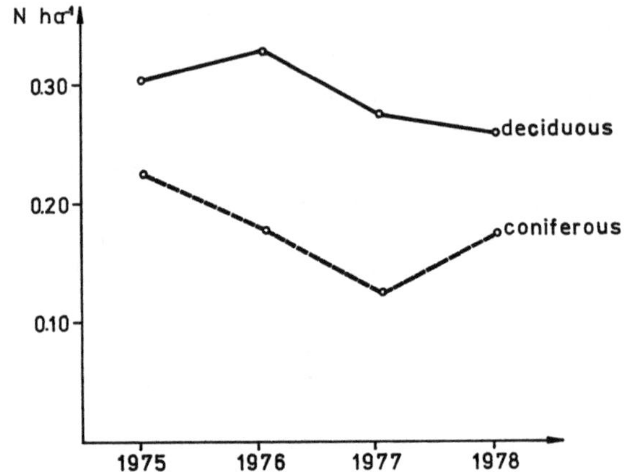

Fig. 4.6. Roe deer abundance in the Niepołomice Forest between 1975 and 1978

ning two broods over an area averaging 400 ha. The tracking results were corrected for underestimation of population size (Pucek et al. 1975).

Red deer number 0.02 individuals ha^{-1} in the deciduous forests, but density is almost three times this in the coniferous forests. The roe deer population, on the contrary, is twice as dense in the deciduous forests where it reaches 0.29 individuals ha^{-1}, with a maximum density noted of 0.33 (Fig. 4.6). The average density of roe deer in 1975–1978 was 0.22 individuals ha^{-1}. The wild boar population is difficult to estimate because of seasonal migration between the two forest complexes. Densities established by tracking range from 0.009 to 0.02 individuals ha^{-1} (0.014 average) (Jelonek 1977).

The standing crop of cervids has been estimated using data on average individual weight (Weiner 1973, Bobek 1974). It is 5 kg ha^{-1} for the roe deer and 5.6 for the red deer. According to Jelonek (1977), it is only 0.75 kg ha^{-1} for the wild boar.

Under the assumption of an annual biomass turnover rate of 0.4 (Bobek 1974), the annual production of roe deer is 2.1 kg ha^{-1}. This is only 1.1 kg ha^{-1} for the red deer. When the bioenergetic parameters of the wild boar are taken into account (Myrcha and Jezierski 1972, Jezierski and Myrcha 1975), the annual production approximates 0.5 kg ha^{-1}. Thus, the total annual productivity of ungulates exceeds 40 t annually.

The average annual energy consumption of a roe deer is 5×10^6 kJ (Weiner 1977 a, b), which is equivalent to the usage of 7.5×10^5 g biomass. The roe deer population consumes 1.1×10^6 kJ ha^{-1} (or 6×10^4 g dry biomass) annually. One red deer requires three to four times more food than a roe deer (Pasławski 1977, Bališ 1980), so that all the cervids of the Niepołomice Forest may be assumed to eat annually 10×10^4 g dry biomass ha^{-1} (or 1.9×10^6 kJ). The food requirements of wild boar are given by Jezierski and Myrcha (1975). From those data, the annual usage of food is calculated to approximate 0.13×10^6 kJ ha^{-1}.

The data given by Pasławski (1977), Weiner (1973), and Jezierski and Myrcha (1975) were used to estimate the energetic costs of the three ungulate species con-

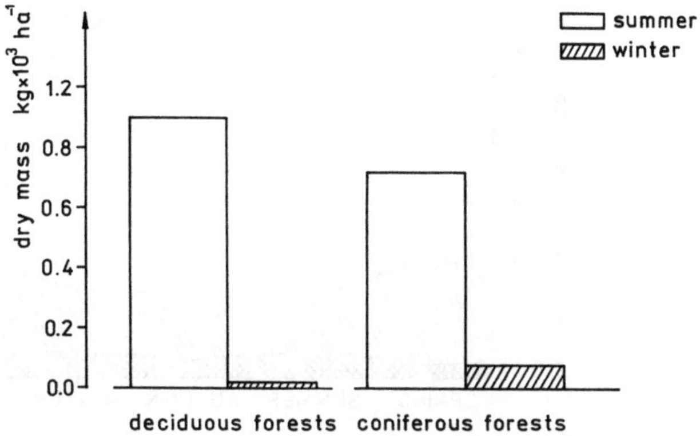

Fig. 4.7. Available food resources of ungulates in the Niepołomice Forest (t dry wt ha^{-1})

sidered. The average annual dissipation of energy (respiration) is 1.24×10^6 kJ ha^{-1}.

In order to estimate the energy flow through big game populations, exploitation by hunters must also be taken into account. In the Niepołomice Forest, this outflow amounts to 1.1 kg biomass ha^{-1}, or 8.3×10^3 kJ ha^{-1}, annually or 20–30% of the annual productivity of ungulates.

The food resources available annually to the cervids in the deciduous forests have been estimated as 890 kg dry biomass ha^{-1} (Bobek 1980, Perzanowski et al. 1982). This value includes the potential productivity of dicotyledonous herbs as well as shoots and leaves of shrubs and trees without economic significance but available for browsing. In the coniferous forests, the food resources of the cervids are 630 kg ha^{-1} annually (Bobek 1980, Perzanowski et al. 1982). These figures are computed assuming that roe deer must consume materials with digestibility not less than 60% in order to maintain energic balance (Perzanowski 1978) (Fig. 4.7).

Seasonal variation of food resources is of great importance because this is the limiting factor in winter. In the deciduous forests, food resources decrease as much as 98% in winter. The greater proportion of shrubs in the understory of the coniferous forests is responsible for the decrease being lower there (91%). An average decline of summer food resources of 720 kg dry biomass ha^{-1} to 0.55 kg ha^{-1} occurs in winter (Figs. 4.7 and 4.8). The exploitation of food resources by cervids is only 14% in the Niepołomice Forest. They are the dominant herbivores with respect to biomass, but eat only 1% of the total primary production.

It is much more difficult to estimate the food resources of wild boar. On the one hand, they often feed in arable fields and meadows surrounding the forest complex, but on the other they are omnivorous to an extent which makes it impossible to provide any reliable estimate of either their food resources or the exploitation rate.

Large herbivorous mammals contain huge amounts of nutrients. The concentration of macroelements in the roe deer is much higher than in the available food. The former exceeds the latter by a factor of 4 for nitrogen, and by almost 50 for sodium (Table 4.14). The turnover rate of macroelements is also accelerated in

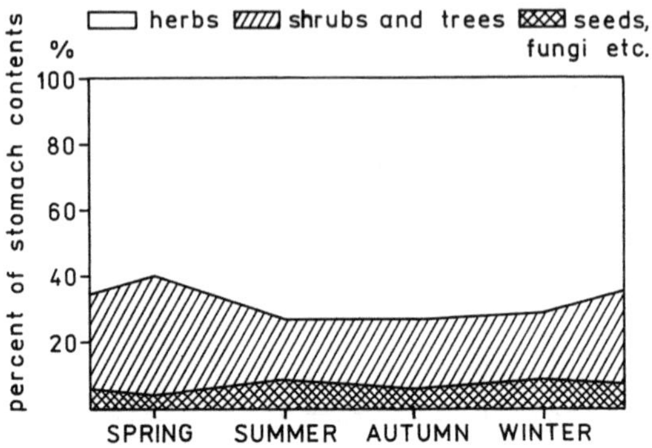

Fig. 4.8. Seasonal variation in diet composition of roe deer. (Data from Siuda et al. 1970, Matuła 1977)

Table 4.14. Concentration of macroelements in some tissues and organs and in the natural food excrement of the roe deer in the Niepołomice Forest

	N	P	K	Ca	Na
	Percent dry weight				ppm
Liver	10.58	1.44	0.96	0.00	3700
Kidney	9.83	0.95	0.93	0.00	1700
Muscles	13.12	0.62	0.97	0.00	1600
Bones	4.35	10.08	0.09	18.00	4200
Fat	0.00	0.02	0.02	0.00	0
Body (weighted mean)	9.15	4.10	0.63	6.05	2500
Food (weighted mean)	2.12	0.22	1.71	0.95	52
Feces	1.54	0.29	0.35	1.60	297

the supply of mammalian feces and urine, and the ungulates contribute 3.0–3.5 kg nitrogen ha^{-1} annually in this way (Table 4.14). In addition, organic matter of the excreta is decomposed much more rapidly than dead plant materials. Furthermore, the occurrence of decaying organic matter accelerates the decomposition rate of neighboring organic matter; hence, mammalian feces contribute also to an increased decomposition rate of the litter.

The significance of cervids in a forest ecosystem cannot be reduced simply to the energy flow through the population. Their impact on the deciduous forests leads to a reduction in standing crop of the herb layer, of production, and of the standing winter crop of browse. These affects considerably exceed the actual amounts of consumption (Bobek et al. 1979). In fact, the biomass of leaf-bearing shoots is twice as high in those areas of deciduous forest which are protected against browsing by cervids (Bobek et al. 1975). The loss in annual wood increment in plantations, caused by cervids, may reach 39% (Zlotin and Hodašova 1974).

The cervids' share of energy flow through forest ecosystems is rather low. Nevertheless, their impact on the dynamics of primary production, especially at the lowest levels in the forest, may be of much importance. This holds true especially for young tree stands (Bobek et al. 1972, Fruziński et al. 1975). The cervids commonly prefer tree species which are the main components of forest ecosystems, hence economically important. In this way, they may affect forestry very disadvantageously (Haber et al. 1975).

Wild boar activities are of high but variable significance, especially on the forest floor. Their mode of feeding contributes to an acceleration of matter cycling and stimulates the development of soil and litter invertebrates (Haber 1969). On the other hand, their activities are destructive of the herb layer (cf. Bratton 1974). An unequivocal evaluation of their impact on forest ecosystems is barely achievable at the moment because of the lack of adequate quantitative data.

4.3.3 Energy and Matter Flow Through Bird Populations

Z. Głowaciński, J. Kozłowski, and J. Weiner

The Niepołomice Forest is a mosaic of different habitats. The bird assemblages are therefore described from a gradient-typological perspective, with age variation among forests taken into account. The main habitats are: alder forests, oak-hornbeam forests, pine forests with oaks and birches, moist pine forests, and meadows.

In all, 180 bird species have been recorded in the Niepołomice Forest, of which 60% are known to breed (Głowaciński 1975, 1978, 1981). Very rare species do not play any significant role in energy and matter flow. This study considers only those species with population densities not less than 0.1 pairs ha^{-1}.

Energy flow through bird populations has been estimated for all the main habitats. The parameters have been analyzed using a computer simulation model (Głowaciński and Weiner 1977, 1980, 1983, Weiner and Głowaciński 1975). To this end, the abundance of juveniles and adults was derived from field observations on population density, phenology, death rate, and breeding success in particular species. Various ecological and bioenergetic data were taken from the available literature.

Of the five main habitats investigated (forests from 75 to 100 years old), the largest bird assemblage occurs in the oak-hornbeam forest; the pine forest is half as rich in birds (Table 4.15). The dominant species are *Fringilla coelebs* and *Anthus trivialis,* associated with *Erithacus rubecula, Parus caeruleus,* and *P. major* (Table 4.15). Meadows are relatively simple in ecological structure, and the bird assemblage there is between 5 and 11 times less rich in the number, but only half as poor in biomass as in the mature forest ecosystems (Table 4.15).

The avifauna is very unstable and largely homogeneous in winter. Only a single winter bird colony has been recognized in the whole Niepołomice Forest. It is dominated by titmice. The bird biomass in winter approximates 200 g ha^{-1}, which is equivalent to 1,600 kJ ha^{-1} (Table 4.16).

The energy flow (or assimilation) through a breeding bird colony in mature forests ranges from 250 to 585 × 10^{3} kJ ha^{-1} in the breeding season depending on the habitat. Ecological efficiency, however, is very even and ranges from 2.1 to

Table 4.15. Abundance (N) and standing crop (Sc) of birds in the main habitats of the Niepołomice Forest in the breeding season. (Głowaciński 1975, Głowaciński and Weiner 1977, 1983). For each habitat three dominant species are listed

Ecosystem	Bird species	N		Sc	
		Pairs per 10 ha	%	g ha^{-1}	kJ ha^{-1}
Meadow	*Alauda arvensis*	3.3	46.7	23.1	184.9
	Anthus pratensis	1.7	23.3	5.8	46.0
	Coturnix coturnix	0.5	7.0	10.0	80.6
	Other species[a]	1.6	23.0	86.8	691.9
	Total	7.1	100.0	125.7	1003.4
Alder forest	*Emberiza citrinella*	8.1	11.0	45.4	361.1
	Fringilla coelebs	8.1	11.0	32.4	257.9
	Erithacus rubecula	6.3	8.5	21.6	171.5
	Other species[a]	51.9	69.5	279.9	2236.1
	Total	74.4	100.0	379.3	3026.6
Oak-hornbeam forest	*Ficedula albicollis*	10.8	13.5	31.3	249.4
	Fringilla coelebs	9.1	11.3	36.4	289.8
	Parus caeruleus	8.2	10.2	20.5	163.2
	Other species[a]	52.2	65.0	464.3	3696.3
	Total	80.3	100.0	552.5	4398.7
Pine mixed forest	*Anthus trivialis*	8.1	12.7	37.3	296.6
	Erithacus rubecula	6.0	9.4	20.5	163.4
	Fringilla coelebs	6.0	9.4	24.0	191.1
	Other species[a]	43.6	68.5	287.8	2541.0
	Total	63.7	100.0	369.6	2942.1
Fresh pine forest	*Fringilla coelebs*	7.5	20.7	30.0	238.8
	Anthus trivialis	4.7	13.0	21.6	172.1
	Phylloscopus collybita	2.3	6.4	4.2	33.0
	Other species[a]	21.7	59.9	177.4	1411.9
	Total	36.2	100.0	233.2	1855.8

[a] The full list is given in the cited publications

2.4% (Table 4.21). Apart from the breeding season, the energy flow is much lower. The total annual energy flow through the birds is 660×10^3 kJ ha^{-1}.

The energy flow through the birds is lower by a factor of 3 to 7 in the meadows, but the ecologic efficiency is considerably higher and reaches 3.5% (Table 4.17). This is due to the greater share of species with higher birth rates and greater population densities. Following cutting in the deciduous forest the ecologic efficiency may be as high as 7% (Głowaciński and Weiner 1977).

The energy balances of bird populations of the deciduous and coniferous forests are widely different from each other in the breeding season. The energy flow is much higher in the former. The standing crop is twice as high there as in the coniferous forest. Furthermore, a maximum is achieved in 15-year-old deciduous

Table 4.16. Abundance (N) and standing crop (Sc) of a winter bird community in the Niepołomice Forest. (Głowaciński and Weiner 1977)

Species	N		Sc	
	Indiv. per 10 ha	%	$g\ ha^{-1}$	$kJ\ ha^{-1}$
Parus major	12.5	20.7	23.6	187.9
Coccothraustes coccotheraustes	10.0	16.6	53.0	421.9
Sitta europaea	9.4	15.6	19.9	158.4
Parus caeruleus	9.2	15.2	11.5	91.5
Certhia sp.	4.1	6.8	3.9	31.1
Parus palustris	3.5	5.8	4.0	31.8
Other species	11.7	19.3	86.9	691.9
Total	60.4	100.0	202.8	1614.5

Table 4.17. Energy budget and ecological efficiency of the main bird communities of the Niepołomice Forest (energy budget of migratory birds from unpublished data of Weiner and Głowaciński) C = consumption, A = assimilation, E = excretion, R = respiration, P = production, P/A = ecological efficiency

Ecosystem	C	A	E	R	P	P/A
Breeding season (150 days) ($kJ \times 10^3\ ha^{-1}\ season^{-1}$)						
Wet meadow	115.5	86.6	28.9	83.6	3.0	3.5
Alder-forest	646.1	484.6	161.5	474.6	10.0	2.1
Oak-hornbeam forest	780.3	585.2	195.1	571.0	14.2	2.4
Mixed pine forest	587.1	440.3	146.8	429.6	10.7	2.4
Fresh pine forest	333.7	250.3	83.4	244.5	5.8	2.3
Weighted mean[a] (without meadow)	443.4	332.6	110.8	324.9	7.7	2.3
Nonbreeding season (215 days) ($kJ \times 10^3\ ha^{-1}\ season^{-1}$)						
Mean (without meadow)	148.6	111.9	36.7	111.9	0.0	
Whole year ($kJ \times 10^3\ ha^{-1}\ yr^{-1}$)						
Mean (without meadow)	592.0	444.5	147.5	436.0	7.7	

[a] Weighted mean of various age classes of various forest communities

forests, but only in mature coniferous forests. In the climax, however, the standing crop of birds is fairly constant in these forest ecosystems.

The energy flow through the birds in the Niepołomice Forest falls to only 50×10^3 kJ ha^{-1} in winter. The standing crop, however, is only slightly lower than in spring and summer in the coniferous forests.

The energy balances of the forest nesting birds are given in Fig. 4.9. Ecological succession from cutting (1 year old) to climax (over 150 years old) is taken into account. The energy flow increases at variable rates during secondary succession (Głowaciński and Weiner 1977, 1980, 1983). Bird productivity is much higher in

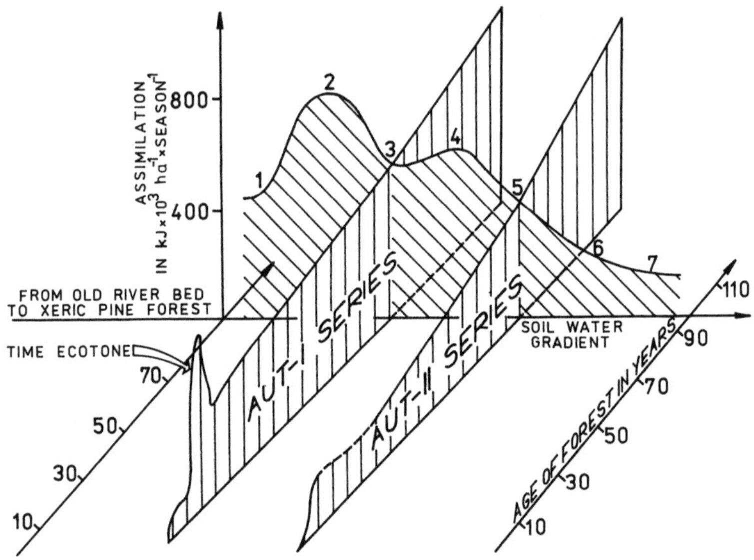

Fig. 4.9. Three-dimensional pattern of energy flow through breeding bird communities in the Niepołomice Forest. *1* moist alder forest; *2* alder forest × oak-hornbeam forest; *3* oak-hornbeam forest; *4* oak-pine forest; *5* mixed pine forest; *6* fresh pine forest; *7* xeric pine forest; *AUT I* ecological succession of oak-hornbeam forest; *AUT II* ecological succession of pine forest

forests developed on moderately humid and transitional sites (alder-oak forest, oak-hornbeam forest, oak-pine forest) than in those on dry (pine forest) and wet sites (alder forest). Energy flow attains a maximum in alder-oak forests (assimilation equal to 851.8×10^3 kJ ha^{-1} in the breeding season), and exceeds by a factor of 5 that in dry pine forests (Głowaciński and Weiner 1983). This is shown in the diagram of energy flow through bird assemblages at various successional stages (Fig. 4.9).

The average energy flow through the avifauna has been estimated taking into account typological variation of the habitats and age variation of the tree stands. The total energy flow averages 440×10^3 kJ ha^{-1} in the breeding season and 150×10^3 kJ ha^{-1} in the nonbreeding season, providing the total annual flow of 600×10^3 kJ ha^{-1}. At least 75% of the bird diets are accounted for by insects (Graczyk 1974, Oko 1975, Sikora 1975, Markiewicz 1978). Consequently, one may assume that the birds consume some 450×10^3 kJ ha^{-1} in the form of insects. Under the additional assumption that 50% of the bird diets are accounted for by butterflies (mostly *Tortrix viridana*), whose production is known (Bandoła-Ciołczyk and Witkowski 1976, and Sect. 4.2), one may conclude that the birds utilize 2–10% of production of the latter, mostly in May and June.

The energetic impact of bird populations on the deciduous and coniferous forests does not exceed 0.5% of primary production.

Having estimated the energy flow through birds, the flow of specific elements can be computed as well. To this end, data on the chemical composition of bird bodies and food are required. The body composition of two small insectivorous

Table 4.18. Body composition in the blackcap (*Sylvia atricapilla*) and the great tit (*Parus major*) (Ambroziak 1979; Białecka 1979). Fat content as percent proportion of dry weight; N, P, K, Ca contents as percent proportions of dry weight without fat

Species and sex	Water	Fat	N	P	K	Ca
Sylvia atricapilla, females	66.5	15.8	9.3	2.8	1.3	3.9
Sylvia atricapilla, males	66.8	16.0	9.6	3.0	1.3	4.2
Parus major, females	69.8	11.6	10.5	2.5	1.1	4.1
Parus major, males	67.5	16.2	11.2	2.8	1.1	5.0
Average	67.5	14.9	10.2	2.8	1.2	4.3

birds, the blackcap (*Sylvia atricapilla* L.) (Ambroziak 1979) and the great tit (*Parus major* L.), is given in Table 4.18. The figures are almost the same for both species. Hence, these values can be extrapolated for the remaining species. Concentrations of nitrogen, phosphorus, potassium, and calcium are given in percentage proportion of dry biomass without fat, because fat contents are highly variable among individuals. Considerable changes in body composition have been observed in great tit females during the period of egg deposition and incubation (Białecka 1979); therefore, the values are averaged over the whole nesting season.

Using these data and also assuming that the feathers (13% nitrogen dry weight) account for 9% of the total biomass, the amounts of particular elements maintained in bird bodies have been computed. There is less than 10 g nitrogen ha^{-1}, 3 g calcium, and even smaller amounts of potassium and phosphorus (Table 4.23). The amounts of those elements assimilated during a single breeding season can be estimated by multiplying standing crop by biomass turnover rate. The latter coefficient approximates 3 yr^{-1}, but the amounts of assimilated elements exceed the average annual level by a factor of 5 because the standing crop of birds is much lower in winter than in the breeding season. This estimate is also biased by nest mortality. However, Bieszczad-Kosch (1979) demonstrated that except for calcium, the concentration of the elements in birds' bodies investigated does not change during postembryonic development. The concentration of calcium in juveniles, especially those 3 to 7 days old, is very low. Even in juveniles leaving the nest it is still much lower than in adults. Calcium, however, is also the main constituent of egg shells. The amounts contained in egg shells counterbalance the difference in calcium concentrations between juvenile and adult birds. Hence, the estimate of element assimilation by birds in the process of production may be sufficiently reliable.

Depending on the age of the forest, the birds assimilate during production 13–90 g nitrogen, 3–17 g phosphorus, 1–7 g potassium, and 4–27 g calcium ha^{-1} in the oak-hornbeam forest. In the coniferous forests these values are still lower. Average values for the whole Niepołomice Forest area: 35 g nitrogen, 7 g phosphorus, 3 g potassium, and 11 g calcium ha^{-1} (Table 4.19). Bird consumption in terms of energy is given in Table 4.17. The estimates provided above follow from

Table 4.19. Flow of elements through bird communities of various ecosystems near maturity in the Niepołomice Forest

Ecosystem	Standing crop Sc				Production P				Consumption C				Excretion E			
	N	P	K	Ca	N	P	K	Ca	N	P	K	Ca	N	P	K	Ca
Breeding season (150 days) (g ha^{-1} season^{-1})																
Meadow	4.7	0.9	0.4	1.4	14.0	2.6	1.1	4.1	711	31	96	10	706	30	95	6
Alder forest	14.0	2.7	1.1	4.2	46.4	8.8	3.8	13.8	3,976	171	535	54	3,962	160	534	40
Oak-hornbeam forest	20.4	3.9	1.7	6.1	66.2	15.5	5.4	19.7	4,802	207	646	65	4,781	203	644	45
Mixed pine forest	13.7	2.6	1.1	4.1	49.7	2.4	4.0	14.8	3,613	156	486	49	3,600	153	485	34
Fresh pine forest	8.6	1.6	0.7	2.6	26.9	5.1	2.2	8.0	2,054	89	276	28	2,045	87	276	20
Weighted mean (without meadows)	10.9	2.1	0.9	3.3	35.5	6.7	2.9	10.6	2,729	118	367	37	2,698	115	364	26
Nonbreeding season (215 days) (g ha^{-1} season^{-1})																
Weighted mean (without meadows)	8.8	1.7	0.7	2.6	0.0	0.0	0.0	0.0	844	36	114	11	844	36	114	11
Weighted mean (without meadows)	9.7	1.9	0.8	2.9	35.5	6.7	2.9	10.6	3,573	154	481	48	3,542	151	478	37

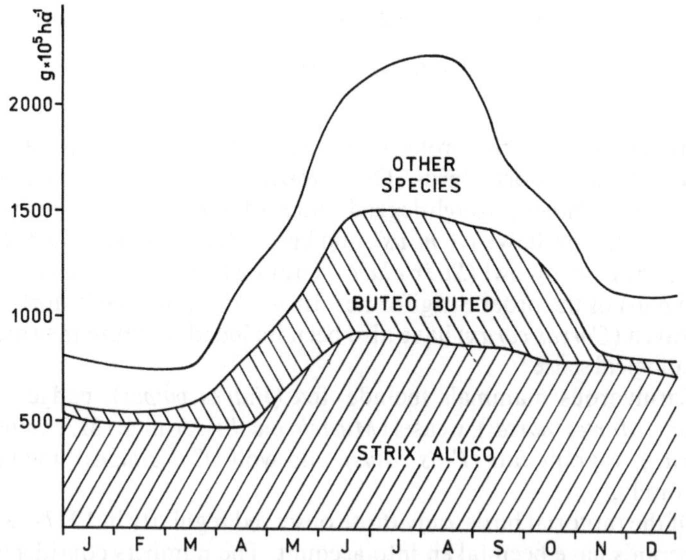

Fig. 4.10. Seasonal variation in standing crop of raptorial birds in the Niepołomice Forest. *Strix aluco* tawny owl; *Buteo buteo* common buzzard

the assumption that the birds feed mostly on insects, especially in the breeding season. The concentrations of specific elements in insects have been assumed as follows: 4.6% of the biomass is accounted for by nitrogen, 0.6% by phosphorus, 0.2% by potassium, and 0.02% by calcium (Spector 1956, Kozłowski unpubl.).

The total consumption of specific elements and the amounts returned to the ecosystem in the form of feces and urine are provided in Fig. 4.16 and Table 4.19. The amounts of elements consumed are strongly dependent on the successional stage of the forest, with a maximum usually achieved in mature forests. Annually, the birds consume 6 kg nitrogen, 260 g phosphorus, 800 g potassium, and 80 g calcium ha^{-1} in mature oak-hornbeam forest (Fig. 4.16). The averages for the whole Niepołomice Forest are much lower and approximate 3.5 kg nitrogen, 150 g phosphorus, 480 g potassium, and 50 g calcium ha^{-1} annually (Table 4.19). Almost the total consumed elements are rapidly evacuated with feces and urine. The birds seem to accelerate considerably the rate of matter, and especially nitrogen, cycling in the ecosystem investigated.

The impact of birds on forest ecosystems exceeds by far that of bioenergetics and matter cycling. It is largely related to food selection, as it depends on which insect species are eaten and what is the role of prey species in the ecosystem. Furthermore, birds eat and/or destroy buds and seeds of trees and other plant species, and as well transport seeds and fruits (e.g., the Turdidae). A quantitative estimate of these effects on the forest ecosystem is difficult and exceeds the scope of this study.

4.3.4 Energy and Matter Flow Through Carnivores

W. Grodziński and J. Wasilewski

Homeotherm carnivorous consumers of the Niepołomice Forest include mammals and raptorial birds. The following raptorials have been recorded: buzzard (*Buteo buteo*), rough-legged buzzard (*B. lagopus*), goshawk (*Accipiter gentilis*), sparrow hawk (*A. nisus*), and kestrel (*Falco tinnunculus*) of the order Falconiformes, tawny owl (*Strix aluco*), Ural owl (*S. uralensis*), and long-eared owl (*Asio otus*) of the order Strigiformes. In addition, the black stork (*Ciconia nigra*) and raven (*Corvus corax*) have also been included, because they are carnivores in an ecological sense.

Carnivorous mammals include: fox (*Vulpes vulpes*), badger (*Meles meles*), marten (*Martes martes*), weasel (*Mustela nivalis*), ermine (*M. erminea*), polcat (*M. outorius*), vagrant cats (*Felis catus*), and wolf (*Canis lupus*), the latter occurring only rarely.

Of the raptorial birds, only those observed regularly in 1977–1978 and for longer periods have been taken into account. The numbers considered in this study relate to the year 1978. They have been obtained by multiple estimates recorded along roads and forest district boundaries over the whole forest. Nesting territories could be established when these estimates were plotted on the map. The populations of the tawny owl were investigated at night on 1,000 ha of deciduous and on 1,500 ha of coniferous forests, and extrapolated for the whole area.

The abundance of specific bird species has been computed from the number of breeding individuals. This number is increased by brood values. Breeding efficiencies of the buzzard and the goshawk were estimated in the field and for the remaining species, literature data were consulted (Belskij 1962, Glutz and Blotzheim 1966, 1971, Gotzman and Jabłoński 1972, Newton 1976, Niethammer 1938). A correction for adult mortality and self-dependent juveniles has been also based on literature data (Glutz and Blotzheim 1971, Huhtala and Sulkava 1976, Newton 1976). This correction has been assumed to be zero for the black stork because the onset of juvenile independence coincides with the beginning of dispersal when mortality is highest.

The abundance of species of raptorial birds is presented in Table 4.20. The buzzard and the tawny owl are clearly numerically dominant. When compared to older records, the recent data indicate a decrease in numbers of species and numbers of individuals. This may either be caused by some disadvantageous environmental changes, or by fluctuations in abundance which commonly occur in raptorial birds. Apart from the nestlings, the rough-legged buzzard is the only raptor that occurs regularly here. In the winter of 1977/78, three to five individuals were observed in the northern part of the forest complex.

The biomass of particular species has been computed from abundance and average weight of adults and juveniles (Belskij 1962, Glutz and Blotzheim 1966, 1971, Gotzmann and Jabłoński 1972, Heinroth, O. and Heinroth, M. 1927, Newton 1976, Niethammer 1938). Seasonal variation is given in Fig. 4.17. The buzzard and the tawny owl jointly account for more than 50% of total biomass. The standing crop per unit area is much greater in the deciduous than in the coniferous

Table 4.20. Number of pairs of raptorial birds breeding in different forest types in the Niepołomice Forest

Species	Coniferous forests		Deciduous forests		Total area	
	Total	per 100 ha	Total	per 100 ha	Total	per 100 ha
Buteo buteo	14	0.15	5	0.28	19	0.17
Accipiter gentilis	3	0.03	1	0.05	4	0.04
Accipiter nisus	2	0.02	–	–	2	0.02
Falco tinnunculus	3	0.03	–	–	3	0.03
Strix aluco	27	0.29	25	1.41	52	0.53
Strix uralensis	1	0.01	–	–	1	0.01
Ciconia nigra	2	0.02	1	0.05	3	0.03
Corvus corax	3	0.03	1	0.05	4	0.04

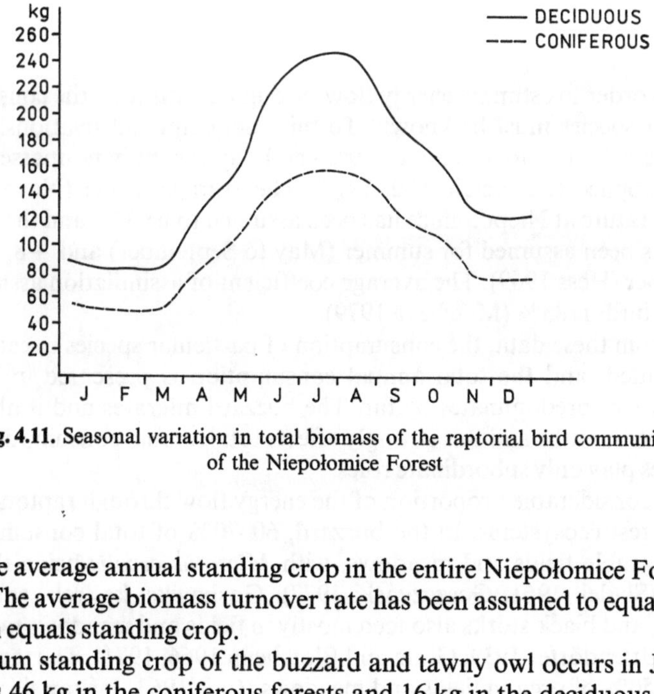

Fig. 4.11. Seasonal variation in total biomass of the raptorial bird community of the Niepołomice Forest

forests. The average annual standing crop in the entire Niepołomice Forest is 1.4 g/100 ha. The average biomass turnover rate has been assumed to equal 1. Hence, production equals standing crop.

Maximum standing crop of the buzzard and tawny owl occurs in July. It approximates 46 kg in the coniferous forests and 16 kg in the deciduous forests for the former, and 52 and 48 kg, respectively, for the latter. The maximum joint standing crop of the remaining raptorial species is achieved in August when it approximates 61 kg in the coniferous and 22 kg in the deciduous forests.

The seasonal variation in biomass of the raptors is given in Fig. 4.18. An increase in biomass occurs in April when migratory species appear and from May to June when the juveniles hatch and grow. The autumn decrease in biomass is due to increased mortality as well as to the departure of migratory species. Maximum total standing crop of the raptors is attained in July when it approximates 158 kg in the coniferous forests and 68 kg in the deciduous forests.

Table 4.21. Total annual consumption of raptorial birds in various forest types in the Niepołomice Forest (kJ ha^{-1})

Species	Consumption, kJ ha^{-1} yr^{-1}		
	Coniferous forests	Deciduous forests	Total area
Buteo buteo	934.6	1772.2	1074.5
Buteo lagopus	64.5	–	54.5
Accipiter gentilis	377.2	657.5	427.5
Accipiter nisus	104.4	–	88.3
Falco tinnunculus	87.6	–	74.1
Strix aluco	2226.5	10,869.6	3626.5
Strix uralensis	116.5	–	98.5
Corvus corax	418.3	672.8	461.6
Ciconia niger	170.4	449.1	216.1
Total consumption	4500.1	14,421.3	6121.6

In order to estimate energy flow through the raptors, the consumption of particular species must be known. To this end, empirical relationships of the daily energetic budget to air temperature and body weight in nonpasseriform birds has been applied (Kendeigh et al. 1977). The average winter (November to March) temperature at Niepołomice has been assumed to be 0 °C and the average of +15 °C has been assumed for summer (May to September) and +8 °C for April and October (Hess 1969). The average coefficient of assimilation assumed for all raptorial birds is 85% (Metcheva 1979).

From these data, the consumption of particular species in each month can be computed, and the total annual consumption is presented in Table 4.21. The tawny owl predominates by far. The buzzard migrates and is absent during the 5 winter months and this strongly affects total annual consumption. All the other species play only subordinate roles.

A considerable proportion of the energy flow through raptors is derived from nonforest ecosystems. In the buzzard, 60–70% of total consumption is derived from arable fields and meadows, with *Microtus arvalis* being the predominant prey (Sladek 1961, Goszczyński 1977). Goshawks, kestrels, rough-legged buzzards, and black storks also feed mostly in fields and meadows rather than in forests (Uttendörfer 1939, Glutz and Blotzheim 1966, 1971). The tawny owl finds almost 50% of prey in fields and meadows (Cajs 1963, Jaśkowski 1956, Kulczycki 1964, Serafiński 1954, Skuratowicz 1950). One may conclude that less than 50% of the total consumption in the coniferous forests is actually derived from forest ecosystems and that this proportion is even lower in the deciduous forests.

Most of the total annual consumption takes place in summer. The food resources of raptorial birds are very variable at that time because in addition to vertebrates a large proportion of insects is included. Up to 40% of the summer diet is accounted for by insects (Sladek 1961). As judged from pellet examination, tawny owls feed to a large extent on insects, especially beetles. The abundance of insects and birds is so high in summer, that it is largely unaffected by predation. Only in winter is it possible for raptorial species to have an impact on the population of small birds. Amphibians account for 10% of the food consumption of the tawny owl, and much more in the black stork. However, the lack of information on amphibian abundance here makes it impossible to provide an estimate of the affect of the raptorial birds on their populations.

The only animal group seriously affected by raptorial birds is the forest rodents, especially the bank vole (*Clethrionomys glareolus*). They account for 5% of the buzzard consumption and almost 10% of that of the tawny owl. Half of this is represented by bank voles. On the average, buzzards consume 300 forest rodents annually in the coniferous forests, and tawny owls as many as 14,000. As indicated in Section 4.3.1, the average annual production of rodents in the coniferous forests is 4,231 kJ ha^{-1}. Buzzards and tawny owls consume 1.8 individuals ha^{-1} (or 260 kJ ha^{-1}) annually, which is equivalent to 6% of total rodent production. In the deciduous forests the raptors may consume 17% of total annual production.

Of the carnivorous mammals referred to only the two largest species, the badger and the fox, were investigated in the field. Both are general carnivores sensu Anderson and Erlinge (1977).

The abundance of badgers and foxes was estimated from counts of inhabited burrows. In addition, tracking in fresh snow was used for fox. Badger burrows were counted from 1975–1978 over the entire forest (Sawicki 1977). Fox dens have been counted from 1978–1980 over a third of the area of coniferous forests (Milczanowski 1980). Each burrow was checked for inhabitants (Kruuk 1978).

In all, 37 inhabited badger burrow systems have been found of which 25 occur in the coniferous and 12 in the deciduous forests. This is equivalent to 2.69 and 6.11 burrows per 1,000 ha respectively. They are unevenly distributed, depending on forest type, distances from margins of the forest complex, and topography (Fig. 4.19). They occur most commonly in oak-hornbeam and in moist oak-pine forests. The majority (63%) are found no nearer than 300 m from the forest margin. Their density increases on dunes and other higher topographic features.

In this period 136 fox dens 46 of which were inhabited, were found in 3,934 ha of the coniferous forests (Milczanowski 1980). If this sample is representative of the whole forest, a total of 130 inhabited fox dens may be expected. They are distributed mostly in dry areas, on hills, and within short distances of the forest margin.

The basic (average) and maximum (breeding season) populations have been estimated from the number of inhabited burrows. For the basic population it has been assumed that each badger burrow is inhabited by two animals and each fox den by a single individual. For maximum population size the following data have been taken from Stubbe (1973a, b): sex ratio 1:1, and average brood 2.4 for badger; sex ratio 1.5♂:1.0♀, and average brood 5 for fox (males inhabit different dens than females with juveniles). The average annual density of fox and badger should be close to the basic because juveniles quickly establish their own burrows or emigrate. The average standing crop computed for basic population size is 0.080 and 0.065 kg ha^{-1} for the badger and the fox respectively (Table 4.22).

The badger diet is well known from analyses of stomach contents and the composition of excrement (cf. Anderson 1954, Lihačev 1956, Skoog 1970). The

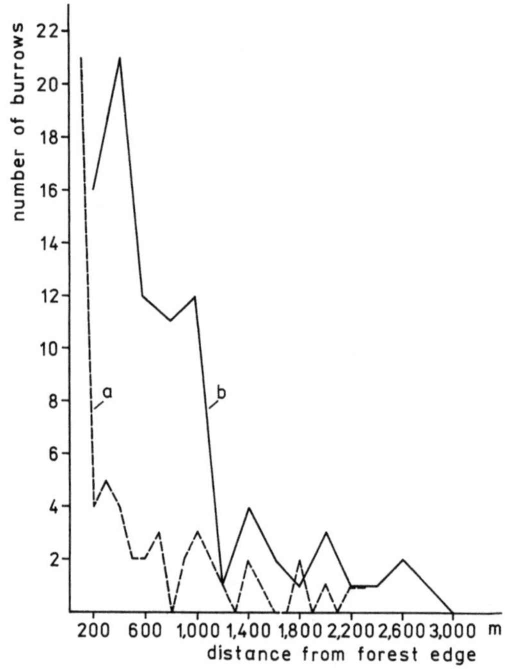

Fig. 4.12. Distribution of badger (**a**) and fox (**b**) burrows in relation to distance from forest edge in the Niepołomice Forest

Table 4.22. Population size and biomass for badger and fox populations in the Niepołomice Forest

	Census of inhabited dens in forest			Estimated population number		Mean standing crop of biomass	
	Coniferous	Deciduous	Niepołomice Forest	Basic (average)	Maximum (breeding season)	Total kg	kg ha^{-1}
Badgers	25	12	37	75	(110)	900	0.080
Foxes	46 (107)[a]	(23)[a]	(130)[a]	130	(300)	750	0.065

[a] Estimates for the whole area

badger is polyphagous and feeds on animals as well as plants. The main diet includes earthworms, insects, plants, and small vertebrates, but in addition, badgers feed on birds, mollusks, amphibians, reptiles, and carrion.

The fox diet has been widely studied in Europe (Jensen and Sequeira 1978), but it was also investigated in detail in the Niepołomice Forest by scat analysis (Ciuła 1979). Fox are largely carnivorous in this area, in terms of both frequency and biomass of various food types. They feed mostly on mammals (87.3% frequency), less commonly on birds (25.2%), insects (43.7%), and plant materials (fruit and seeds – 16.8%). Mammalian prey is dominated by small rodents

Table 4.23. Energy flow and consumption of badger and fox populations in the Niepołomice Forest ($kJ \times 10^6\ yr^{-1}$); the share of various food types in total consumption is also given

	Assimilation	Production[a]	Respiration[b]	Total	Consumption		
					Animal foods	Small rodents	Woodland rodents
Badgers	59.9	3.7	56.2	70.5	52.8	10.5	3.0
Foxes	108.9	5.5	103.4	136.2	129.4	32.5	10.0
Total in Niepołomice Forest	168.8	9.2	159.6	206.7	182.2	33.0	13.0
Per ha yr^{-1}	0.0150	0.0008	0.0142	0.0183	0.0161	0.0029	0.0013

[a] Computed from $P_n = \bar{B} \times \Theta_N$
[b] Computed from $R_t = A_t - P_n$

(77.3%). The contribution of various foods to fox diet is better illustrated by share in dry biomass of total diet. Annually, rodents account for 42% of the dry biomass eaten, hares 24%, cervids 12%, birds 7%, insects 6%, and others 9%. Over half of the rodents are *Microtus arvalis* and *Mus musculus*, while only one third are typical forest species (*Clethrionomys glareolus, Apodemus flavicollis, Pitymys subterraneus*). Fox diet varies seasonally, as well as from year to year, due to the variation in rodent populations.

The assimilation of badgers has been estimated from the daily consumption rate (full stomach contents; Skoog 1970) as well as from the general metabolic equation given by Kleiber (1961). The active period (early April to mid-November – 230 days) and the wintering period (mid-November to late March – 135 days) have been distinguished. A daily consumption of 550 g or 1.5 BMR level has been accepted for the active period, and 250 g or 0.7 BMR level for the winter. Both estimates, the metabolic one having been modified (15% correction for rejects) (Grodziński and Wunder 1975), are close to each other, and are 2,480 and 2,670 kJ per day calculated by the consumption and metabolic techniques, respectively. If an average of 2,575 kJ per day is accepted, the total annual consumption of a badger is 940,000 kJ, of which 75% is accounted for by animal prey. This is equivalent to 130 kg biomass of earthworms, rodents, and insects.

Fox consumption has been estimated from feeding experiments performed in fox farms (Pierielolik, after Sławoń and Woliński 1975). The assimilation of an adult varies from 1,800 to 2,680 kJ per day, and 2,300 kJ on the average. Taking the coefficient of digestibility in wild fox (Vogtsberger and Barret 1973) and the production of urine in carnivores (Moors 1977) into account, the average daily consumption of an adult approximates 2,870 kJ, hence, the consumption amounts to 1,047,500 kJ annually. This is equivalent to 146 kg biomass. In the Niepołomice Forest this includes 35 kg rodents (1,720 microtines, voles, and mice), 43 kg hares, 50 kg cervids, 9 kg small birds, 1 kg insects, 1.5 kg insectivorous mammals, and 1 kg plants.

The consumption rates given above have been used to estimate the total assimilation by badger and fox (Table 4.23). These carnivores assimilate almost

Table 4.24. Nutrient analysis (N, P, K, Ca, Na) in fox caracasses, and estimates of nutrient contents in total biomass and annual net production of fox and badger populations of the Niepołomice Forest

Concentration in fox carcasses (percent of wet weight)	N	P	K	Ca	Na
	3.040	1.591	0.206	2.307	0.058
Fox and badger populations					
Total kg/Niepołomice Forest					
Standing crop of biomass (750+900) 1650	50.16	26.25	3.40	38.07	0.96
Net production (705+405) 1110	33.74	17.66	2.29	25.61	0.64
Per area unit g ha^{-1}					
Standing crop of biomass 146.4	4.45	2.33	0.30	3.38	0.08
Net production 98.5	2.99	1.57	0.20	2.27	0.06

20×10^6 kJ annually, of which 182×10^6 kJ are derived from various animals. Their impact on rodents seems to be enormous. They eat 33×10^6 kJ in the form of rodent biomass, of which 13×10^6 kJ is represented by forest rodents. This is equivalent to 188 g of rodent biomass (or 8–9 individuals) ha^{-1} of forest, and it approaches the total production of rodents in the entire Niepołomice Forest (Sect. 4.3.1). This seems to be an overestimate. In fact, the exact number of badger and fox inhabiting a single burrow is unknown. On the other hand, the other carnivores (marten, weasel, ermine, polecat, and vagrant cats) also feed on rodents (for comparison see Pelikán and Vačkar 1978, Ryszkowski 1982).

At the moment, the energy flow through badger and fox can be estimated only very approximately (Table 4.23). The total assimilation rate has been computed with digestion efficiency being assumed to be 85% for the badger, and 80% for the fox. The net production has been calculated by multiplying average standing crop (Table 4.22) by biomass turnover rate. To this end, the age structures given by Stubbe (1973 a, b) have been used to produce life tables which indicate that average longevity is 2.2 years for the badger and 1.6 for the fox. Consequently, the turnover rate is 0.45 and 0.84, respectively. As indicated by Table 4.23, the average total annual consumption of badger and fox exceeds 18×10^3 kJ ha^{-1}, of which almost 15×10^3 kJ are assimilated but only 816 kJ (5.45%) are used for production. Some 10–30% of fox production is exploited by hunters, who killed 12 to 45 individuals annually in the last decade.

The concentration of nutrients (N, P, K, Ca, Na) was analyzed in fox foods and feces and also in a few fox caracasses (Prajsnar 1981). When these data are extrapolated for the populations of fox and badger, the amounts of nutrients in biomass and production can be estimated (Table 4.24).

4.4 Flow of Heavy Metals Through Selected Homeotherm Consumers

K. SAWICKA-KAPUSTA and J. KOZŁOWSKI

The recent bioindication studies have demonstrated that the Niepołomice Forest is polluted not only with sulfur dioxide (Grodzińska 1977), but also with heavy metals (Sawicka-Kapusta 1979, Makomaska 1978, Sawicka-Kapusta et al. 1981, Grodzińska et al. 1983). Therefore, an attempt has been made to estimate the flow of heavy metals through selected groups of homeotherm consumers.

Cadmium, lead, zinc, and iron have been considered in this study. Small insectivorous birds are represented by the great tit (*Parus major*). The concentrations of heavy metals in a dozen other small bird species are very similar to those recorded in the great tit, which can therefore be considered representative. When estimating the flow of heavy metals through bird populations, total bird consumption, production, and excretion have been taken into account. The dominant rodents are the bank vole (*Clethrionomys glareolus*) and the yellow-necked field mouse (*Apodemus flavicollis*) which account jointly for some 95% of the total number of rodents. Hence, only these two species have been studied. Due to the lack of other data, only one ungulate species, the roe deer (*Capreolus capreolus*), and one carnivorous species, the fox (*Vulpes vulpes*), could be investigated in respect to the flow of heavy metals.

The concentration of heavy metals was analyzed by atomic absorption spectrophotometry in muscle, liver, kidney, bone, hair, and whole carcasses, as well as in food and excrement. The results are presented in Table 4.25. The elements considered are very unevenly distributed. It seems important to recognize the pat-

Table 4.25. Concentration of heavy metals in homeotherm consumers in the Niepołomice Forest (ppm dry weight)

	n	Tissue	Cd	Pb	Zn	Fe
Small birds	19	Carcass	0.60	15.0	115.0	391.9
	19	Feathers	0.87	28.5	173.0	363.3
	19	Bones	0.34	41.4	283.6	228.9
	19	Liver	2.08	5.8	162.9	1947.7
Rodents	20	Carcass	1.10	16.6	102.3	264.9
	20	Hair	0.00	1.7	27.8	39.4
	67	Liver	1.11	1.5	97.8	514.6
Roe deer	4	Carcass	0.00	0.0	52.3	143.1
	4	Hair	0.56	11.2	301.8	1223.4
	4	Bones	0.00	26.9	100.7	42.2
	4	Liver	1.59	0.0	215.7	1075.9
	4	Kidneys	11.05	0.0	303.1	261.6
Fox	2	Muscle	0.00	15.9	155.2	148.3
	2	Bones	0.00	30.3	152.7	90.7
	2	Liver	1.56	5.2	23.8	954.6
	2	Kidneys	1.28	0.0	41.8	250.6

tern of distribution because this may permit an estimate of the impact on organisms, as well as give an insight into possible detoxification mechanisms.

The lowest concentrations are those of cadmium (0–11 ppm), however, this is an extremely toxic element (Dulka and Risby 1976, Luckey and Venugopal 1977). Such concentrations are the lowest in bone and muscle (less than 1 ppm), and the highest in kidney and liver (1.5 to 11 ppm). Notably, cadmium does not appear in mammalian hair, while its concentration is fairly high in bird feathers.

Lead occurs more commonly than cadmium in the investigated animals. It is concentrated primarily in bones where concentration averages 41.6 ppm in birds, 26.9 ppm in roe deer, and 30.3 ppm in fox. It occurs also in considerable amounts in hair and feathers. In turn, its concentration is low in liver and it is indiscernible in kidney of roe deer and fox (Table 4.25).

The concentration of zinc ranges from 40 to 300 ppm. When compared to the other elements, it is relatively evenly distributed over the whole animal body (Table 4.25).

The highest concentrations (90 to 1,900 ppm) are those of iron, which is natural due to its physiological significance in vertebrates. The lowest concentration of iron occurs in bone, and the highest in liver. In fact, the liver is able to accumulate large amounts of iron when the concentration exceeds the physiological capacity of the animal (Harper 1969).

These data have been compared to those taken in the Białowieża Forest. The latter is widely accepted in Poland as a nonpolluted control site (Grodzińska 1977, Sawicka-Kapusta 1979). The concentration of all the elements considered in virtually all animal tissues exceed those recorded in this control area at the rate of a dozen to several hundred percent (Sawicka-Kapusta 1979, Zawołek 1978). The Białowieża Forest is not totally free from pollution by heavy metals (Grodzińska 1978 b). The present data indicate therefore that the level of pollution is very high indeed at Niepołomice. The investigated consumers represent various trophic levels. Nevertheless, no significant differences in the concentrations of the heavy metals considered have been observed. This indicates that these metals do not tend to accumulate along the trophic chain in forest ecosystems. This is also corroborated in that concentrations are almost the same in the body of animals as in the food.

The amounts of heavy metals contained in various consumers have been computed from estimates of abundance and biomass of the animal populations (Table 4.26). These amounts are very small indeed, much less than 1 g ha^{-1} for iron and only a fraction of a milligram ha^{-1} for cadmium. Very small amounts of heavy metals are used in the process of production (Table 4.27).

Given the assimilation and the concentration of heavy metals in food, the flow of heavy metals through homeotherm consumers can be estimated (Table 4.28). The largest amounts of cadmium are assimilated by birds (90 mg ha^{-1} annually); the largest amounts of the other elements by roe deer (804 mg lead, 6 g zinc, 21 g iron ha^{-1} annually). Certainly, this flow is very small relative to the input from the air. However, the consumers investigated participate in cycling only a small portion of the mobile pool of the elements considered (Chap. 2).

The net production of the consumers considered is low. There is also little accumulation of heavy metals in ontogeny. Hence, almost the total amounts of

Table 4.26. Amounts of heavy metals in biomass of some homeotherm consumer populations in the Niepołomice Forest

	Average concentration in body				Biomass standing crop g ha^{-1}	Total amounts (mg ha^{-1})			
	Cd	Pb	Zn	Fe		Cd	Pb	Zn	Fe
Small birds	0.57	14.8	107.9	355.1	262	0.05	1.27	9.2	30.2
Rodents	1.06	15.6	99.5	264.6	184	0.08	1.1	7.1	19.0
Roe deer	0.20	9.7	103.7	358.8	5,000	0.36	17.3	185.1	640.5
Fox	0.05	21.5	150.1	144.7	65	0.00	0.56	3.9	3.8

Table 4.27. Amounts of heavy metals involved in annual production of some homeotherm consumer populations in the Niepołomice Forest

	Production (biomass) g ha^{-1} yr^{-1}	Cd	Pb	Zn	Fe
		mg ha^{-1} yr^{-1}			
Small birds	1013.1	0.19	4.9	35.6	116.9
Rodents	632.3	0.25	3.8	23.9	63.6
Roe deer	2100.0	0.15	7.3	77.7	269.0
Fox	97.1	0.00	0.9	5.8	14.0

heavy metals assimilated should be evacuated with excrement and urine. The amounts of evacuated heavy metals have been computed on the basis of the amounts of excrement (excreta in birds) and the concentration of heavy metals therein (Table 4.28). A complete balance with assimilation can hardly be expected even experimentally, nevertheless, the results are consistent with estimates made on the basis of consumption. This is especially true for birds, because all their excreta could be analyzed, whereas mammalian urine has remained beyond the scope of this analysis. The loss in excrement is much lower than consumption in rodents and fox, but this difference may be balanced in urine. In turn, the evacuation of heavy metals is much higher than consumption in roe deer, especially for zinc and iron. It is not possible to determine at the moment whether this is due to errors in estimation, or to the supplementary input of heavy metals in water.

It should be emphasized that the amounts for heavy metals which are evacuated with excrement are small, but they are partly transformed into insoluble compounds and therefore removed from the matter cycle in the ecosystem.

Table 4.28. Flow of heavy metals through homeotherm consumer populations in the Niepołomice Forest

	Annual consumption mg ha^{-1} yr^{-1}				Excretion with feces[a] mg ha^{-1} yr^{-1}				C_m–P_m [c]			
	Cd	Pb	Zn	Fe	Cd	Pb	Zn	Fe	Cd	Pb	Zn	Fe
Small insectivorous birds	90	136	5,361	6,704	39	149	1,585	6,644	89.8	131	5,325	6,587
Rodents	47	160	2,264	6,825	3	34	874	1,601	46.7	156	2,240	6,761
Roe deer[b]	60	804	6,204	20,928	26	1,310	12,878	46,322	59.8	797	6,126	20,659
Fox	0.5	9	58	173	0.2	2	24	23	0.5	8	52	167

[a] Feces plus urine in birds
[b] Figures extrapolated from data in winter (November to March)
[c] C_m: Amounts consumed, P_m: amounts involved in production

4.5 Impact of Consumers on the Forest Ecosystems of the Niepołomice Forest

A. Górecki

The impact of consumers on ecosystems was recently considered by several students. Most of these studies have focused on the energy flow through populations or trophic levels (Medwecka-Kornaś et al. 1974). Several authors have also considered the nonenergetic affects of consumers on various ecosystems, especially with respect to the cycle of matter (Petrusewicz and Grodziński 1975, Lee and Inman 1975, Wiener 1975, Batzli 1978, Hayward and Phillipson 1979). Such studies, however, are very time-consuming and difficult and generally incomplete.

Bioenergetic parameters of the investigated poikilo- and homeotherm consumers of the Niepołomice Forest are presented in Table 4.29. What is striking is the preponderant share of insects in the total energy flow, relative to homeotherm animals. Even though the investigations covered only a part of the insect fauna, the production of insects exceeds that of homeotherms by two orders of magnitude and this holds true also for consumption. The smallest differences occur in respiration, which is due to the absence of energetic costs of thermoregulation in insects. Therefore, the ratio of production to consumption is relatively high in insects ranging from 4.5 to 37%. This coefficient is even higher, exceeding 61%, in phyllophagous insects. In turn, these ratios only slightly exceed 2% in homeotherms (Table 4.29) (Grodziński and French 1983).

Table 4.29. Bioenergetic parameters of consumer populations investigated in the Niepołomice Forest ($g\ ha^{-1}$ for standing crop, $kJ\ ha^{-1}\ yr^{-1}$ for the others)

	Sc	P	C	R	A	FU	P/C %
Rodents	197.6	4,336	227,700	185,630	189,860	37,840	1.9
Small birds	202.8	7,700	592,000	436,000	444,500	147,500	1.3
Ungulates	11,350.0	29,430	1,930,000	1,240,000	1,270,000	660,000	1.5
Carnivorous mammals	145.0	816	18,340	14,170	14,980	4,170	4.4
Raptorial birds	14.2	90	6,120	5,410	5,500	620	1.5
Homeothermic consumers	11,909.6	42,372	2,774,160	1,881,210	1,924,840	850,130	2.1
Foliophagous insects	630,000[a]	63,000,000	17,100,000	3,900,000	10,300,000	6,800,000	36.8
Rhizophagous insects	11,430[a]	114,300	570,800	199,600	313,900	256,900	20.0
Xylophagous insects	850						
Cambiophagous insects		8,500	188,400				4.5
Poikilothermic consumers	642,280	64,222,800	17,859,200				20.4

[a] $Sc = 0.1 \times P$ (Bandoła-Ciołczyk and Witkowski 1976)

Table 4.30. Consumption of the consumers studied at various trophic levels (kJ ha^{-1} yr^{-1})

	Producers	Primary consumers	Secondary consumers
Rodents	223,146	4,154	400
Small birds	150,000	405,000	45,000
Ungulates	1,930,000	–	–
Avian and mammalian carnivores	–	22,000	2,700
Homeothermic consumers	2,303,146	431,554	47,700
Phyllophagous insects	17,100,000		
Rhizophagous insects	114,300		
Cambio-xylophagous insects	8,500		
Totals	20,162,346	431,554	47,700

It is interesting to compare the consumption of the consumers investigated as ordered according to their trophic position (Table 4.30). The values are based on average annual diets. The disproportion between poikilo- and homeotherms is due to the consumption of plant production by insects. The consumption relationships between successive trophic levels are 47:9:1. One should, however, keep in mind that various consumers have been omitted from the present study (for instance, diverse carnivorous insects), and this could strongly affect the relationships presented above. On the other hand, simple trophic chains can hardly be distinguished among consumers, since their interrelationships present commonly complex webs.

Many problems must be overcome to estimate the proportion of primary production eliminated from the ecosystem by consumers. It is not easy to calculate even the ratio of consumption to primary production. Variable herbivore populations consume only a small fraction of the primary production (Golley et al. 1975, Pętrusewicz and Grodziński 1975), less than 10% as a rule. However, primary production cannot be consumed entirely. Therefore, the term "food available" has been introduced (Grodziński 1968), defined as "food which is easy to find, is being chosen and being eaten by these animals" (see also Grodziński 1975).

Petrusewicz and Grodziński (1975) compiled some data on the ratio of consumption to food available for herbivore consumers in some terrestrial and aquatic ecosystems. It happens that some populations can exploit from a few percent to some tens of percent of the food available. This proportion is lower in terrestrial than in aquatic ecosystems, as a rule. For instance, Drożdż (1966) provided data indicating that rodents consume 4.5% of the food available in a beech wood. The data given by Medwecka-Kornaś et al. (1974) for the Niepołomice Forest indicate that mammals, birds, and *Tortrix* and other lepidopteran larvae consume some 7.7% of available food. Certainly, the latter estimates are very incomplete, just as those presented in the present study. Furthermore, matter flow through consumers was not investigated by those authors.

Table 4.31. Amounts of nutrients in the annual production of the consumers investigated (g $ha^{-1} yr^{-1}$)

	N	P	K	Ca
Rodents	19.9	4.5	2.1	5.9
Small birds	35.5	6.7	2.9	10.6
Ungulates	340.0	150.0	20.9	220.6
Carnivores	4.0	2.1	0.2	3.0
Phyllophagous insects	46,864.0	6262.0	2020.0	–
Rhizophagous insects	29.4	3.3	1.0	–
Cambio-xylophagous insects	814.7	108.5	35.1	–

The major role in matter flow in the ecosystems investigated is by poikilotherm consumers. This is clearly indicated by the amounts of nutrients in the biomass produced by particular consumer groups (Table 4.31). The amounts of nitrogen in the biomass produced annually by insects in the deciduous forests of the Niepołomice Forest approach 48 kg ha^{-1}, whereas they approximate only 0.4 kg ha^{-1} in homeotherms. Similar relationships have been obtained for the other nutrients (Table 4.31). This is only slightly influenced by the fact that phyllophagous insects are almost entirely confined to the deciduous forests in the area considered. Even when extrapolated over the whole forest, the amounts of nitrogen in the annual production of phyllophagous insects approximate 6 kg ha^{-1} and exceed by an order of magnitude those used by the homeotherms.

Much more constant among the consumer populations investigated are the amounts of nutrients evacuated with excrement. In fact, the phyllophagous insects evacuate almost 7 kg ha^{-1} annually, but the homeotherm consumers evacuate jointly some 7.3 kg ha^{-1}. The greatest share in the latter value is that of the small birds (almost 3.6 kg ha^{-1}). The total amount of evacuated nutrients is very high. It approximates 22 kg ha^{-1} annually, which is comparable to the amounts of N, P, K, and Ca delivered by aircraft to fertilize the ecosystem.

The impact of various consumers (insects, rodents, ungulates) on primary production, as well as on its decomposition through consumption and nutrient dissipation in ecosystems has been presented by Batzli (1978) from his work and that of others. His conclusions are here supplemented with considerations concerning small birds (Table 4.33).

The impact of consumers on ecosystems is highly complex and multifaceted, and an unequivocal evaluation is hardly possible at the moment. Most authors are of the opinion that consumers only slightly affect the majority of ecosystems. It is commonly assumed (Golley et al. 1975, Wiener 1975) that the impact of consumers is of significance only in simple ecosystems, whereas it is negligible in such complex ecosystems as forests. The majority of relevant data have been produced from grasslands and agrocoenoses where the impact of consumers can be easily traced and directly affects human economics.

The affects of consumers activities on ecosystems are multifaceted and variable. It is well known that primary production is affected, and most commonly decreased, by grazing. Several authors, however, point out that grazing may stimulate plant growth, while damage to leaves can only slightly affect primary production (Chew 1974, Mattson and Addy 1975, Petrusewicz and Grodziński 1975). For instance, in most tree species wood increment remains unaffected by grazing of even up to 50% of the leaves. The impact of grazing may be considerable, however, if the growing points are seriously damaged (Chew 1974). Some authors refer to the consumption of plant seeds in this context, since it may lead to a decrease in primary production and even to inhibition of the process of self-rejuvenation of the forest (Golley et al. 1975). Most commonly, however, these herbivore activities (especially those stocking food) seem to contribute to an increase in plant diversity and to the dissemination of some plant species (Chew 1974).

Burrowing activities of consumers change the physicochemical properties of the soil and stimulate plant growth, as well as increase diversity and primary production (Golley et al. 1975, Chew 1974). Recently, the impact of consumers on matter or nutrient cycling has been repeatedly emphasized (Lee and Inman 1975, Petrusewicz and Grodziński 1975, Batzli 1978). This may be due to various processes which lead, generally, to a shortening and simplification of trophic chains. The reducers get a continuous supply of matter that is grazed but not consumed, as well as excrement and urine (Lee and Inman 1975, Hayward and Phillipson 1979). In some ecosystems, especially those with a small pool of mineral resources, this process decreases oscillations in primary production. However, this mechanism seems to attain some significance only in simple and nutrient-poor ecosystems. Concerning the Niepołomice Forest, only a shortage of sodium may represent a mechanism limiting primary production (Lee and Inman 1975), but even this has not been empirically corroborated.

Two consumer groups should be distinguished when considering their excrement as a contribution to the nutrient supply of the ecosystem (Petrusewicz and Grodziński 1975, Hayward and Phillipson 1979). One of these includes animals with a very low digestibility coefficient, which supply very large amounts of excretions with low nutrient contents. The other includes consumers with high digestibility supplying smaller amounts of excretions but with more concentrated nutrients. In particular, nitrogen in urine is of paramount importance for plant growth, because it acts not only directly but also by its impact on an increased rate of litter decomposition (Batzli 1978). The difference between the two consumer groups mentioned above is well exemplified in the Niepołomice Forest. When the amounts of excrement per area unit are compared to the nutrient content (Tables 4.31 and 4.32), the amounts of excrement are by an order of magnitude larger in insects than in homeotherms, the latter showing much higher digesting efficiency (Petrusewicz and Grodziński 1975). Nevertheless, the amounts of nutrients supplied by the insects and by the homeotherms are almost equal to each other, which indicates clearly that the concentration of nutrients in excrement is much lower in insects than in homeotherms.

The impact of consumers on ecosystems, especially on such complex ecosystems as the Niepołomice Forest, can scarcely be determined in detail. It is widely different between the deciduous and coniferous forests. This difference is even greater with respect to the meadows within and around the forest complex. Consumer density is much higher in the deciduous than in the coniferous forests. The

Table 4.32. Evacuation of nutrients with excrement ($g\,ha^{-1}\,yr^{-1}$)

	N	P	K	Ca
Rodents	750	22	27	12
Small birds	3589	153	484	–
Ungulates	3000	100	100	650
Insects:				
phyllophagous	6877	2085	5436	–
rhizophagous	211	64	167	–

Table 4.33. Impact of herbivores on mineral cycling (Batzli 1978, slightly emended)

Herbivore	Effects of primary production (nutrient uptake)	Effects on decompositon (nutrient release)	Effects on dispersion of nutrients
Insects	Consumption reduces production of most plants but it may be compensated by change in vegetational composition	Deposition of frass stimulates decomposition and damage to leaves increases leaching of nutrients	Nutrients concentrated at nest sites by ants and termites
Rodents	Consumption reduces short-term production but it may increase long-term production because of more rapid nutrient cycling	Nutrients released more rapidly through urine and feces than from plant litter	Nutrients concentrated at burrows and nest sites
Ungulates	Consumption usually reduces production but selective feeding may compensate by reducing senescence or changing vegetational composition	Nutrients usually released more rapidly through urine and feces, whose decomposition also stimulates decomposition of litter	Nutrients concentrated on loafing grounds and near water sources
Birds	Seed consumption only slightly reduces production, contributes to plant sowing; feeding on insects reduces their abundance and contributes to increased primary production	Nutrients usually released more rapidly through excrement than through decomposition; very high concentration of nitrogen in excrement (a dozen percent or so) accelerates litter decomposition	Nutrients distributed rather evenly in ecosystem, slightly concentrated near nests

phyllophagous insects, which rank first in consumption among the investigated consumer groups, are virtually absent from the coniferous forests (Table 4.5). Consequently, the consumption as well as the supply of nutrients per unit area, is much higher in the deciduous forests. Density of consumers, especially the homeotherms, is much lower in the meadows than in the forest; however, the impact of grazers is much more serious in the former ecosystems (Petrusewicz and

Grodziński 1975). In general, the detrital chain of energy flow exceeds the grazing chain in all the ecosystems considered. The impact of consumers on the meadows cannot be reliably estimated because they are economically exploited and virtually the whole primary production is removed from the ecosystem.

For the Niepołomice Forest, one can only attempt to predict what would happen to the ecosystem after elimination of the consumers. To this end, the effects of air pollution (Chaps. 1–2) must be kept firmly in mind. Under the influence of industrial air pollution, decomposition of the litter can be considerably inhibited (Chap. 5). Any increase in amounts of the litter has not been recorded, but nevertheless the turnover rate of various nutrients in the litter is decreased (Chap. 5).

No doubt an absence of consumers would have some effect on the primary production. The stimulation of plant growth by grazing seems to be overestimated. As reported by Bobek et al. (1979), production is greater on those plots in the Niepołomice Forest which have been protected from grazing by ungulates. Similar relationships relate also to the impact of rodents (Sect. 4.3.1) and insects (Sect. 4.2). One may therefore expect that the primary production would increase in the absence of consumers.

Considerable changes could be effected in the species composition of forest communities. Certainly, those species would increase their production which are preferably eaten by particular consumers. Lack of burrowing and zoochory would contribute to unification and simplification of the species composition.

The most important effects would be those on reducers. Perhaps rapid oscillations in primary production would not occur, but the decomposition rate of the litter would certainly be decreased. Absence of easily decomposable excrement and much smaller amounts of nitrogen would modify the process of decay. The Niepołomice Forest is relatively poor in microorganisms, especially in the humus layer of the soil, and therefore the absence of stimulation of litter decomposition would lead finally to the carpet effect or an inhibition of the decay process. This is why matter flow through consumers contributes to increased homeostasis of the ecosystem, which is especially important in this case because of the strong pressure of industrial air pollution (Chap. 1).

The carnivores limit to a considerable extent the herbivores. This holds especially for the affects of carnivores on rodents (Sect. 4.3.4) and birds on insects (Sect. 4.3.3). Were the carnivores eliminated, the impact of herbivores on the primary production would increase considerably. The extermination of carnivores might effect not only an increase in herbivore abundance, but also some changes in species composition. Dominance would be achieved by those herbivores which are highly efficient in production but with weak defence mechanisms against carnivores. This would certainly have enormous effects on the plant communities. One may conclude that the carnivores may help to preserve the forests of Niepołomice until the tree stands can be restructured, even under the existing level of air pollution.

Chapter 5

Decomposition in the Pine Forests of Niepołomice

Chapter Author: J. ZIELIŃSKI

5.1 Introduction

Litter is one of the most important factors in nutrient flow through forest ecosystems. It comprises a large pool of elements derived from decomposed organic matter and transported to deeper soil layers where they may be reassimilated by plants. This pool may include more than 60% of the total nutrients in an ecosystem (Weaver 1975). Decomposition of dead organic matter and activation of nutrients are the most important functions of forest litter and in this way have an influence on all living conditions in the ecosystem. Inhibition of these processes may result in a deficiency of mineral nutrients indispensable for autotrophic life, whereas excessive acceleration may cause the disappearance of a large pool of mineral elements, thus disturbing the balance of the ecosystem. Disturbance of the decomposition rate of forest litter may also strongly affect water and air relations in specific soil layers.

Estimates of elemental flow through litter may therefore assist in the ecosystem balance. This is particularly important for forest ecosystems affected by industrial air pollution (Smith 1974, 1981). The Niepołomice Forest provides a significant example because it occurs within range of the industrial emissions from the Cracow steel complex (Grodziński 1978, Grodziński and Yorks 1981). This ecosystem is much more strongly polluted with heavy metals than the Białowieża Forest, which has served as a control area in Poland (Grodzińska 1977, 1980; Makomaska 1978). A preliminary survey of organic matter decomposition in the Niepołomice Forest demonstrates that, in spite of considerable variation in heavy metal pollution, the rate of organic matter decomposition remains reasonably constant throughout the whole forest system (Derlaga 1981). One may therefore postulate that the actual pollution level, even though very high (cf. Chap. 1), does not dramatically affect the activity of soil microorganisms responsible for organic matter decomposition. Inhibition of decomposition processes is in fact observed mostly at higher concentrations of heavy metals (Jackson and Watson 1977, Strojan 1978, Smith 1981). An increase in heavy metal concentration in the humus layer, just below fresh litter, was observed in Sweden (Tyler 1972). The decomposition process is generally inhibited in areas with heavy industrial emission, possibly due to the blocking of organic colloids by heavy metal ions that replace those of potassium, magnesium, and calcium (Rühling and Tyler 1973). In forest ecosystems influenced by heavy metal emissions microorganisms decrease in abundance in the A_1 layer (Jordan and Lechevalier 1975). This documents the damaging effects of heavy metals on microorganisms responsible for decomposition of dead organic matter. Inman and Parker (1978) record similar reactions in forest ecosystems near large urban complexes.

The present study assumes that in a mature forest (70-year-old temperate pine forest) the litter should not increase in weight. Annual production of dead organic matter reaching the forest floor should be in balance with annual respiration at that level and outflow of organic matter from the ecosystem [See general studies on ecological succession in forest ecosystems (Crocker and Major 1955, Odum 1969)]. A disruption of balance may therefore be interpreted as a disturbance in the forest ecosystem.

Various methods may be applied to determine the rate of flow and the balance of matter in ecosystems. They fall into two general categories, direct and radioisotope methods. Classic investigations of matter flow in forest ecosystems using direct methods have been carried out on the Hubbard Brook watershed in the United States (Likens et al. 1977), and in northern Ontario, Canada (Foster and Morrison 1976). Radioisotopes have been widely used in the Oak Ridge National Laboratory in the United States (Auerbach 1969).

In the present study direct methods have been applied. Matter flow through litter has been determined from annual input and outflow of matter. Organic matter and the elements N, P, K, Ca, S, Mg, Fe, Zn, Ni, Cu, Pb, Cd which were specifically measured reach the forest floor in the rain passing through the canopy, as well as in organic matter and dust derived from the canopy or introduced from outside. The only significant pathway for outflow is water passing downward through the litter.

Some assumptions or simplifications have been made. The input of elements by trunk flow has been neglected. In fact, element inflow from trunks amounts to 3% of that contributed by rain passing through the canopy (Foster and Gessel 1972, Patterson 1975). The effects of the herb layer on the balance of elements has also been neglected because this represents only the so-called "small cycle of matter flow" that makes up a part of the total balance of the litter. The cycle is closed and the annual production of herbs comes back to the litter in the form of dead organic matter. In respect to the balance of nitrogen, denitrification as well as nitrification with atmospheric nitrogen have been omitted because the two processes are assumed to be in balance. In coniferous forests they are dominated by nitrogen derived from tree canopies (Todd et al. 1975).

5.2 Study Areas

The study areas have been located in pine-oak forests (Pino-Quercetum) typical of those of the Niepołomice Forest (cf. Chap. 1.1). They are on podzolic soils on uncompacted fluviglacial sands. The A_0 layer is well developed, having a pH ranging from 3.7 to 3.9. At a depth of approximately 8 cm, there is a rapid transition to the A_{1+2} layer, the latter consisting of uncompacted sands with small amounts of humic compounds (0.8% organic carbon) and with a pH of about 3.6. The B_f consists of more compacted sands with a pH of 4.1. At depths of 23 to 200 cm, the layers B, BC, and C occur. They consist of light-colored sands with some humic compounds, and pH ranging from 4.3 to 4.4 (cf. Chap. 1).

The A_0 contains 81% of the total organic materials of the soil, 73% of the nitrogen, 6% of the phosphorus, 31% of the potassium, 10% of the calcium, and

Table 5.1. Proportion of organic carbon (C) to nitrogen (N), phosphorus (P), potassium (K), calcium (Ca), and magnesium (Mg) in various litter layers and mineral soil of pine forests, Niepołomice, Poland

Horizon	C/N	C/P	C/K	C/Ca	C/Mg
A_0L	55	3067	547	702	8871
A_0F	25	5960	1710	252	4635
A_0H	23	8800	1895	198	3623
A_0	32	4463	905	322	5422
A_{1+2}, B_f	28	331	346	41	201

3% of the magnesium. The remaining amounts occur in the layers A_{1+2} to C. Organic carbon has been found in the A_{1+2} and B. The other elements extend down to the bottom of the investigated soil profile.

The litter is of the mor type consisting of three clearly distinct horizons. These are the A_0L, which includes pine needles and twigs and herb remains, the A_0F, which includes brown, fresh to partly decomposed rotted materials and the A_0H, including brown to black, fresh but decomposed and nonstructured rotted materials intercalated with sand particles. These three horizons of litter contain 75 t of organic matter (dry weight) per hectare, of which 45% is contributed by the A_0L, 35% by the A_0F, and 20% by the A_0H. The three horizons differ in mineral composition. Phosphorus and potassium occur in largest amounts in the upper horizon, while their concentration decreases downward. The same holds for organic carbon. However, nitrogen, calcium, and magnesium occur in more or less constant concentration throughout the litter. This variability of particular elements obviously is proportionate to carbon (see Table 5.1).

The soil and litter of the study areas exhibit moderately strong biological activity[1]. Soil enzymatic activity has been measured using the saccharase, B-glucosidase, urease, and acid phosphatase techniques. Samples for the enzymatic and microbiologic analyses were taken in autumn 1978. Soil enzymatic activity and abundance of microorganisms in the horizons A_0L and A_0F–H are given in Tables 5.2 and 5.3.

Decomposition of organic matter proceeds intensively in the A_0L horizon, while it ceases in the A_0H horizon. Enzymatic activity is in fact higher in the former horizon than in those below. This reflects the activities of microorganisms, especially those contributing to mineralization of organic nitrogen. This variation between specific horizons holds true also for the incidence of microorganisms involved in the decomposition of organic matter. In the study area the latter are represented mostly by fungi of the genera *Penicillium, Verticillium, Mucor,* and others, which predominate over the bacteria *Bacillus, Streptomyces,* and *Arthrobacter.*

The herb cover of the study area is dominated by *Vaccinium myrtillus, Molinia coerulea, Calamagrostis arundinacea, Agrostis vulgaris,* and *Trientalis europaea.*

1 Enzymatic and microbiologic analyses courtesy of Dr. K. Marczyńska-Gałkowska and Dr. J. Olszewski

Table 5.2. Enzymatic activity of the litter in pine forests Niepołomice, Poland (in ml $Na_2S_20_3$ per 10 g of litter for sacharose and B-glucosidase, ml NH_3 per 10 g of the litter for urease, and mg P_20_5 per 10 g of the litter for acidic phosphatase)

Horizon	Saccharase	B-glucosidase	Urease	Acid phosphatase
x̄ A_0L	39.5	32.5	10.80	2.7
A_0F+H	9.2	13.7	6.92	1.8
SE A_0L	10.3	7.2	3.7	0.4
A_0F+H	2.8	1.6	1.8	0.4
Cv A_0L	52.1	44.3	68.5	29.6
% A_0F+H	60.9	23.4	52.0	44.4

Table 5.3. Abundance of microorganisms in the litter of pine forests Niepołomice, Poland (thousands g^{-1} litter)

Horizon	Bacteria and Actinomycetes	Fungi	Microorganisms using $N–NO_3$
x̄ A_0L	88,250	188,000	245,000
A_0F+H	20,400	31,650	26,000
SE A_0L	72,300	39,550	53,000
A_0F+H	11,350	6,150	6,400
Cv A_0L	164	42	44
% A_0F+H	111	39	49

These species contribute jointly 97% of the kg dry weight per hectare for the above materials. Production of the herb layer in this type of forest reaches on average 662 kg dry weight per hectare (Chap. 3; cf. Table 3.11).

The tree stand is dominated by Scotch pine (*Pinus silvestris*), approximately 70 years old. Standing crop is about 161 t dry wt ha^{-1}. The undergrowth includes mostly oak (*Quercus sessilis*), and buckthorn (*Frangula alnus*). Standing crop amounts to 2.4 t dry wt ha^{-1}. The annual increase in weight of all these three species averages 2.0 t of dry wt per ha (Rieger; cf. also Chap. 3). When loss due to the fall of organic matter including needles, leaves, twigs, and bark, totalling 5 t dry wt ha^{-1} annually, is taken into account, the primary production of above-ground parts of the tree layer approximates 4% of standing crop.

In total, the ecosystem under investigation contains some 310 t of organic matter dry wt ha^{-1}, of which 52% is accounted for by the tree stand, 1% by the herb layer, and 47% by the soil with the litter.

5.3 Matter Dynamics and Balance in the Litter

During a 2-year period (April 1, 1976 to March 31, 1978) an average annual rain and snow fall of 666±16 mm was recorded, with pH of 5.7±0.2. Of this amount, 439±22 mm reaches the forest floor where pH decreases to 5.0±0.2. A proportion of the fall is retained in the litter and hence the mineral soil receives

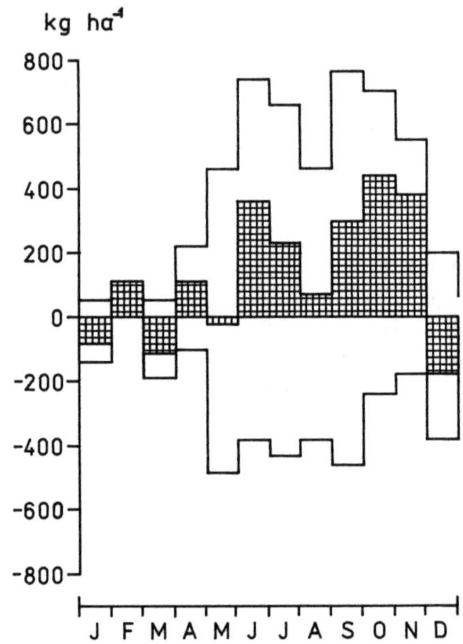

Fig. 5.1. Balance of organic matter in pine forest litter in the Niepołomice Forest. *Upper curve* input; *lower curve* outflow; *cross-hatched area* balance

Table 5.4. Concentration of elements in rain water reaching the canopy, the litter, and the mineral soil of pine forests in ppm, Niepołomice, Poland

Element	Concentration ppm in the rain water reaching:		
	Canopy	Litter	Soil
N (total)	3.463 ± 0.289	4.341 ± 0.325	15.349 ± 0.885
P	0.310 ± 0.066	0.286 ± 0.072	0.811 ± 0.084
K	1.399 ± 0.126	5.163 ± 0.458	8.830 ± 0.478
Ca	4.325 ± 0.361	9.448 ± 0.827	17.593 ± 1.430
S (total)	7.894 ± 1.592	14.275 ± 1.753	22.239 ± 3.258
Mg	0.429 ± 0.028	1.870 ± 0.110	3.357 ± 0.290
Fe	0.001 ± 0.001	0.107 ± 0.044	0.399 ± 0.064
Zn	0.084 ± 0.012	0.272 ± 0.028	0.866 ± 0.094
Ni	0.000 ± 0.000	0.005 ± 0.002	0.050 ± 0.023
Cu	0.046 ± 0.018	0.052 ± 0.021	0.039 ± 0.020
Pb	0.032 ± 0.013	0.064 ± 0.027	0.030 ± 0.014
Cd	0.000 ± 0.000	0.005 ± 0.003	0.003 ± 0.002

only 290 ± 38 mm annually, with a lowered pH of 4.1 ± 0.3. Thus, interception by the canopy is 34 ± 5%, and interception by the litter is 34 ± 12%. These levels retain 56 ± 7% of the total infall reaching the forest.

Contrary to the condition in deciduous forests, coniferous forests show a relatively constant annual fall of dead parts from the canopy. A minimum occurs only in winter, and two weak maxima take place early in summer and in autumn

Table 5.5. Concentration of elements in recently deposited organic matter, litter, and herb layers of pine forests Niepołomice, Poland. Computed as ash-free dry weight (in percent for N, P, K, Ca, S, Mg, Fe, and ash, and in ppm for Zn, Ni, Cu, Pb, Cd)

Element	Litter fall	Litter	Herb layer
N (total) %	1.009 ± 0.038	1.443 ± 0.140	1.358 ± 0.069
P	0.085 ± 0.006	0.102 ± 0.011	0.129 ± 0.005
K	0.252 ± 0.010	0.188 ± 0.024	0.487 ± 0.037
Ca	0.300 ± 0.024	0.097 ± 0.022	0.405 ± 0.150
S (total)	0.173 ± 0.016	0.230 ± 0.020	0.209 ± 0.019
Mg	0.047 ± 0.003	0.045 ± 0.004	0.078 ± 0.003
Fe	0.304 ± 0.095	0.500 ± 0.067	0.024 ± 0.013
Zn ppm	86.496 ± 3.294	225.145 ± 19.739	156.354 ± 5.977
Ni	9.808 ± 2.374	7.542 ± 0.492	1.953 ± 0.120
Cu	10.249 ± 1.249	16.650 ± 1.316	5.250 ± 0.747
Pb	32.512 ± 4.435	191.206 ± 16.408	20.405 ± 8.265
Cd	1.054 ± 0.079	3.849 ± 0.318	5.233 ± 0.687
Ash %	5.017 ± 0.296	43.122 ± 4.931	6.188 ± 0.287

Table 5.6. Annual input of elements to the canopy, the litter, and the mineral soil of pine forests Niepołomice, Poland (in kg ha^{-1}). Fractions introduced in solution in brackets

Element	Annual input (kg ha^{-1})		
	Canopy	Litter	Soil
N (total)	20.80	64.88 (28%)	47.65
P	1.83	3.99 (11%)	2.50
K	8.59	31.53 (65%)	24.13
Ca	26.27	54.56 (68%)	45.02
S (total)	44.10	44.62 (87%)	41.70
Mg	2.54	8.61 (73%)	7.57
Fe	0.01	4.75 (9%)	0.85
Zn	0.50	1.36 (73%)	1.25
Ni	0.00	0.04 (30%)	0.04
Cu	0.25	0.17 (85%)	0.13
Pb	0.16	0.29 (67%)	0.10
Cd	0.00	0.02 (78%)	0.00

(Fig. 5.1). In the course of the 2-year study, this fall averaged 5,038 ± 596 kg dry wt ha^{-1}.

Given the weight of the media, water, and organic matter, and given the concentrations of the elements considered (Tables 5.4 and 5.5), the annual input of those elements to the canopy, the litter, and the mineral soil could be determined (Table 5.6). This provided the basis for determining the balance of all of the major elements in the litter.

In these pine forests the litter is very rich in iron but very poor in calcium. The remaining nutrient elements show concentrations similar to those summarized by

Table 5.7. Balance of matter in the litter of pine forests Niepołomice, Poland, P – pool, I – annual input, 0 – annual outflow, B – balance ($kg\,ha^{-1}$)

Element	P	I	O	B
Organic matter	73,122	5,038	3,433	1,595
N (total)	1,108.0	64.9	47.7	17.2
P	78.5	4.0	2.5	1.5
K	143.5	31.5	24.1	7.4
Ca	74.9	54.6	45.0	9.6
S (total)	171.8	44.6	41.7	2.9
Mg	34.8	8.6	7.6	1.0
Fe	367.0	4.7	0.8	3.9
Zn	16.6	1.4	1.2	0.2
Ni	0.58	0.04	0.04	0.00
Cu	1.23	0.17	0.13	0.04
Pb	13.90	0.29	0.10	0.19
Cd	0.29	0.02	0.00	0.02

Rodin and Bazilevich (1967). Of the heavy metals which are here considered to be pollutants, the most abundant are zinc and lead (Table 5.7).

The elements measured reach the litter in the form of minerals as well as in solution (Table 5.6). The form in which elements appear controls their mobility in the ecosystem. The more abundant an element is in solution, the more rapid its turnover rate. Turnover rates of elements (Θ) are computed from their actual pool and annual input (Olson 1963), assuming that the latter is equal to the outflow. The formula is

$$\Theta_I = \frac{P}{I}, \tag{1}$$

where P denotes the pool or the amounts of the element in the part of the exosystem analyzed, and I denotes the input or the amounts of element introduced into the ecosystem per unit of time.

In the forest ecosystem the balance was positive for all considered elements, input exceeded the outflow (Table 5.7). Therefore, two turnover rates have been computed (Chap. 5), one using formula (1), the other according to the formula:

$$\Theta_0 = \frac{P}{O}, \tag{2}$$

where P is the pool, and Θ is the outflow or the amounts of element removed from the part of ecosystem controlled per unit of time.

In this way, a minimum turnover rate (Θ_I) has been distinguished from a maximum turnover rate (Θ_0), and the actual figure is considered to be intermediate (Table 5.8).

The specific elements vary in rate of turnover and can be divided into three groups: weakly mobile with turnover rates of several scores of years including iron, lead, and cadmium; moderately mobile with turnover rates of a dozen years or more, including phosphorus, nitrogen, nickel, and zinc; and highly mobile ele-

Table 5.8. Turnover rate and total input of matter for pine forests Niepołomice, Poland. B/P – percent proportion of balance of the actual pool, or reverse of the period of accumulation, B/I – accumulation rate (in percent), Θ_I – minimum turnover rate (in years), Θ_0 – maximum turnover rate (in years), I_{tot} – total input (as the factor of difference from the actual pool)

Element	B/P	B/I	Θ_I	Θ_0	I_{tot}
Organic matter	2.18	31.6	15	21	3.2
N (total)	1.55	26.5	17	23	3.8
P	1.91	37.5	20	32	2.7
K	5.16	23.5	4	6	4.3
Ca	12.68	17.5	1	2	5.7
S (total)	1.70	6.5	4	4	15.2
Mg	2.99	12.1	4	5	8.3
Fe	1.06	82.1	77	432	1.2
Zn	0.68	8.2	12	13	12.1
Ni	0.17	2.5	15	15	39.8
Cu	3.00	21.9	7	9	4.6
Pb	1.39	65.8	47	138	1.5
Cd	5.27	84.4	16	103	1.2

ments with turnover rates of less than 10 years, including copper, potassium, magnesium, sulfur, and calcium. Iron is least mobile with a turnover rate which falls within the interval of 77 to 432 years. The most mobile is calcium, with a turnover rate of 1 to 2 years (Table 5.8). Potassium is generally considered to be very mobile, but it ranks fourth after calcium, sulfur, and magnesium in this ecosystem. Nutrient elements are either moderately or highly mobile. Nitrogen and phosphorus are very strongly related to the organic matter and in fact, behave similarly. The minimum turnover rate for organic matter is 14.5 years. A maximum turnover rate could not be computed because the annual outflow is unknown.

It is notable that elements as toxic as lead and cadmium have very low turnover rates in this system. This is dangerous because it indicates that their accumulation in the litter is very high. Iron exhibits a still lower turnover rate but it is almost nontoxic and occurs naturally and in large amounts throughout the area.

Given the basic components of the balance (pool, input, and outflow), total input can be computed for each element, according to the formula introduced by Zieliński (unpubl.)

$$I_{tot} = \frac{\frac{P}{B} I}{P} = \frac{I}{B}, \tag{3}$$

where I_{tot} is the total input necessary to attain the actual pool of an element, B is the balance or the difference between the annual input and outflow, P is the pool or the amounts in the litter and I is the annual input. The total input is computed by using the period of accumulation $\left(\frac{P}{B}\right)$ of the actual pool. This figure is estimated

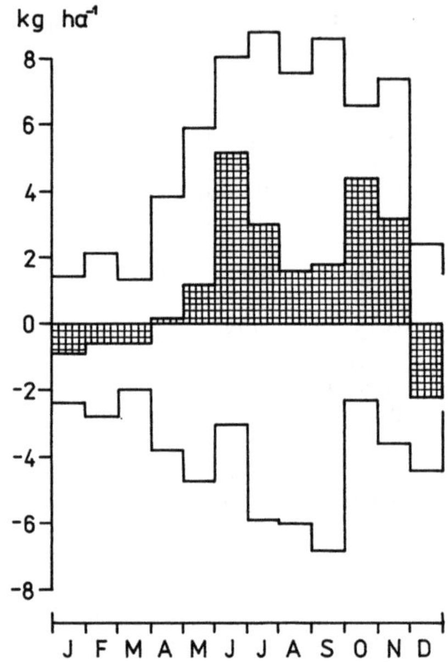

Fig. 5.2. Balance of nitrogen in the litter of pine forests Niepołomice, Poland; *upper curve* input; *lower curve* outflow; *cross-hatched area* balance

assuming that the conditions of accumulation have been constant. Application of the formula assumes that a constant fraction of the annual input has been accumulated.

The highest rate of accumulation is that for cadmium for which the balance accounts for 84.4% of the annual input, and the total input exceeds the actual pool by a factor of only 1.18. On the contrary, the balance of nickel accounts for merely 2.5% of the annual input, while the total input exceeds the actual pool by a factor of 39.8. Nutrient elements show a moderate rate of accumulation.

With respect to rate of accumulation in the forest, three groups of elements can be distinguished. These include those with high rates of accumulation – cadmium, iron, and lead; those with moderate rates of accumulation – phosphorus, nitrogen, copper, calcium, and magnesium; and those with a low rate of accumulation – zinc, sulfur, and nickel. This grouping is related to that based on turnover rates because the higher the mobility of an element, the lower the rate of accumulation. The correlation coefficient of turnover rate versus total input is -0.58 or -0.61, depending on whether the minimum or maximum turnover rate is considered.

The components of the balance show some seasonal variation. This is presented in Figs 5.2–5.5 for nitrogen, phosphorus, potassium, and calcium. The balance is highly negative in winter and in early spring, while it is positive in summer and particularly so in autumn. The shape of the curve suggests that this variation is due to the pattern of organic matter deposition. This is especially true for nitrogen and phosphorus because these elements are introduced into the litter mainly by the newly fallen organic matter. The balance of calcium is much more evenly

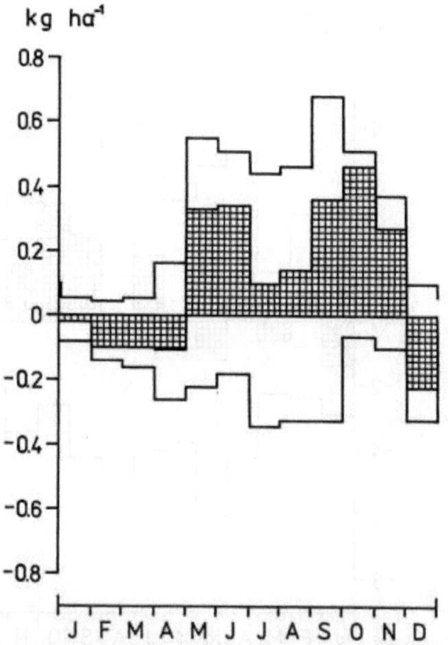

Fig. 5.3. Balance of phosphorus in the litter of pine forests Niepołomice, Poland; *upper curve* input; *lower curve* outflow; *cross-hatched area* balance

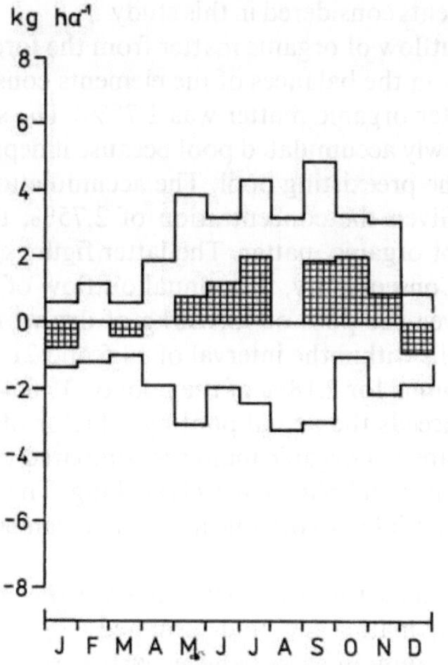

Fig. 5.4. Balance of potassium in the litter of pine forests Niepołomice, Poland; *upper curve* input; *lower curve* outflow; *cross-hatched area* balance

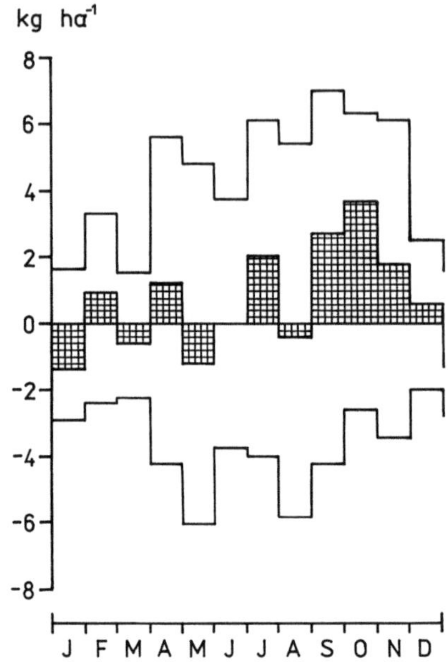

Fig. 5.5. Balance of calcium in the litter of pine forests Niepołomice, Poland; *upper curve* input; *lower curve* outflow; *cross-hatched area* balance

distributed seasonally, except for a positive peak in autumn. The same holds true for the other elements considered in this study.

The annual outflow of organic matter from the forest is unknown. It has only been estimated from the balances of the elements considered. The concentration of these in the litter organic matter was 2.75%. The same concentration can be expected in the newly accumulated pool because it depends not only on the recent fall but also on the preexisting pool. The accumulation rate of these elements is 43.95 kg ha^{-1}. Given the concentration of 2.75%, this figure is equivalent to 1.598 kg dry wt of organic matter. The latter figure gives the annual balance of organic matter. Consequently, the annual outflow of organic matter is 3,439 kg dry wt ha^{-1}. Given the pool of 73,122 kg of dry wt ha^{-1}, the turnover rate of organic matter falls within the interval of 14.5 and 21.3 years. The balance of organic matter accounts for 2.18% of the pool or 31.6% of the annual input, while the total input exceeds the actual pool by a factor of 3.2 (Table 5.8).

When the balance of organic matter is compared to the nutrient elements, the similarity to nitrogen and phosphorus is striking. This is natural, as the latter elements are strongly related to organic matter and can be used to indicate its behavior.

Using similar calculations for particular seasons, the annual dynamics of the balance of organic matter can be determined (Fig. 5.1). The outflow of organic matter is at a minimum in winter and in early spring. A weak maximum extends over a relatively long period in summer. The outflow is also high in early winter. The dynamics of balance depend on variation in the deposition of organic matter.

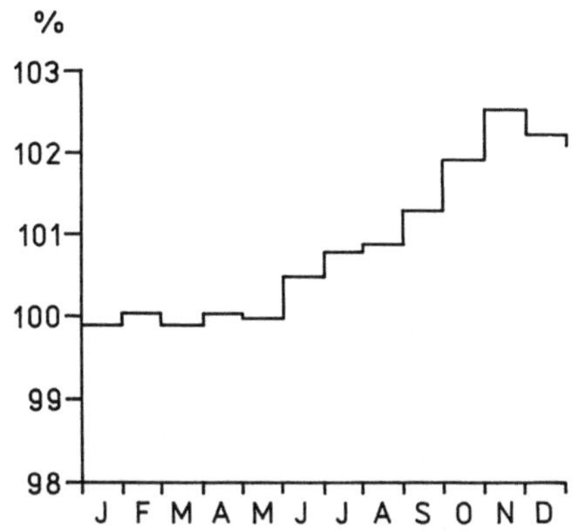

Fig. 5.6. Accumulation rate of organic matter in the litter of pine forests Niepołomice, Poland

Hence, there are two positive peaks in balance, early in summer, and in autumn, and two negative ones, in winter, and in early spring.

Given the input and outflow of organic matter in specific seasons, the dynamics of accumulation of organic matter in the litter can be determined (Fig. 5.6). A starting point of 100% is assumed. The highest accumulation rate is in summer and autumn and the minimum in winter and spring. Late in autumn, the accumulation rate is negative, which indicates that the outflow exceeds the input.

Having thus determined all the components of the balance of organic matter, the turnover rate in particular months may be computed (Fig. 5.7). This is very slow in winter, and most rapid in summer. This pattern correlates well with variation in temperature and precipitation in the study area (Hess 1969, Klein 1978). Decomposition is most rapid with the highest turnover rate under high temperature and heavy deposition in summer, while it attains a minimum with the lowest turnover rate under low temperature and deposition in winter. This pattern the dynamics of decomposition is consistent with experimental observations on the decomposition process (cf. Sect. 5.5), as well as data provided by other authors (Douglas and Tedrow 1959, Mikola 1960, Wood 1974).

5.4 Turnover Rate of Matter in Forest Ecosystems

According to Foster and Morrison (1976), the turnover rate of organic matter in litter in forests dominated by Jack pine (*Pinus banksiana*) is approximately 5 years. Lamb and Florence (1975) presented data on forests dominated by Monterey pine (*Pinus radiata*) indicating that the turnover rate ranges from 3 to 8 years, depending on soil conditions. Tappeiner and Alm (1975) reported a turnover rate of 5 years in stands dominated by red pine (*Pinus resinosa*), the rapidity of turnover increasing with greater amounts of deciduous species (*Corylus cornuta, Betula papyrifera*). Van Lear and Goebel (1976) investigated pathways of organic matter in forests of *Pinus taeda* and their data show that the turnover rate falls within an interval of 4.7 to 5.1 years, regardless of forest age.

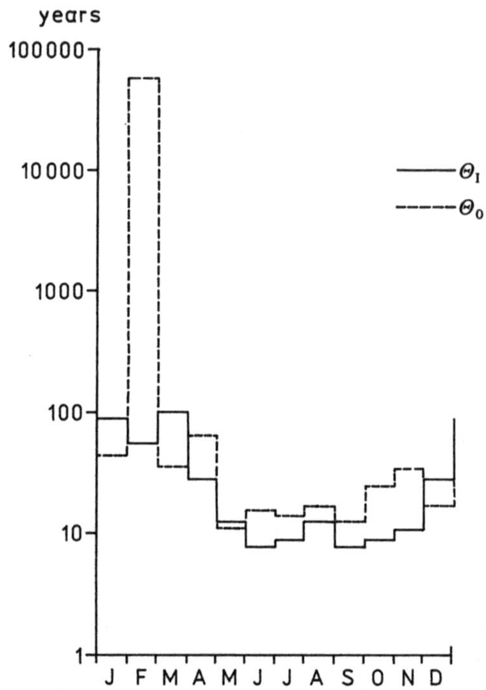

Fig. 5.7. Turnover rate of organic matter in the litter of pine forests Niepołomice, Poland; Θ_I minimum turnover rate; Θ_O maximum turnover rate

Similar studies on various deciduous forests (Cromack and Monk 1975, Phillipson et al 1975, Rochow 1975, Tappeiner and Alm 1975, Yount 1975, Gosz et al. 1976) indicate that the turnover rate of organic matter ranges from 2 to 6 years.

When compared with the data compiled above, the turnover rate of organic matter in the Niepołomice Forest is much slower. Computed from the actual pool and the input, it is approximately 15 years. It takes three times as long for comparable decomposition as in other coniferous forests. The maximum turnover rate is considerably slower and approximates 21 years. Consequently, organic matter is accumulated at the surface of the mineral soil. Amounts of 73 t dry wt of organic matter per hectare observed in this forest exceed the levels normally recorded in coniferous forests of the temperate zone (Rodin and Bazilevich 1967).

Similar patterns are indicated by observations of the flow of most of the elements tested in this ecosystem. Exceptions are calcium and iron, as their turnover rate is more rapid at Niepołomice than in forests studied elsewhere. This is especially true for calcium, where rapid turnover may be related to the high soil acidity (Grodzińska 1980) and lower concentrations of this element in the litter. Concerning the turnover of iron, the relative rapidity is due to a very high input to the forest understory. Of all the elements considered, potassium is the only one to exhibit no deviation from the normal pattern of turnover. Its turnover rate ranks among the most rapid and approximates the values reported from other forest ecosystems.

What seems to be most dangerous in the Niepołomice ecosystem is the intense accumulation of lead and cadmium. The actual pool in the litter approximates 14 kg Pb and 0.3 kg Cd per hectare (Table 5.7), which is equivalent to a concentration of 200 and 4 ppm, respectively. Higher concentrations of these two elements have been recorded in forests, but only in close proximity to nonferrous smelters (Jackson and Watson 1977).

Cadmium not only shows a very high accumulation rate, but its concentration in herbs has exceeded that in the litter (Table 5.5). Conversely, lead and iron are only weakly incorporated by the herbs. This is especially important in the case of lead, as it is among the highly toxic elements. In fact, the concentration of lead and iron in the herbs is very low when compared to the content in organic matter in litter (Table 5.5).

It is noteworthy that sulfur, zinc, and nickel show rapid turnover rates and very low accumulation rates. Sulfur and its various compounds which contribute to acidification of ecosystems is rapidly leached downward from the litter and perhaps even carried outside the ecosystem (Swoboda and Thomas 1965, Karkanis 1976).

5.5 Decomposition Rate of Dead Organic Matter

Parallel with observations on the balance of organic matter in the litter, decomposition rates of dead organic matter were investigated by a modification of the litter-bag technique (Bocock et al. 1960). Precisely determined amounts of fresh and dried litter were exposed in litter bags scattered on the forest floor and analyzed after 1.5, 3, 6, 12, and 24 months (Fig. 5.8). Six such experiments were carried out, each one being started 2 months after the preceding one.

After a year of exposure in the litter bags, 51.1% of dry wt of the organic matter was still not decomposed. After 2 years, this proportion was reduced to 43.4% (Table 5.9). Nutrient elements behaved very differently from heavy metal pollutants in this experiment. The former were removed from decomposing litter, the latter derived from the surrounding and accumulated in the litter bags. Of the nutrient elements, sulfur was removed most slowly, after 2 years of exposure, 91.3% still remained in the litter bags. This apparently contradicts the previous observations on the rapid turnover rate and low accumulation rate of this element. Perhaps, however, this is caused by a high concentration of sulfur in precipitation in this area and by saturation of the litter with sulfur. The samples analyzed were taken from the most recently deposited organic matter. The results of this analysis thus may indicate that the upper layers of litter become rapidly saturated with sulfur, and the excess of sulfur introduced by wet and dry fall must be almost totally leached downward, into the soil. If this is so, these forest ecosystems are indeed highly contaminated with sulfur and strongly acid from its compounds (cf. Grodzińska 1980).

The most rapidly removed element was potassium. After 2 years of exposure, only 9.9% of it still remained in the litter bags. The concentration of all of the other nutrients remained constant or was even slightly increased. The heavy

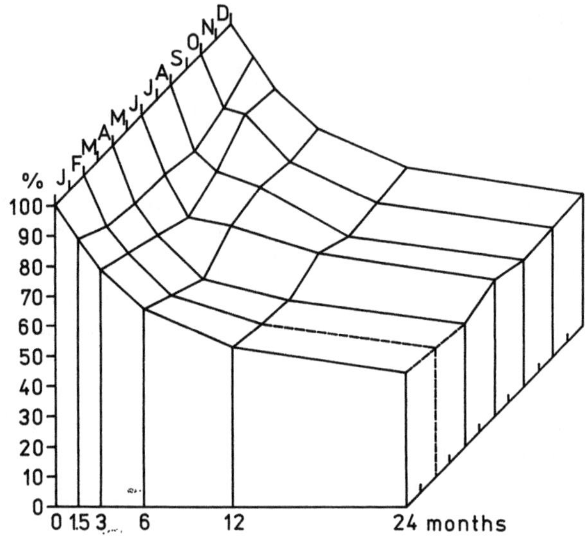

Fig. 5.8. Decomposition rate of recently deposited litter in pine forests Niepołomice, Poland; axes represent: *x* time of exposition (in months); *y* proportion of nondecomposed organic matter (in per cent): *z* timing of the beginning of experiment

Table 5.9. Decomposition of recently deposited litter in pine forests Niepołimice, Poland. Amounts of organic matter and concentrations of elements (in percent for N, P, K, Ca, S, Mg, Fe, and in ppm for Zn, Ni, Cu, Pb, Cd) after periods of exposure

Element		Exposure period in months					
		0.0	1.5	3.0	6.0	12.0	24.0
Organic matter remaining %		100.0	82.6	73.6	62.6	51.1	43.4
N (total)	%	0.898	1.054	1.091	1.257	1.308	1.401
P		0.084	0.082	0.078	0.090	0.103	0.115
K		0.329	0.170	0.141	0.136	0.092	0.075
Ca		0.335	0.315	0.310	0.312	0.296	0.248
S (total)		0.113	0.140	0.148	0.192	0.208	0.238
Mg		0.047	0.047	0.047	0.046	0.039	0.035
Fe		0.250	0.473	0.498	0.535	0.787	0.832
Zn	ppm	91.5	101.0	123.4	152.7	338.2	393.3
Ni		3.27	5.55	5.81	6.43	10.39	10.31
Cu		8.35	10.57	10.28	11.73	15.03	18.37
Pb		26.9	41.5	48.0	62.5	86.7	128.6
Cd		1.211	1.417	1.840	2.908	4.320	6.346

metals were accumulated in the litter bags, with the exception of copper, the concentration of which remained constant (Table 5.9). In accordance with previous observations, lead and cadmium became highly concentrated in the litter bags. After 2 years of exposure, their concentrations increased by a factor of 5 (from

1.21 to 6.35 ppm for cadmium, and from 26.9 to 128.6 ppm for lead). The increase for the other heavy metals was by a factor of 2–4.

The seasonal variation in the decomposition rate of dead organic matter is presented in Fig. 5.8, with the axes representing months of exposure, percent non-decomposed organic matter, and timing of initiation of the experiment. The graph shows clearly that the decomposition rate attains a maximum in summer and autumn, whereas the process is strongly inhibited in winter. This is entirely consistent with previous observations on turnover rate of organic matter in the litter. The pattern is most distinct at the onset of the decomposition process, whereas the range of seasonal variation decreases with increased time of exposure of a sample.

The experimental data permitted computations of the period of half-decay of the fresh litter, which approximates 1.4 year ($x_t/x_0 = e^{-0.50\,t}$, where t is time in years). Similar experimental data provided by Karkanis (1975) for deciduous forests of the Niepołomice Forest give the period of half-decay of 0.6 years ($x_t/x_0 = e^{-1.21\,t}$). These two figures indicate that after a year of exposure, 60% of the fresh litter still resists decomposition in coniferous forests, whereas in deciduous forests this fraction in only 30%.

Karkanis (1975) investigated also the behavior of nitrogen, phosphorus, potassium, calcium, magnesium, and copper during the decomposition of the litter in deciduous forests at Niepołomice. Comparing this study with Karkanis' data it is possible to show that after a year of decay, the pine forest litter contains much higher concentrations of elements than that of the deciduous forest. The difference varies by a factor of 1.5 for calcium to 2.6 for copper. After 2 years of decay, this difference increased on the average to a factor of 3.7. The mobility of matter in the litter is thus twice as high in pine forests here than in deciduous forests. Such relationships result from varying concentrations of nitrogen in pine and deciduous litter. Nitrogen is the basic element which governs the rate of decomposition of forest organic matter (Berg et al. 1982). Organic matter rich in nitrogen undergoes more rapid decomposition than a similar type of organic matter containing less nitrogen (Zieliński 1980). This suggests that the deciduous forests are less endangered by pollution with heavy metals, despite the accumulation rate in their soils (cf. Chap. 8).

Smith (1974, 1981) has distinguished three levels of effect of air pollution on forest ecosystems. At the lowest level, the ecosystem acts as a sink for pollutants, but functioning may not be discernibly disturbed. At the second level, some functions of the ecosystem are disturbed in a subtle manner. At the final level, air pollution causes dramatic stress to the ecosystem.

The observations presented in this chapter demonstrate that the pine forest ecosystems of the Niepołomice Forest are at the second level of pollution-induced disturbances. The turnover rate of matter is very slow. Some elements have undergone considerable accumulation, especially lead and cadmium but the behavior of sulfur may also indicate that the litter is close to saturation with this element. The decomposition process of organic matter in the litter is discernibly, although not dramatically, inhibited. Considerable inhibition of that process has been reported from various forest ecosystems with comparable concentrations of heavy metals similar to that measured here (cf. Smith 1981).

Chapter 6

The Outflow of Macro- and Microelements from the Niepołomice Forest Watersheds

Chapter Editor: M. RECZYŃSKA-DUTKA

Contributors:

T. BEDNARZ, M. BOMBÓWNA, A. LEWIŃSKI, K. PASTERNAK, M. RECZYŃSKA-DUTKA, J. SZLAGA, K. TRELA, S. WRÓBEL, and J. ZYGMUNTOWA

6.1 The Watershed and Typical Streams of the Niepołomice Forest

K. Pasternak and J. Szlaga

6.1.1 Watershed of Selected Streams

A watershed has a considerable effect on the structure of a stream because it influences the natural retention of water, the water influx to streams, variability in the water level and balance, the chemical composition of water, and consequently also the existence of aquatic communities (Pasternak 1976).

The water cycle of forest watersheds generally follows the rules valid for other types of watersheds. However, the controlling factors undergo certain modification due to local environmental conditions. Consequently, the rainfall, evaporation, and condensation, as well as the surface flow and underground run-off show distinctive characteristics in forest watersheds (Obmiński 1977). It is noteworthy that surface flow is commonly transformed into underground run-off in forests due to the high retention of litter and mosses, the slow rate of snow thaw, and the high level of soil freezing. The water retained in litter passes slowly to the subsurface and therefore water outflow from a forest watershed becames rather evenly distributed in time.

The Niepołomice Forest is located in lowlands with only slight relief; the area slopes slightly northward, and only at the southernmost edge of the forest complex does a low sandy elevation occur, this being the springs of Traczówka Creek.

Traczówka Creek, which is the main object of this study, is a tributary of the Drwinka River that crosses the Niepołomice Forest eastwards and divides in into two complexes. The southern is dominated by mixed oak-pine forests, and the central part is the watershed of Traczówka Creek. The northern complex consists mostly of deciduous forests drained by Śmigne Creek.

The substrates of the southern forest complex consist of Pleistocene sands and gravels underlain by Miocene clays. In places, the sands and gravels are intercalated with clays and silts. The sands and gravels, which represent the main water-bearing bed in this area, are quite variable in thickness. The thickness ranges from 1 to 1.5 m in the south, near the springs of Traczówka Creek, up to 12 or more m in the north. Consequently, the groundwater level is very shallow, actually less than 3 m. In places, especially in depressions, this level may reach the surface. Locally, groundwater may appear at the surface due to the overlying impermeable strata being cut through by drainage features. The groundwater run-off from the watershed is north to northeastward, conforming with the general slope of the area. The slope of the groundwater level, however, does not exceed 1% (Pietryga 1969). Groundwater is often locally stagnant, poorly oxygenated, and acidic.

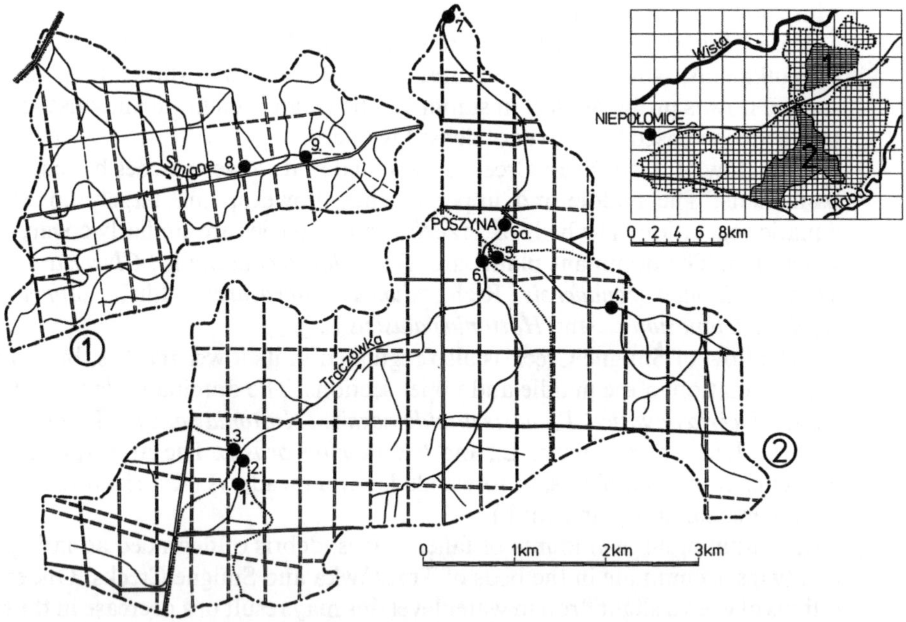

Fig. 6.1. Hydrographic network of Traczówka and Śmigne Creeks and location of stations where hydrological and hydrochemical measurements were taken

Under such hydrogeologic conditions, most of the area is covered with loamy, gleyified sandy soils of the podzol type with patches of mostly oligotrophic half bog-peat-gleyey soils. The sandy soils are generally poor in bases, trophic, and colloidal components, and are acidic, as are the half bog and peat soils. The latter soils have developed mostly in depressions and in situations where the groundwater covers loamy patches scattered in sandy soils (Chap. 1). A low concentration of loamy and colloidal components in the sandy soils results in poor capillarity, which, in turn, causes severe drying of litter and accumulation horizons, especially in periods of drought.

The substrate of the northern deciduous forest complex consists of alluvia from the Vistula River. This substrate has permitted the development of mesoeutrophic, heavy to moderate gleyified loam soils of the old gleyified alluvial type. These weakly acidic soils include sandy, silty, and clayey intercalations (Karkanis 1973). The occurrence of those impermeable beds in the soil profile hinders precipitation water from penetrating deep into the subsurface, and the groundwater from rising upwards to the surface horizons. Consequently, the water flow in Śmigne Creek may decrease considerably and even dry totally in the upper reaches when there is a lack of precipitation in summer (August) or in winter. The watershed of Śmigne Creek is 10.92 km^2 in area (Fig. 6.1).

The watershed of Traczówka Creek, which has been selected for detailed hydrologic studies, is approximately 21 km^2 in area (Fig. 6.1). This is equivalent to 25% of the total area of the southern forest complex. It receives several tributaries, and only Baczków Creek with a watershed area of 4 km^2 is of significance.

Most other tributaries are just main drainage ditches. Smaller ditches constitute a dense network covering the entire watershed. They are delimited by earthern embankments, which makes the direct flow of rain water difficult. In addition, surface flow is problematical in summer, when water retention and dust accumulation are increased by a cover of dense high herbs.

The flow of Traczówka Creek is regulated in its lower reaches by fascine embankments. The middle and upper reaches, however, are largely covered by aquatic vegetation. The beds of Traczówka tributaries are similarly covered with vegetation. The dominant plants are: *Batrachium aquatile* L., *Glyceria aquatica* Wahlb., *Elodea canadensis* Rich., *Juncus caprocarpus* Ehrb., *Juncus* sp., *Equisetum limosum* L., and *Hottonia palustris* L.

The flow of Śmigne Creek is also regulated in its lower reaches, but covered with vegetation in the middle and upper sections. The dominant plant species include *Glyceria aquatica, Phalaris arundinacea* L., *Hottonia palustris* L., *Oenanthae aquatica* Lam., *Lemna minor* L., and *Iris pseudoacorus* L. The latter species occur particularly in open places, in forest glades and meadows (Krzeczkowska-Wołoszyn, personal communication).

In autumn, large amounts of fallen leaves, debris of degraded aquatic plants and twigs accumulate in the beds of Traczówka and Śmigne Creeks. Under conditions of even a slight drop in water level this may result in a decrease in the speed of water flow and cause severe silting. These creeks may additionally freeze throughout their entire lengths, particularly Śmigne Creek during unusually cold winters. These processes affect local dams and cause considerable variations in the speed of water flow; thus classic methods of flow estimation are unreliable.

6.1.2 Analysis of Water Flow in Traczówka Creek

To the present time no creek in the Niepołomice Forest has been mentioned for continuous flow. In order to determine the water flow in periods when hydrochemical and biological observations were being taken, hydrological measurements were also made. For this, a set of control points were established along Traczówka Creek and tributaries and also along Śmigne Creek.

Measurements of water flow and level were taken in 1977–1978, and these are given in Table 6.1. The values are highly variable because the measurements were taken under variable climatic conditions to ensure compatibility with hydrochemical and biological investigations.

Except for the section at Poszyna, no correlation has been found between water level and water flow in Traczówka Creek (Fig. 6.2). One may be led to suppose that this is caused by variable vegetation cover and varying rates of organic debris accumulation in the main stream and tributaries.

Therefore, additional data on the groundwater levels in the Traczówka watershed have also been employed to estimate the pattern of water flow in the stream. These data have been collected from 1954 by the Institute of Meteorology and Water Management. Those collected in 1966–1975 have been selected for additional analysis. They indicate a mean level of groundwater of 264 cm in that period. The maximum level was achieved in spring (March to June), and the mini-

Table 6.1. Water flow ($m^3 s^{-1}$) in Traczówka Creek and tributaries, and in Śmigne Creek. h = height of level from constant; H = water level on gauge; Q = water flow; H_G = level of ground water

ta- ons	Hydrological sections		Dates of measurements						
			1977			1978			
			3.05.	27.07.	28.10.	27.04.	1.07.	19.08	23.10.
	Right source stream	h	41	29	43	50	55	37	32
		Q	0.021	0.0002	0.0006	0.031	0.020	0.034	0.022
	Left source stream	h	36	24	37	46	55	54	48
		Q	0.008	0.0	0.0	0.007	0.0035	0.012	0.009
	Traczówka stream – 10 m below the fusion of source streams	Q	0.029	0.0002	0.0006	0.040	0.024	0.050	0.031
a	Traczówka stream – gauge 15 m below the footbridge at the road to Poszyna	H	–	–	–	227	262	279	266
		Q	–	0.0003	0.019	0.103	0.060	0.159	0.065
	Baczków stream at Poszyna	h	45	27	48	50	62	72	68
		Q	0.026	0.0004	0.0007	0.029	0.008	0.070	0.025
	Traczówka stream – mouth	h	21	1	16	28	37	48	–
		Q	0.078	0.0001	0.0045	0.124	0.093	0.277	0.110
	Śmigne stream – mouth	h	42	21	34	50	60	29	35
		Q	0.078	0.0001	0.0045	0.087	0.020	0.007	0.016
	Drwinka river – 10 m below the mouth of Traczówka stream	Q	–	–	–	–	0.343	–	–
	Groundwater level at poszyna (IMuWM station)	H_G	235	292	257	227	236	262	236

mum in fall (October to December). The annual amplitude was approximately 140 cm.

The levels of groundwater in 1977–1978 are presented in Fig. 6.3. The graph is supplemented with daily precipitation measured in the same period at a nearby station. The mean groundwater level was 244 cm in 1977 and 240 cm in 1978, thus higher than observed between 1966 and 1975.

In 1977, the highest groundwater was observed late in February and at the beginning of March, and the lowest occurred in July and early August. The pattern of groundwater level and water flow was strongly influenced by a long period of more than a month with no precipitation. Because of the absence of any surface flow, the streams were fed only by groundwater in that period.

In 1978, the groundwater level pattern was less variable. There was little snow in winter and thus the spring inflow to groundwater was only slightly increased. Additionally, the summer and autumn had much rain which counteracted the usual drop in water level.

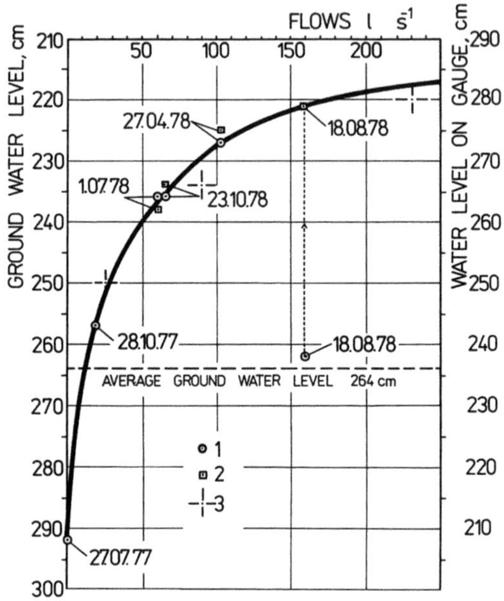

Fig. 6.2. Relation of flow to creek level and to groundwater level in the watershed, as exemplified by Traczówka Creek at Poszyna

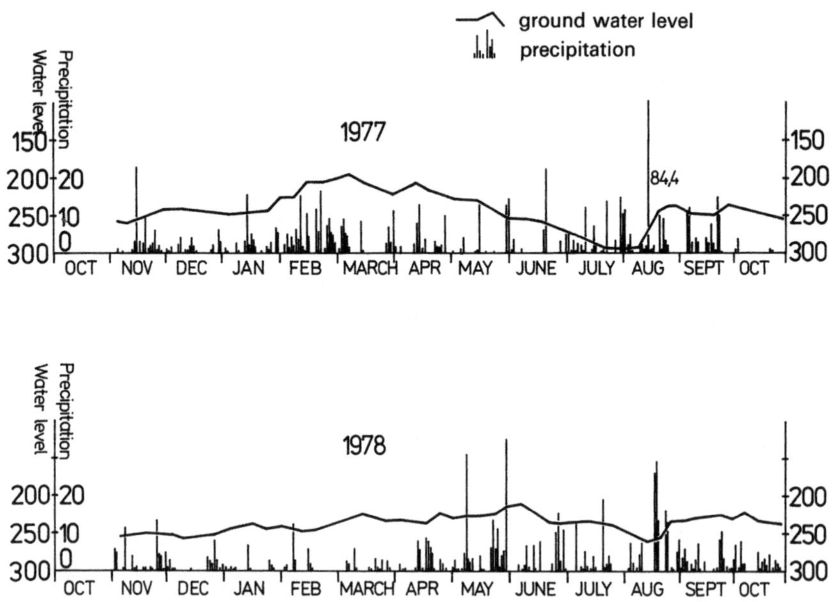

Fig. 6.3. Seasonal variation in precipitation and groundwater level in Traczówka watershed at Poszyna in 1977–1978

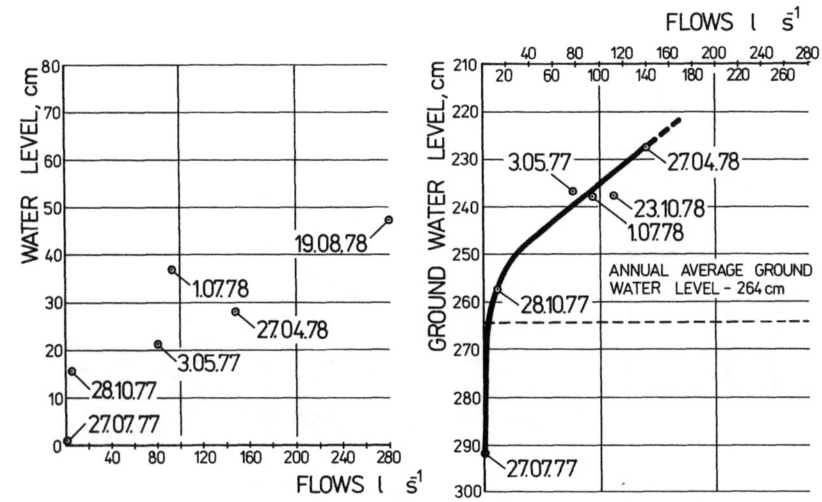

Fig. 6.4. Relationship of flow to groundwater level in the watershed and to level at the mouth of Traczówka Creek

As demonstrated by Table 6.1, flow in Traczówka Creek is highly variable from season to season. A minimum is achieved in summer-autumn, it is higher in the latter, but a maximum occurs in the spring. This variation is caused mostly by an oscillation in groundwater level (Fig. 6.3). The lower the groundwater level, the less the flow in streams. This correlation is consistent with data on flow in other lowland streams on sandy substrates. According to Dynowska (1971), spring run-off from such streams accounts for some 40% of the total annual run-off. Kiciński (1964) claims that in lowland rivers with slight slope the share of spring run-off is greater, even as much as 65%.

The results of this study suggest also that under conditions of predominant groundwater supply to creeks, and when there is considerable vegetation cover and heavy accumulation of organic debris, the most reliable estimates of flow are provided by direct measurement supplemented by observations on groundwater level. This method has been used to estimate the flow in Traczówka Creek (Fig. 6.4). It is also used, for other purposes, to predict low outflow periods from rivers (Lambor 1962).

As indicated by the graphs of groundwater level and water flow in Traczówka Creek, the flow is maintained at more or less constant rate when the groundwater level drops below its mean position. This is due primarily to a continuous supply of groundwater to Traczówka Creek. Given a higher level of groundwater and greater flow in the streams, the relationship is controlled mainly by other factors (e.g., flow conditions as controlled by bed shape and size).

Variation in flow through the bed of Traczówka Creek is shown in Fig. 6.5. At low water levels, flow is decreased in the lower reaches relative to the upper. This indicates an increased proportion of Traczówka flow in the lower reaches in the form of groundwater (through sandy-gravely beds) directly to the Drwinka River, the bed of which has recently been deepened. With higher water levels, flow in Traczówka Creek does not approximate the normal pattern.

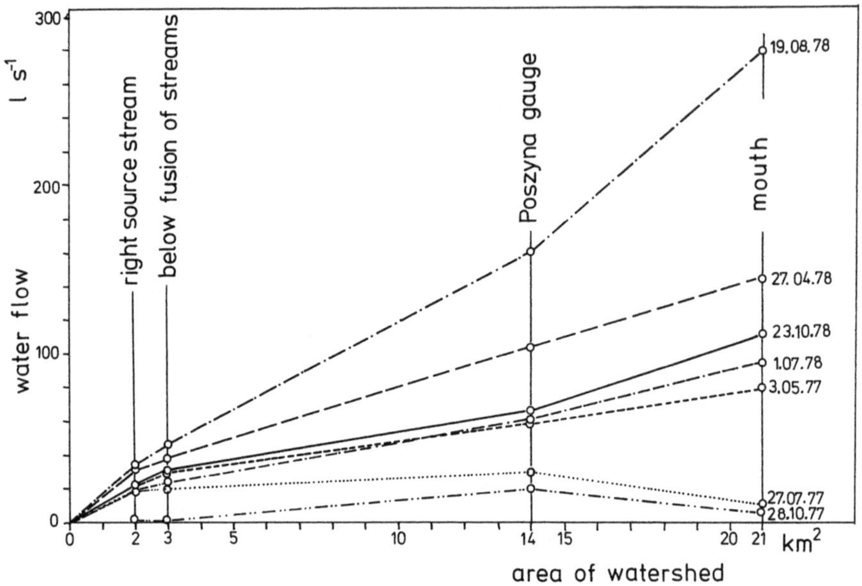

Fig. 6.5. Water flow along the course of Traczówka Creek

6.1.3 Average Annual Flow in Traczówka and Śmigne Creeks

The average annual flow in Traczówka and Śmigne Creeks could not be determined from observations on water flow because the sampling period was too short. Comparative hydrological data on the Drwinka River were unavailable, and the measurements did not cover the maximum and minimum levels to be expected in these creeks. Therefore, average annual flow has been computed using the formula given by Punzet (1959, 1975). On this basis the average annual flow toward the mouth is $0.14\ m^3\ s^{-1}$ in Traczówka and $0.072\ m^3\ s^{-1}$ in Śmigne Creek.

The results of this study indicate that variation in the hydrologic regime of creeks in the southern part of the Niepołomice Forest is controlled mainly by groundwater contribution. In the northern forest complex, however, the substrate contains impermeable silt and clay horizons which complicate the interaction of ground and surface waters and as a result, streams are fed mostly by surface flow there. Consequently, creek beds in the north may become totally dry in the upper reaches during drought.

The hydrologic regime is also strongly affected by drainage systems, commonly built by widening and deepening natural streams. These ditches contribute to a fall in the level of groundwater and facilitate run-off from the watershed. The hydrologic regime is also influenced by the road network with associated embankments and ditches, which may significantly change the directions of surface flow, as well as the size and shape of natural watersheds.

6.2 Effect of Drainage Systems on the Hydrological Regime of the Niepołomice Forest

A. Lewiński

Groundwater within the reach of tree roots is of crucial importance for forests because it supplies water, as well as easily assimilable nutrients. Therefore, forest management is largely dependent on adjustments, both vertical and lateral, of groundwater. These movements are most easily promoted by drainage systems. Unfortunately, systems developed here have not always and everywhere been well suited for their purpose.

To be useful, drainage networks must be adequate to local soil-hydrologic conditions and also be responsive to the effects of drainage devices themselves. Therefore, a knowledge of the hydrologic regime and dynamics of the groundwater in the area considered is a precondition to amelioration projects.

It has been emphasized in the preceding section (6.1) that two distinct hydrologic regimes occur in the Niepołomice Forest. A hydrologic regime of the terrestrial type (Obmiński 1977) prevails in the southern complex where groundwater results mainly from underground flow northward. In contrast, a hydrologic regime of the ombrophilous type (Benecke and Van der Ploeg 1978) is found in the northern areas where precipitation contributes the greatest share to groundwater.

Underground flow in the substrate in the south is blocked by each folding of water-bearing beds, which causes a decrease in the cross-section of flow. The free surface of groundwater undergoes tension and consequently groundwater comes very close to, or even occurs at, the surface. Under impermeable overlying strata, however, groundwater persists under continuous, although low, pressure.

The groundwater pattern is quite different in the northern forest complex. The beds bearing pressured water occur at a greater depth and are overlain by 1.5–3.0 m of heavy alluvial soil. Furthermore, the form of the bottom, as well as the limitations to drainage of the Vistula River caused by embankments built to counteract inundations, produces a surface tension of the ground water. As a result, drainage ditches that do not cut across the alluvial soil merely drain the excess of meteoric water. Additionally, those ditches which do cut across, or at least disturb, the alluvial soil, are filled with pressured water and in such instances, the effect is the reverse of that intended.

Most drainage devices in the Niepołomice Forest were constructed in the 1960's. Ditches as long as 120 km came into existence at that time, equivalent to an average density of 1 km of drainage network km^{-2}. Planning assumed that the surface groundwater is free, and expected the open ditches to drain excessive waters gravitationally. For this purpose a nonsystematic network was developed, mostly in local depressions, to drain rain water supplied to those depressions by surface flow. This network does not trap water flowing along natural slopes and hence the effects on retention of groundwater, and by implication on the natural water management of fall-retention type, are negligible (Ogijewski 1951, Janert 1961). Amelioration actually consisted in rejuvenation of old ditches and their linkage to the new network. In various ditches regulatory devices have been built.

The actual drainage network only partly meets expectations, except for localities such as the glade in the vicinity of Szarów where swamps have been transformed into meadows. The regulation of Traczówka Creek has also been partly successful.

Low efficiency of the drainage system is due mostly to the occurrence of pressured waters over extensive areas, which results in very limited reach of depression curves in the ditches despite the wide distribution of sandy soils in the southern forest complex. The slopes of ditches are very low and perpendicular to the natural slope of the area, which results in low speed flow. The latter enables phytoplanktonic organisms to flourish, which further contributes to a blockage of the flow and consequently the water in ditches becomes stagnant, with all the associated chemical and hydrobiological implications. In addition there is no gravitational run-off in periods of high water level in the recipient rivers (Vistula, Raba, Drwinka). Finally the drainage system is inadequately exploited.

It is necessary to improve the functioning of the existing drainage network by rejuvenating ditches and other devices. Moreover, systematic hydrologic and phytosociological studies are essential to recognize the affects of such devices on plant communities.

6.3 Hydrobiologic Characteristics of Selected Streams

T. Bednarz, K. Trela, and J. Zygmuntowa

Study was devoted to a bacteriological analysis of the water of Traczówka Creek and tributaries. The following bacteriological and hydrobiological characteristics were investigated; the number of all heterotrophic bacteria; the numbers of ammonifying, decomposing protein compounds, desulfurizing, iron, and denitrifying bacteria, as well as those decomposing protein compounds and producing H_2S; the trophic value of the water (by growth test on the alga *Chlorella pyrenoidosa*); phytoplankton abundance; and concentration of dissolved proteins and free amino acids.

The results indicate that the creeks are representative of two hydrobiological groups. The water is richer in nutrients in the northern and southern parts of the forest complex than in the central. In the former areas, the algae are more prolific, the phytoplankton is more abundant, the bacteria are by two orders of magnitude more numerous, and the concentration of free amino acids in water is lower. In turn, acidic waters of the central part of the forest are rich in acid humic matter and free amino acids that have not been used by consumers.

6.3.1 Bacteriological Characteristics of the Streams

The abundance of heterotrophic bacteria is relatively constant throughout. It averages 10^3 cells ml^{-1}, with a maximum (10^4 cells ml^{-1}) being achieved in the upper reaches of Traczówka and in Śmigne Creek. However, there is distinct sea-

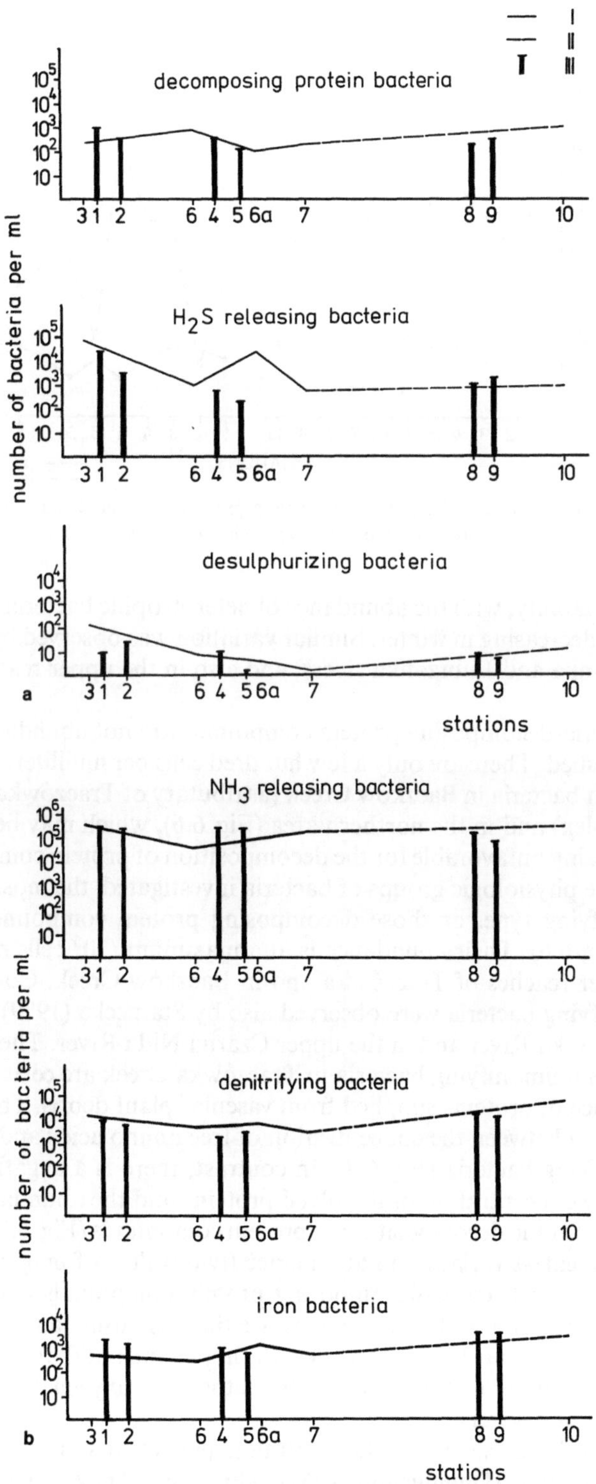

Fig. 6.6. Average abundance of physiological groups of bacteria investigated *I* Traczówka Creek; *II* Drwinka River; *III* tributaries

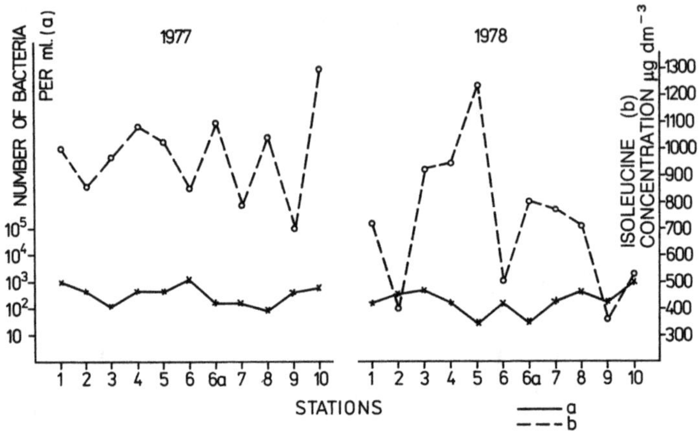

Fig. 6.7a, b. Abundance of bacteria decomposing protein compounds and the concentration of dissolved protein compounds. **a** protein bacteria; **b** protein compounds

sonal variability, with the abundance of heterotrophic bacteria increasing in summer and decreasing in winter. Similar variation was observed by Starzecka (1979) in Belnianka and Lubrzanka rivers, and also in the upper reaches of the Czarna Nida.

Bacteria-decomposing protein compounds are not abundant in the Traczówka watershed. There are only a few hundred cells per milliliter. These are the least abundant bacteria in Baczków Creek (a tributary of Traczówka in the central forest complex) and in the northern area (Fig. 6.6), which may be due to local conditions being unfavorable for the decomposition of protein compounds.

Of the physiologic groups of bacteria investigated, the most abundant are the ammonifying type, or those decomposing protein compounds and producing NH_3 (Fig. 6.6). Their abundance is at a maximum (10^5 cells ml^{-1}) in the upper and lower reaches of Traczówka and in Baczków Creek. Comparable levels of ammonifying bacteria were observed also by Starzecka (1979) in clean waters of the Belnianka River and in the upper Czarna Nida River. The flourishing populations of ammonifying bacteria in Traczówka Creek are certainly favored by an abundance of proteins supplied from vascular plant debris. There is a significant correlation between the concentration of free amino acids and the abundance of ammonifying bacteria (Fig. 6.8). In contrast, there is a negative correlation between the concentration of dissolved proteins and the abundance of bacteria responsible for the decomposition of protein compounds (Fig. 6.7).

The greatest variation in abundance (two orders of magnitude; Fig. 6.6) has been found in bacteria-decomposing protein compounds and those producing H_2S. The bacteria in Baczków Creek are the least abundant (10^2 cells ml^{-1}) and the most abundant (10^5 cells ml^{-1}) in the upper and middle sections of Traczówka Creek. The upper section of the latter is the least oxygenated, and the central is swampy flood waters.

Denitrifying bacteria reduce nitrates, particularly in presence of dissolved organic matter (Duvigneaud 1967), (Fig. 6.6). They reach maximum levels (10^4 cells ml^{-1}) in Śmigne and in the upper section of Traczówka Creek (Fig. 6.6).

Iron bacteria are not abundant and constant in frequency (10^2–10^3 cells ml^{-1}) in the waters investigated (Fig. 6.6). Their abundance is maximum in tributaries of Traczówka Creek at the margins of the forest complex, whereas Traczówka itself is poor in these bacteria. However, iron bacteria increase to a few thousand cells per milliliter in the swampy waters in the central sections, probably owing to flourishing populations of species able to exploit iron compounds of humic and fulvic acids (Kononowa 1963). Nevertheless, the generally low abundance of iron bacteria in the Traczówka watershed, and in tests conducted, indicate that humic matter is favorable for the development of heterotrophic iron bacteria. This may be so because the fixation of iron by humic compounds restricts the environment for microbiological reduction, which is of crucial importance for the mobility of Fe^{3+} ions (Janota-Bassalik and Kermen 1978).

In spite of the high concentration of sulfate compounds in these waters, desulfurizing bacteria are very rare, with concentrations of only a few tens of cells per milliliter (Fig. 6.6). This may be due to high acidity and high oxygen content of water, and to similar concentrations of humic acid matter.

6.3.2 Concentration of Dissolved Proteins and Free Amino Acids in the Streams

The concentration of dissolved organic substances with α-amino nitrogen was highly variable during the time interval of the sampling. This is indicated by a coefficient of variation of 1.02 for dissolved proteins, and 1.27 for free amino acids. In 1977, the average concentration of dissolved proteins varied from 700 to 1.289 µg dm^{-3} in Traczówka Creek and tributaries (Fig. 6.7). In 1978, the concentration of dissolved proteins decreased, probably due to the higher meteoric fall and ranged from 391 to 1,052 µg dm^{-3} (Fig. 6.7).Concerning the amounts of free amino acids, the concentration of α-amino nitrogen varied from 75 to 372 and from 10 to 557 µg dm^{-3} in 1977 and 1978, respectively (Fig. 6.8).

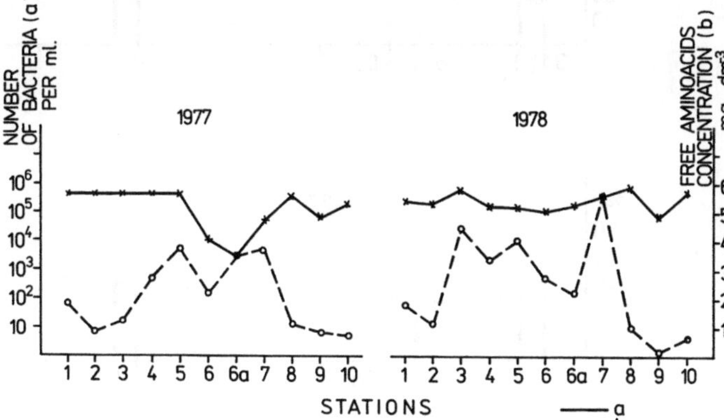

Fig. 6.8 a, b. Abundance of ammonifying bacteria and concentration of free amino acids in water. **a** ammonifying bacteria; **b** free amino acids

The greatest constancy in the concentration of dissolved proteins occurs in forest areas in the central part of the forest where the coefficient of variation ranges from 0.40 to 0.50. This relative constancy may be due to the formation of stable compounds of proteins and humic substances (Steinberg 1977). Furthermore, the low pH (4.0–5.0) may hinder the decomposition of protein compounds (Kabawa et al. 1978).

6.3.3 Variation in the Trophic Value of Water

The trophic value of water was determined by a 12-day growth test with the alga *Chlorella pyrenoidosa*. It is highest in the southernmost and northernmost parts of the forest complex, while it decreases in the central areas. There is considerable seasonal variation. The alga grows best in autumn. Any components toxic to *C. pyrenoidosa* have not been observed in autumn, winter, or in summer at high water levels. However, the rate of development of the alga decreased in spring and summer at normal and low water levels. In the central forest, the water contains relatively high concentrations of humic and iron compounds and is toxic to this alga. In 1978, a rainy year, the alga growth was more prolific. This may be indicative of an increased flow of biogens out of the watershed. Average annual rates of growth are shown in Fig. 6.9.

Recently, the trophic value of water is more and more frequently estimated using growth tests on various algae (Skulberg 1964, Bombówna et al. 1978, Claesson and Forsberg 1978). The tropic value of water, according to Liebig's rule,

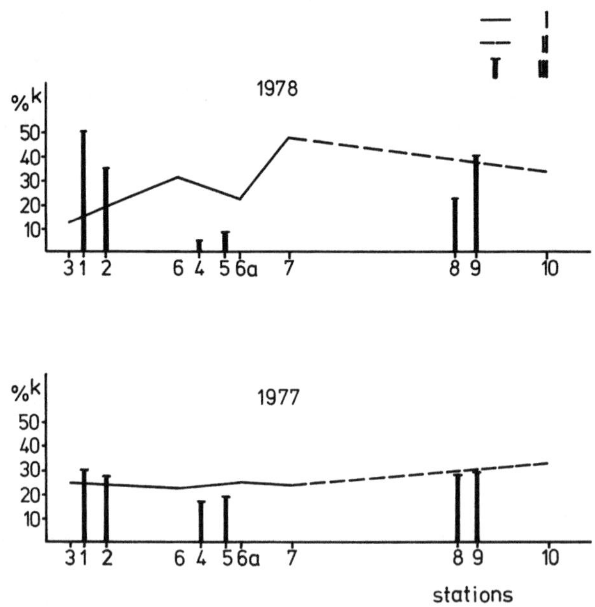

Fig. 6.9. Growth tests on the alga *Chlorella pyrenoidosa*. *I* Traczówka Creek; *II* Drwinka River; *III* tributaries

is dependent on the most limiting factor (Matusiak et al. 1976, Rhode 1976, Marvan 1979). When considered in such terms, the poor development of *C. pyrenoidosa* in water removed from the central part of the forest indicates a scarcity of nutrients.

6.3.4 Characteristics of Phytoplankton

A similar pattern has been obtained in analyses of phytoplankton in the Traczówka watershed. The abundance of phytoplanktonic organisms is much higher in the Drwinka River than in Traczówka Creek or its tributaries (Fig. 6.10). In the latter, the phytoplankton is very poor and strongly dominated by diatoms (87%). Green algae account for 7%, euglenoids for 2%, chrysophytes for 1%, and other groups for 3% (Fig. 6.10). Diatoms usually account for 60–70% of the phytoplankton in other Polish rivers (Cabejszek 1959, Siemińska 1956, Kawecka 1964, Wasylik 1971).

Of the diatoms, the following are the most abundant in the watershed: *Eunotia* (*E. arcus, E. praerupta, E. bigibba, E. exigus*), *Achnanthes, Gomphonema* (*G. angustatum, G. olivaceum*), *Fragillaria, Navicula,* (*N. cryptocephala*), and *Synedra* (*S. ulna*). The predominance of *Eunotia* may be caused by the high concentration of iron hydroxide (1.3–5.7 ppm) (Margalef 1960).

Among the green algae, *Closterium kützingii* and *C. ehrenbergi* are notable because they are usually confined to acidic oligotrophic waters (Kosinskaja 1960).

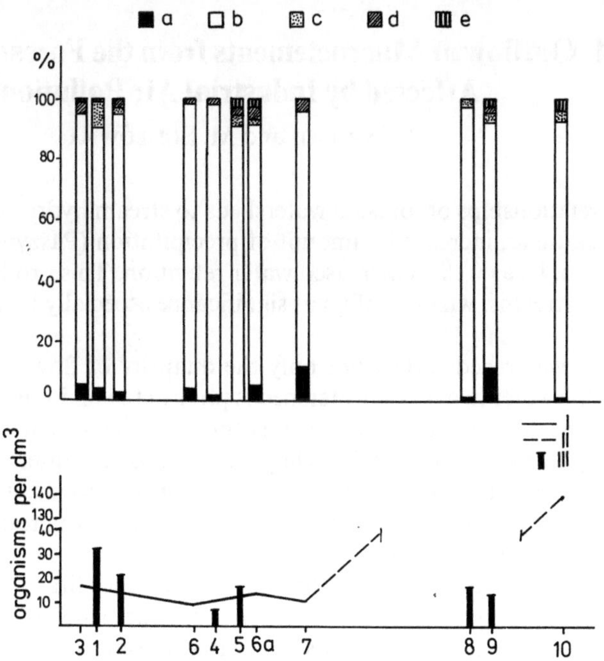

Fig. 6.10. Abundance (1 dm^3 water) and percent composition of algae *a* greens; *b* diatoms; *c* euglenoids; *d* chrysophytes; *e* others. *I* Traczówka Creek; *II* Drwinka River; *III* tributaries

The dominant euglenoides are *Trachelomonas hispida* and *T. volvocina*, both species reported from acidic waters (Huber-Pestalozzi 1955). There are also the mixotrophic species *Euglena acus* and *E. granulata*, both confined in the Niepołomice Forest to water with a decreased concentration of amino acids.

The observed seasonal variation in phytoplanktonic distribution is concerned mostly with the relative abundances of particular groups. Maximum abundance of phytoplankton occurs in autumn, and minimum in winter.

6.3.5 Hydrobiological Characteristics of the Streams

Because of the scarcity of phytoplankton, it may be supposed that most proteins dissolved in the water are derived from the decomposition of aquatic vegetation and fallen leaves. The increased concentration of organic matter of plant origin favors the development of ammonifying bacteria.

Bacteria and algae are more prolific at high water levels than at normal or low levels, owing to a higher influx of mineral and organic compounds from the substrate under the former conditions. A decrease in the concentration of dissolved protein compounds and a parallel increase in the concentration of free amino acids appears to be caused by an increase in the rate of decomposition and exploitation of proteins by aquatic organisms.

6.4 Outflow of Macroelements from the Forest Watershed Affected by Industrial Air Pollution

S. Wróbel and M. Bombówna

The relationships of forested watersheds to stream hydrology are well known. Forests cause an increase in amount of precipitation (Paszyński quoted in Bac 1968; Chap. 1), as well as increased water retention. This produces a levelling-off of flow in streams, which is of great significance especially in areas with irregular rainfall.

Forest watersheds affect not only the quantity of flow, but also quality. In general, the water in a forest watershed is poor in the major ions, such as calcium, magnesium, chlorides, and sulfates. It is poor also in the main nutrients – nitrogen and phosphorus compounds. This happens because the litter and high content of humic organic substances counteract the run-off of cations, while rapid mineralization and oxygen depletion contribute to denitrification, restricting nitrate supply to the water (Gosz 1978).

Forest soils strongly retain nutrients. This applies firstly to forest areas only slightly affected by human activities. In turn, the water flowing from forested watersheds contains high amounts of organic substances, both dissolved and aggregated. This is not only humic matter, but also peptides and amino acids produced by the decomposition and leaching of leaves.

The retentive ability of forest soils is disturbed in areas affected by industrial air pollution with high levels of sulfur oxides. These pollutants cause an increase in soil acidity and disturb the sorption complex. Consequently, the mobility of most elements and the rate of their flow through the ecosystem are increased (Norton 1976).

To study the effects of industrial air pollution on water composition and macroelement transport, two creeks in the Niepołomice Forest were studied.

6.4.1 Characteristics of the Creeks

The streams investigated were Traczówka in the southern forest complex and the other drainage ditch in the northern forest complex. Traczówka creek is the largest of ten streams or ditches in the southern forest complex (cf. Sect. 6.1.1). Drwinka River splits the Niepołomice Forest into two pedologic areas. The northern is dominated by brown soils of alluvial origin, while rot soils and peat-bog-gley soils prevail in the southern (Chap. 1). The latter soils occur in the lower reaches of Traczówka Creek, whereas podzolic soils occur in the upper. The northern pedologic area is richer in nutrients, but both are wet and only weakly affected by winds. The northern forest complex includes mostly deciduous forests, the southern is dominated by mixed oak-pine forests. Thus, the streams investigated occur under different soil and forest conditions.

6.4.2 Physicochemical Characteristics of the Water

Water temperature is lower in forest streams than in lowland rivers running through open areas. This difference is 2–5 °C between Traczówka Creek and the Drwinka River (Table 6.2).

The streams studied show high concentrations of carbon dioxide and have weakly acidic to acidic (station 5) water. Water conductance is relatively high in the forest streams. In this respect, there is variation not only between Traczówka Creek and the drainage ditch investigated and the Drwinka River, but even along Traczówka Creek itself. Water conductance is higher in the upper reaches than in the right tributary (station 5). At each particular station, however, water conductance is rather constant, as demonstrated by the low coefficient of variation (0.09–0.18).

Oxygen content in the forested streams is close to the average. The only supersaturation measurement was recorded in the Drwinka River. Low water temperature, an unfavorable concentration of hydrogen ions, and resistance of organic compounds, prevent the total loss of oxygen, even though saturation decreases to as low as 5% (stations 3 and 4) under the ice. Low bioactivity of the water is indicated also by a low biochemical oxygen demand (Table 6.2).

Dry residue and residue after ignition are also constant in the streams studied. With the constancy of conductance, this indicates considerable stability in water composition despite a variation in water flow.

Table 6.2. Physicochemical properties of water in Traczówka Creek (stations 2, 5, 6, 7), northern drainage ditch (station 9), and the Drwinka River (station 10). $\bar{x}$ = mean; V = coefficient of variation

	2		5		6		7		9		10	
	$\bar{x}$	V	$\bar{x}$	V	$\bar{x}$	V	$\bar{x}$	V	$\bar{x}$	V	$\bar{x}$	V
Temperature °C	0.5 –14.5	–	0 –15.1	–	0.1 –16.8	–	0.1 –18.1	–	0.2 –19.5	–	0.3 –20.1	–
pH value	5.8 – 6.6	–	4.2 – 5.6	–	5.2 – 6.4	–	4.4 – 6.6	–	6.0 – 6.8	–	6.5 – 7.4	–
Conductivity uS 18 °C	207.4	0.18	141.9	0.11	162.9	0.10	165.8	0.10	270.7	0.09	335.4	0.12
Carbon dioxide CO_2 l^{-1}	26.0	0.73	25.5	0.35	19.6	0.31	15.9	0.26	27.1	0.44	12.8	0.71
Oxygen dissolved mg O_2 l^{-1}	7.12	0.28	7.64	0.24	8.57	0.17	9.31	0.18	7.15	0.40	9.65	0.17
BOD_5 mg O_2 l^{-1}	2.14	0.42	2.35	0.44	2.57	0.47	2.52	0.46	2.42	0.31	3.15	0.52
Total residue mg l^{-1}	232.0	0.36	149.3	0.10	169.9	0.12	162.2	0.11	243.2	0.16	288.4	0.14
Loss on ignition mg l^{-1}	82.6	0.48	47.2	0.40	57.0	0.38	52.6	0.32	68.4	0.32	82.8	0.27
Total fixed residue mg l^{-1}	149.5	0.33	102.1	0.14	112.9	0.14	111.2	0.14	174.9	0.17	205.5	0.14

6.4.3 Ionic Composition of the Water

The ionic composition of the water (Table 6.3) deviates considerably from the average for surface waters and this is especially true where Traczówka Creek flows over poorer and more acidic soils. Surface waters usually are of the calcium bicarbonate type in the temperate zone but Traczówka water is of the calcium sulfate type (Fig. 6.11). Sulfates prevail over bicarbonates at all seven stations along the creek, and prevalence increases downstream. Under less acidic soil conditions in the north, the ionic composition of the drainage ditch studied, as well as of the Drwinka, continues to be of the calcium bicarbonate type.

It is notable that sulfates increased during high water levels in February 1977, and considerably decreased at the exceptionally low level in July 1977. The most drastic change took place in the northern ditch, where water was distinctly of the calcium sulfate type in February 1977 (Fig. 6.12).

The concentrations of calcium and magnesium are relatively low in Traczówka Creek, but much higher in the northern ditch and in the Drwinka. Variability with time is fairly low.

The concentration of potassium is highly variable with time (coefficient of variation up to 0.96), although it never falls below 1 mg l^{-1}. This variability is due to changes in the usage of potasium by aquatic plants growing abundantly in certain areas of the streams.

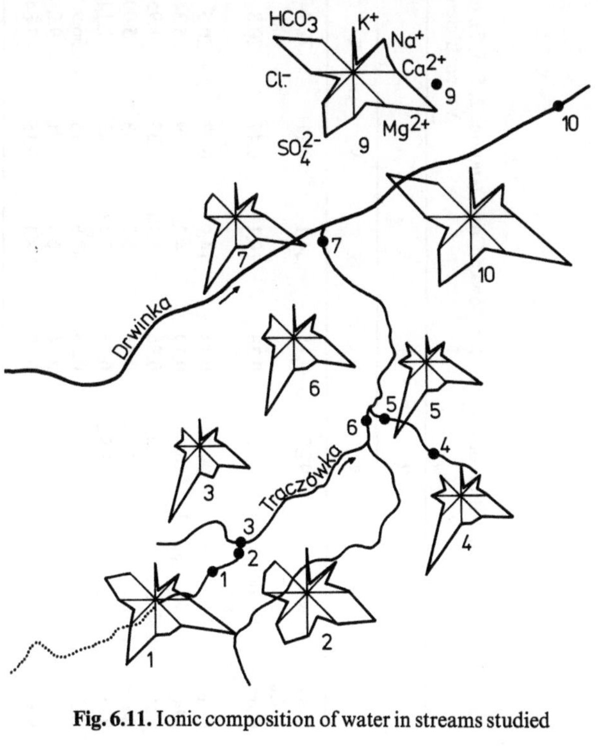

Fig. 6.11. Ionic composition of water in streams studied

Table 6.3. Ionic composition of water in Traczówka Creek (stations 2, 5, 6, 7), northern drainage ditch (station 9), and the Drwinka River (station 10). $\bar{x}$ = mean; V = coefficient of variation

	2		5		6		7		9		10	
	$\bar{x}$	V	$\bar{x}$	V	$\bar{x}$	V	$\bar{x}$	V	$\bar{x}$	V	$\bar{x}$	V
Total hardness $CaCO_3$ mg l^{-1}	97.9	0.15	54.1	0.11	72.7	0.10	75.2	0.11	121.5	0.08	158.6	0.12
Carbonate hardness $CaCO_3$ mg l^{-1}	43.0	0.38	10.1	0.47	20.8	0.10	22.7	0.38	84.6	0.31	99.2	0.24
Calcium Ca mg l^{-1}	29.2	0.18	14.9	0.17	20.5	0.15	21.4	0.14	36.3	0.12	48.8	0.14
Magnesium Mg mg l^{-1}	6.14	0.19	4.20	0.19	5.34	0.10	5.30	0.16	7.57	0.24	8.95	0.13
Potassium K mg l^{-1}	3.32	0.96	1.10	0.33	1.95	0.45	1.52	0.34	1.03	0.61	3.87	0.21
Sodium Na mg l^{-1}	7.05	0.46	5.26	0.33	5.90	0.27	5.50	0.25	17.10	0.35	17.50	0.25
Total iron Fe mg l^{-1}	1.92	0.87	2.43	0.89	2.14	0.70	1.97	0.58	3.49	1.13	1.39	0.67
Sulfate SO_4 mg l^{-1}	55.0	0.23	48.9	0.14	56.9	0.12	56.1	0.11	64.0	0.26	74.6	0.06
Chloride Cl mg l^{-1}	12.4	0.34	6.7	0.21	9.2	0.18	9.2	0.33	8.4	0.27	17.9	0.22
Silica Si mg l^{-1}	8.16	0.17	8.14	0.15	7.84	0.16	7.63	0.15	6.56	0.26	5.87	0.15

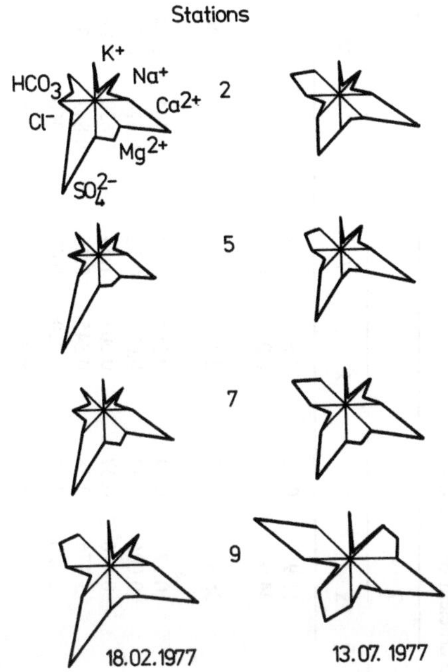

Fig. 6.12. Ionic composition of water at high (February 18, 1977) and low (July 13, 1977) water levels

The concentration of total iron (0.35 to 14.4 mg l^{-1}) is also highly variable. Concentration increases during low water levels and in periods of strong surface flow.

The concentration of dissolved silica is high, usually more than 6 mg l^{-1}.

6.4.4 Organic Matter and Nitrogen and Phosphorus Compounds in the Water

The concentration of organic matter in the streams studied is high. This is indicated by high oxidability and chemical oxygen demand (Table 6.4), because oxidability over 10 mg l^{-1} and chemical oxygen demand over 30 mg l^{-1} are only rarely found in surface waters. Higher concentrations of organic matter are usually confined to summer at low water levels as well as following heavy rains. In September 1976, the water was sampled at station 5 prior to and immediately following a heavy rain, and the chemical oxygen demand increased from 27 to 128 mg l^{-1}. This increase was caused partly by surface flow, and partly by increased water flow and speed which resulted in a resuspension and run-off of organic matter accumulated on the bottom. It is notable that water samples taken in February 1978 from beneath the ice have very low concentrations of organic matter, because the supply from deeper parts of the soil section is very low.

In an effort to characterize organic matter in the water, the percent ratio of oxidability to chemical oxygen demand has been computed. This index reflects

Table 6.4. Organic matter, nitrogen and phosphorus compounds in water of Traczówka Creek (stations 2, 5, 6, 7), the northern drainage ditch (station 9), and the Drwinka River (station 10). $\bar{x}$ = mean; V = coefficient or variation

	2		5		6		7		9		10	
	$\bar{x}$	V	$\bar{x}$	V	$\bar{x}$	V	$\bar{x}$	V	$\bar{x}$	V	$\bar{x}$	V
Oxydability O_2 mg l^{-1}	12.3	0.16	14.1	0.42	13.1	0.29	11.9	0.22	9.6	0.42	9.2	0.18
Chemical oxygen demand C.O.D. O_2 mg l^{-1}	40.2	0.26	40.7	0.40	36.0	0.25	34.0	0.21	30.9	0.40	33.1	0.15
Oxydability/C.O.D. in percent	31.4	0.24	36.5	0.30	37.9	0.29	35.8	0.21	33.1	0.41	27.6	0.26
Ammonia N–NH_4^- mgl^{-1}	0.427	0.76	0.361	0.36	0.375	0.43	0.372	0.34	0.362	0.78	0.464	0.64
Nitrite N–NO_2^- mg l^{-1}	0.006	0.67	0.007	0.43	0.005	0.40	0.006	0.33	0.009	0.56	0.014	0.36
Nitrate N–NO_3^- mg l^{-1}	0.687	0.45	0.352	0.58	0.392	0.45	0.476	0.67	0.503	1.34	0.639	0.88
Organic nitrogen N mg l^{-1}	1.328	0.26	0.935	0.15	0.899	0.27	1.176	0.28	0.845	0.26	0.842	0.35
Total nitrogen N mg l^{-1}	2.470	0.22	1.660	0.16	1.670	0.15	2.020	0.23	1.710	0.47	1.960	0.21
Phosphate P–PO_4 mg l^{-1}	0.025	0.64	0.024	0.62	0.019	0.68	0.018	0.61	0.093	0.83	0.040	0.85
Organic phosphorus P mg l^{-1}	0.079	0.25	0.087	0.37	0.071	0.28	0.075	0.32	0.129	0.68	0.087	0.54
Total phosphorus P mg l^{-1}	0.108	0.31	0.111	0.36	0.091	0.19	0.092	0.24	0.222	0.68	0.127	0.62

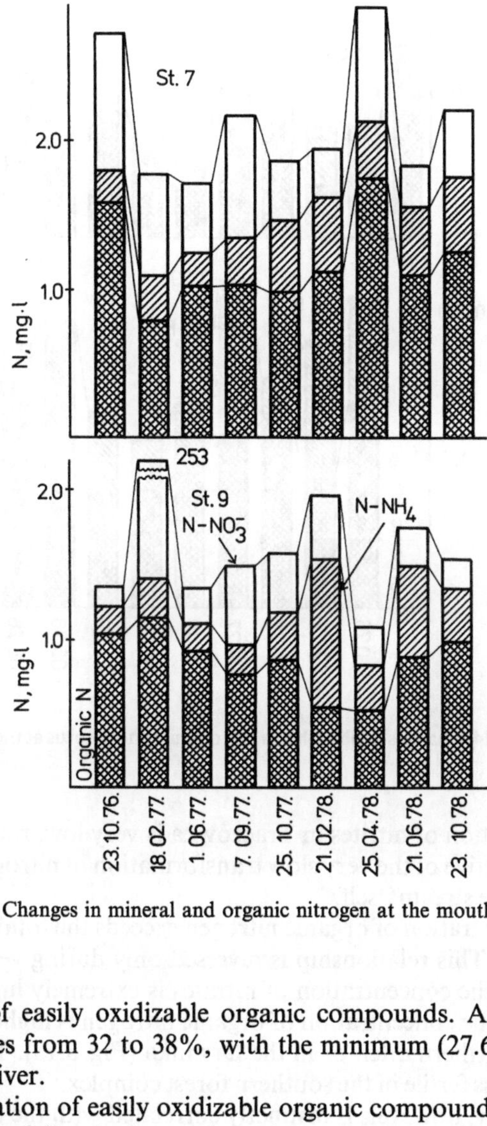

Fig. 6.13. Changes in mineral and organic nitrogen at the mouths of creeks

the proportion of easily oxidizable organic compounds. As demonstrated by Table 6.4, it varies from 32 to 38%, with the minimum (27.6%) being achieved in the Drwinka River.

The concentration of easily oxidizable organic compounds is correlated with the concentration of dissolved organic carbon, as measured by the method of Lewis and Canfield (1977). In fact, the proportion of dissolved organic carbon attains a maximum at the mouth of Traczówka Creek and a minimum in the Drwinka River.

The streams studied contain relatively large amounts of mineral nitrogen, especially in the form of ammonium. Concentration of the latter varies from 0.18 to 1.38 mg l^{-1} (Fig. 6.13), with the lowest coefficient of variation of 0.34 at the mouth of Traczówka Creek. Nitrates are always in the waters, usually at concentrations higher than ammonium. This rule is violated in samples taken from beneath the ice and at very high water levels in summer. Under such conditions, the concentration of ammonium may exceed 1.0 mg l^{-1}.

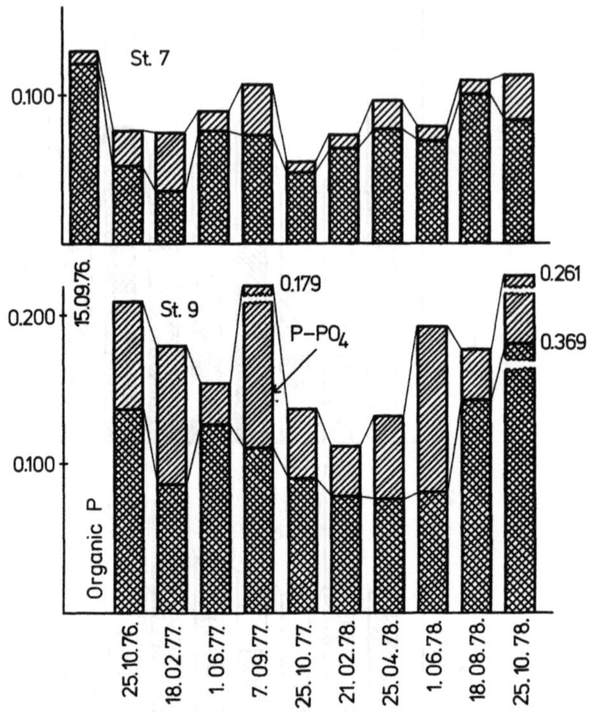

Fig. 6.14. Changes of phosphates and organic phosphorus at the mouths of creeks

Concentration of nitrites in Traczówka is very low, rarely over 0.010 mg l^{-1}, which is indicative of the very slow transformation of nitrogen compounds in soils as well as in the stream itself.

The concentration of organic nitrogen exceeds the total concentration of mineral nitrogen. This relationship is reversed only during very high water levels in spring, when the concentration of nitrates is extremely high. It is worthwhile to mention that the concentration of organic nitrogen is higher in Traczówka Creek than in the northern ditch or in the Drwinka (Fig. 6.13). This is striking because the soils are less fertile in the southern forest complex.

Soil fertility, however, is distinctly correlated with the concentration of phosphates and organic phosphorus. The concentration is much higher in the northern drainage ditch than in Traczówka Creek (Fig. 6.14). Organic phosphorus accounts for about 80% of the total phosphorus in the latter.

6.4.5 Chemical Erosion by the Streams

The nature and rates of outflow of macroelements are highly variable both in the watersheds studied and in the Traczówka Creek watershed. During relatively high water levels in April 1977, much larger amounts of calcium, magnesium, sodium, iron, sulfates, and phosphorus were lost from the deciduous than from

Table 6.5. Daily transport of macroelements (in g ha^{-1}) as measured on April 25, 1978

Components		Station	
		7	9
Calcium	Ca	143.0	359.0
Magnesium	Mg	31.0	73.0
Potassium	K	9.9	10.5
Sodium	Na	39.0	186.0
Iron	Fe	0.3	3.3
Sulfate	$S\text{-}SO_4$	110.0	265.0
Chloride	Cl	107.0	80.0
Ammonia	$N\text{-}NH_4$	2.4	2.8
Nitrate	$N\text{-}NO_3$	4.2	3.3
Total nitrogen	N	17.0	16.3
Phosphate	$P\text{-}PO_4$	0.09	1.05
Total phosphorus	P	0.58	1.80
Organic carbon	C	59.0	68.0

Table 6.6. Daily chemical erosion in Traczówka watershed (in g ha^{-1}), as measured on August 18, 1978)

Components		Station		
		2	5	7
Calcium	Ca	326.0	271.0	196.0
Magnesium	Mg	90.0	73.0	59.0
Potassium	K	51.0	24.0	19.0
Sodium	Na	62.0	82.0	66.0
Iron	Fe	68.0	36.0	38.0
Sulfate	$S\text{-}SO_4$	236.0	199.0	169.0
Chloride	Cl	120.0	109.0	82.0
Ammonia	$N\text{-}NH_4$	9.82	8.32	5.52
Nitrate	$N\text{-}NO_3$	2.04	0.56	1.42
Total nitrogen	N	30.4	22.4	19.7
Phosphate	$P\text{-}PO_4$	0.15	0.16	0.10
Total phosphorus	P	1.32	1.62	1.25
Organic carbon	C	219.0	141.0	171.0

the coniferous forests (Table 6.5). Total run-off of these macroelements was almost equal in both, although that of Traczówka is nearly 2.5 times the area of the other. However, the rate of outflow of nitrogen compounds and potassium per unit area is relatively constant between the watersheds investigated.

Traczówka loses more macroelements during high water levels in summer (Table 6.6). This pertains to sulfates, calcium, iron, phosphorus, and organic matter. In summer, nitrogen run-off from the southern forest complex is confined mainly to ammonium, which is consistent with pedologic data (Chap. 1).

Most macroelements are fairly constant in concentration. Hence, given the average water flow in Traczówka Creek, run-off of the main macroelements can be estimated.

Average annual water flow in Traczówka Creek, as computed by Lewiński (unpublished) after Jankowski's formula (Lambor 1971), is approximately 0.120 $m^3 s^{-1}$. Thus, the run-off of the main macroelements is 40.9 kg Ca ha^{-1}, 9.5 kg Mg ha^{-1}, 2.7 kg K ha^{-1}, 33.3 kg S (in SO_4) ha^{-1}, 3.64 kg N ha^{-1}, 0.165 kg P ha^{-1}, and 23 kg organic carbon ha^{-1}. If the average annual flow in Traczówka Creek is 0.140 $m^3 s^{-1}$, as estimated by Pasternak and Szlaga (Sect. 6.1.3), these values would be 17% higher. In any case, the run-off is quite low when compared to that from arable fields (Borowiec et al. 1978) or even from mountain forests (Płużański 1976)

6.4.6 Mechanisms of Macroelement Run-Off from the Traczówka Watershed

Among the most difficult problems in studies on transport of chemical compounds, both dissolved and aggregated in water, is that of the relative significance of underground and surface flow. This is true especially for watersheds with considerable relief and slope. The amount and nature of transported compounds depends upon whether the precipitation reaches the streams by infiltration through the soil, or by more or less rapid surface flow. In areas affected by industrial air pollution this is not only important for the chemical composition of water, but also for the rate of pollutant accumulation.

The Niepołomice Forest is on a flat plain. Thus the Traczówka watershed shows only slight topographic relief. Furthermore, most of the tributaries are drainage ditches with high embankments making surface flow difficult after rains. Traczówka Creek, therefore, receives water mostly from infiltration through litter and soil. This may account for the independence of constant chemical composition from highly variable water flow. A similar phenomenon was recorded by Likens et al. (1967).

Chemical composition and macroelement content of infiltrated rainfall water are largely controlled by the organic-accumulation horizon and the deeper strata of the soil. As shown by Zieliński (Chap. 5), the organic-accumulation horizon, which is rich in organic matter and nutrients (Gosz et al. 1976), delivers water that is much richer in macroelements than that of the stream. Thus, run-off of macro elements is to a large extent dependent upon mineral features of the soil. The simple relationship between the chemical composition of infiltrated and stream water is further disturbed during the vegetative season if vascular plants grow in the stream bed. The vegetation not only causes a decrease in the concentration of nutrients but also of some other macroelements (e.g., iron and humic matter). The concentration of organic and mineral compounds is affected also by sorption by bottom sediments and photooxidation (Klotz and Matson 1978).

Low saturation of the sorption complex with bases, decomposition of organic matter, and acidity of soils and waters restrain the process of denitrification. This is the reason for the prevalence of ammonium over nitrates in summer. This phenomenon is very characteristic of coniferous forests (Chap. 1), but rarely observed in surface waters. In general, the outflow of macroelements from the watersheds investigated is low. It is greater by an order of magnitude in mountain watersheds covered with coniferous forests (Płużański 1976). In the latter case, maximum daily run-off of nitrogen, phosphorus, and potassium compounds is associated with very large amounts of minerals being transported in suspension, which demonstrates that run-off is controlled chiefly by surface flow.

Run-off of the main cations (Ca, Mg, K, Na) is greater in the Niepołomice Forest than in various other watersheds (Likens et al. 1967). This run-off is controlled mainly by geological structure and composition of the source rocks, but nevertheless the increase appears to be due to the sulfur dioxides in air pollution. This hypothesis is corroborated by the outflow of sulfates being higher here than in strongly polluted Scandinavian areas (Oden 1976). Infiltration of rainfall acidified on the way through tree canopies (Hornbeck et al. 1976, Mayer and Ulrich 1976, Zieliński 1980) and along the bark (Grodzińska 1976) results in an increased rate of run-off of alkaline elements. Ecological stress (Barret et al. 1976) caused by influx of sulfur dioxides is manifested by the increased outflow of calcium, magnesium, potassium, and sodium. Because of the buffering abilities of even poor forest soils (Greszta 1975), the affects of industrial air pollution are weak at the outset. However, once the reserve of alkaline elements is used up, the soils and forest communities may undergo rapid degradation.

6.5 Outflow and Accumulation of Heavy Metals in Streams of the Niepołomice Forest

K. Pasternak and M. Reczyńska-Dutka

A considerable degree of contamination of the Niepołomice Forest by industrial air pollution is indicated by changes in the natural forest environment (Chap. 1). Measurements of the influx of pollutants indicate considerable concentrations of heavy metals (Chap. 2), and this is among the most important reasons for changes in the ecosystem.

Heavy metals are among the most dangerous pollutants. This is due to the ability to form complex organic compounds. Iron, copper, and zinc catalyze metabolism (Williams 1967). Heavy metals are active centers of enzymes and therefore their effects on living organisms may be great in spite of low concentrations (Vahrenkamp 1973). Replacement of a metal in an enzyme with another metal results in the loss of, or changes in catalytic abilities of the enzyme. Heavy metals (Cd, Zn, Pb, Hg, Cu, Ni, Co) react very easily with organic compounds. In small amounts they are indispensable for life, but in higher concentrations they restrain metabolism (Bowen 1966) or are even toxic. Various organisms respond variably to pollution with heavy metals. Copper, for instance, is highly toxic to blue-green algae and fungi. Some fungi, however, are able to survive even in a saturated solution of copper (Bowen 1966). Pollution with heavy metals generally increases with time. Hence, the deteriorating affects are of a long-term nature. Their effects may be reinforced by other factors. Co-occurrence of some metals (nickel and zinc, copper and zinc, copper and cadmium) may be much more toxic than the simple sum of effects would suggest (Khangarot et al. 1981, Wong and Beaver 1981). Heavy metals accumulate in bottom sediments but they may be released from there, and this endangers aquatic life. Heavy metals can undergo biological accumulation and concentration along the trophic chain.

Much attention has been paid recently to the pollution of surface waters with heavy metals transported by air in areas affected by industrial emissions. Generally, the main source of pollution is the rain of industrial dusts and their transport by surface flow. Winchester and Nifong (1971) demonstrated that influx from the air is the main source of Zn, Cu, and Ni in Lake Michigan. This is also the source of anthropogenic Pb, Zn, Cu, and Cd in sediments of Lake Ontario (Kemp and Thomas 1976), or Zn, Pb, and Cd in sediments of the Reservoir Kozłowa Góra (Reczyńska-Dutka 1984).

The availability of toxic compounds for biological processes is largely dependent upon physicochemical properties. In general, heavy metals are more active in solution than in mineral forms. Changes in hydrochemical properties effected by pollutants influence water-sediment interaction. Thus, the influx of sulfur dioxide contributes to changes in water acidity. Under low pH, the solubility of heavy metal compounds is increased, while their sorption by clay minerals, humic matter, and hydroxides is decreased (Förstner 1976). Release of heavy metals from the sediment is facilitated by large amounts of natural or synthetic chelating substances (Zitko and Carson 1972). Metals form soluble but less toxic complex compounds with natural organic matter in water (Sunda and Lewis 1978, Khangarot 1981). Streams flowing out of the Niepołomice Forest contain large amounts or organic matter. It is well known that organic matter (humic acids, fulvic acids) in water is able to release Pb, Zn, Ni, and Cu ions from compounds (Pauli 1975, Götz 1978, 1979).

Air pollutants transported to a watershed may accumulate there, but may also get into streams and flow to areas beyond the watershed.

6.5.1 Seasonal Variation in Water Flow in the Southern Forest Complex

The concentration of heavy metals transported by Traczówka Creek (watershed 21 km^2) has been computed from measurements taken at station 7, near the mouth (Fig. 6.1). For comparative purposes the measurements were also taken at station 5, near the mouth of Baczków Creek (watershed 4 km^2) which is the largest tributary. Flow in both streams has been estimated from data on groundwater levels and meteoric fall (Sect. 6.1). In fact, the hydrological regime of the southern forest complex is controlled mainly by groundwater.

The highest groundwater level was reached in 1977 late in February and persisted to the beginning of March due to the thawing of deep snow. The lowest level occurred in July and the beginning of August. Therefore, maximum flow occurred in February, was still considerable in April but decreased to a minimum in July. It increased again in September but decreased once more in October (Fig. 6.15).

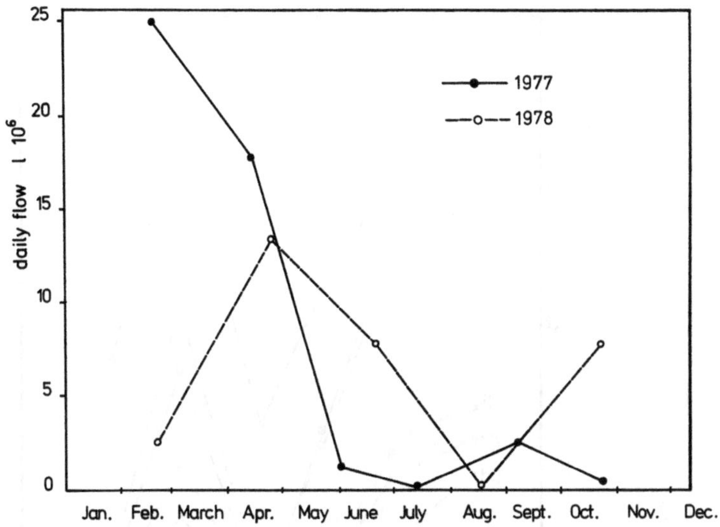

Fig. 6.15. Seasonal changes in daily water flow of Traczówka Creek

The level of groundwater was less variable in 1978. There was little snow in winter, but persistent and heavy rains fell in April, May, and in mid-August. Consequently, flow was almost the same in April, June, and October, whereas it declined to a minimum in February and August (Fig. 6.15). The seasonal variability in daily flow was very similar at stations 5 and 7 in both 1977 and 1978.

6.5.2 Concentration of Heavy Metals

In 1977 there was some variation in total concentrations of Pb, Ni, Co, Cd, Cu, and Mn in Traczówka Creek. Concentrations were higher in winter than in autumn, were very low in April and July, but later increased rapidly. This happened because the concentration of pollutants on plants and on the watershed surface was very high after a long summer without rainfall. When the heavy and acidic (Chap. 2) rainfall came, it washed the pollutants off the surfaces, and this was reflected in a rapid increase in the concentration of heavy metals.

There was some seasonal variation in the total concentrations of heavy metals also in 1978, but any consistent pattern is only with difficulty discernible. This could be due to variation in industrial pollution as well as the frequency of short but intensive rainfall.

When compared to a forest stream in the vicinity of Wierzchosławice, the metal pollutants in the streams of the Niepołomice Forest are zinc and manganese, periodically also accompanied by lead, copper, and nickel. Iron is a significant component of the industrial dusts falling on the watershed investigated. In this study, however, it is not possible to ultimately determine the share contributed naturally from the environment and that from air pollution to the high concentration of iron.

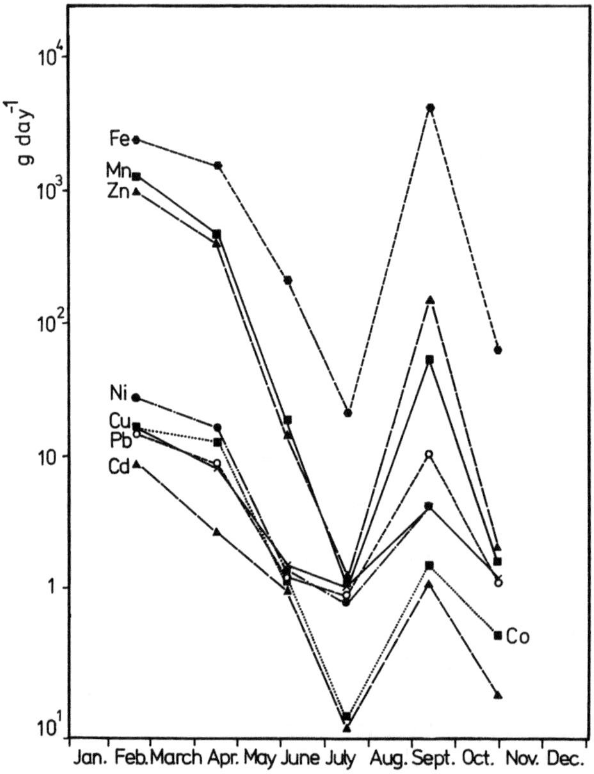

Fig. 6.16. Seasonal changes in daily outflow of heavy metals from the Traczówka Creek watershed in 1977

Air pollution contributes also to an increase in the concentration of mercury in the Niepołomice Forest. The highest concentrations of this element (0.76 μg Hg dm^{-3}) measured are eight times those recorded in forested streams in the vicinity of Wierzchosławice where industrial air pollution is negligible (Pasternak and Reczyńska-Dutka 1979).

6.5.3 Seasonal Variation in the Daily Outflow of Heavy Metals

Using the total concentrations of heavy metals and the daily water flow, the daily outflow of heavy metals from the watershed of Traczówka Creek could be computed. The seasonal variation resembles the pattern of variation in water flow. However, the maxima of the two curves are disparate.

In 1977, the daily outflow of heavy metals (Fig. 6.16) was at a maximum in February. It was very high also in September, when a maximum for iron occurred. The minimum was in July, with another in October. The range of variation was larger in 1977 than in 1978 (Fig. 6.17). In 1978, daily outflow was more constant, except in August when it fell to a minimum, coincident with a minimum for water flow.

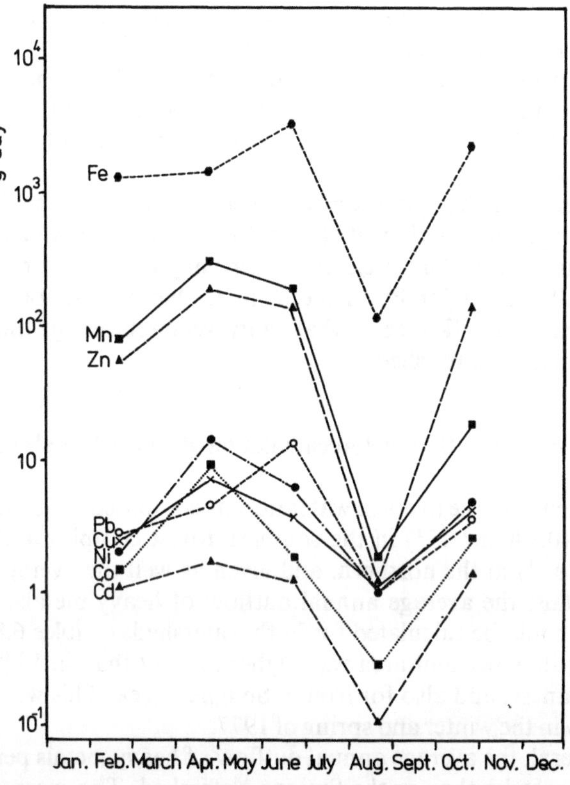

Fig. 6.17. Seasonal changes in daily outflow of heavy metals from the Traczówka Creek watershed in 1978

6.5.4 Spatial Variation in the Outflow of Heavy Metals from the Southern Forest Complex

In 1977, a year more typical of regional climatic conditions than 1978, the outflow of heavy metals was considerably greater from the central than from the southern part of the forest. Average daily outflow of copper, lead, and cobalt per unit area was 50% higher in Baczków than in Traczówka Creek (Table 6.7). The out-

Table 6.7. Average daily outflow of heavy metals g day^{-1} km^{-2} from Baczków (a) and Traczówka (b) watersheds in characteristic periods of years 1977–1978

Year		Cu	Zn	Pb	Cd	Co	Ni	Mn	Fe
1977	a	3.94	144.60	2.36	1.03	4.08	5.81	165.59	852.35
	b	2.48	126.82	2.39	1.01	2.16	4.31	125.24	849.42
	a/b	1.59	1.14	1.66	1.02	1.89	1.35	1.32	1.00
1978	a	2.69	46.01	1.01	0.38	1.86	3.03	100.32	1,034.96
	b	2.06	54.86	2.39	0.39	1.51	2.75	95.17	957.26
	a/b	1.31	0.84	0.44	0.97	1.23	1.10	1.05	1.07

flow was the same in both streams for iron and cadmium only. This difference between the two areas may be controlled perhaps by water acidity because an increase contributes to solubility and desorption of heavy metals (Förstner 1976). In fact, the highest stream acidity in the forest has been recorded at the mouth of Baczków (4.2 to 5.6), while it is lower at the mouth of Traczówka Creek (4.4 to 6.6; Sect. 6.4).

The trends of this variation remained largely the same, but less distinct, in 1978. Average daily outflow of Ni, Co, Cu, Mn, and Fe was a little higher in Baczków than in Traczówka Creek. It was virtually the same for Cd, a little lower for Zn, and 50% lower for Pb in the former. One may suppose that surface flow gained greater significance in that rainy year relative to the outflow of heavy metals from these watersheds.

6.5.5 Outflow and Accumulation of Heavy Metals in Streams

Provided with the average water flow as computed in Section 6.1 for Traczówka Creek (0.140 $m^3\ s^{-1}$) in the southern forest complex and for Śmigne Creek (0.072 $m^3\ s^{-1}$) in the northern, and given as well the average concentrations of heavy metals, the average annual outflow of heavy metals, both total and per unit area, could be calculated for both watersheds (Tables 6.8 and 6.9). Average annual outflow per unit area was higher in 1977 than in 1978, except for cobalt and manganese, and also for iron in Śmigne Creek. This was caused by the high water flow in the winter and spring of 1977.

In general, the average annual outflow of heavy metals per unit area is higher in the Traczówka than in the Śmigne watershed. The exceptions are nickel and manganese, and in 1978 also cobalt and iron. When the input of pollutants is similar in both watersheds, this may be indicative that the accumulation of heavy metals is higher in the northern forest complex than in the southern.

Table 6.8. Average annual outflow, total (a) and per km^2 (b) of heavy metals from Traczówka watershed (kg)

Year		Cu	Zn	Pb	Cd	Co	Ni	Mn	Fe
1977	a	43.83	1,116.12	47.24	6.62	18.10	49.89	867,56	25,514.96
	b	2.04	53.15	2.25	0.32	0.86	2.38	41.31	1,215.00
1978	a	39.29	712.59	39.74	4.86	18.98	36.64	1,194.27	17,120.64
	b	1.87	33.93	1.89	0.23	0.90	1.74	56.87	815.27

Table 6.9. Average annual outflow, total (a) and per km^2 (b), from Śmigne watershed (kg)

Year		Cu	Zn	Pb	Cd	Co	Ni	Mn	Fe
1977	a	17.94	242.95	19.30	2.27	7.27	28.84	886.89	5,674.20
	b	1.63	22.09	1.76	0.21	0.66	2.62	80.63	515.84
1978	a	13.62	91.05	14.76	1.82	12.03	20.21	2,448.60	11,538.00
	b	1.24	8.28	1.34	0.17	1.09	1.84	222.60	1,048.91

In 1977, the average concentration of Cu, Ni, Cd, Co, and Zn was almost the same in bottom sediments at stations 5 and 7, but the concentration of Fe and Mn was much lower at station 5. However, the concentration of all these elements in bottom sediments in Śmigne Creek (station 9) was more than twice as high as in Traczówka and Baczków Creeks. This would corroborate the fact that accumulation of heavy metals is higher in the northern forest complex. It is notable that water acidity is lower in Śmigne (6.0 to 6.8) than in Baczków (4.2 to 5.6) and Traczówka Creeks (4.4 to 6.6 Sect. 6.4), and therefore the solubility of heavy metals is decreased and their adsorption by hydroxides and other compounds is increased. On the other hand, Śmigne Creek becomes temporarily dry and this restrains sediment transport considerably.

When the average annual outflow of heavy metals per unit area is compared to average annual input from dust fall (Chap. 2 and Table 6.10), it turns out that outflow in 1977 did not exceed 7.3%, 10.7%, and 20.0% of the input of lead, copper, and cadmium, respectively. These are the elements which are the most intensively accumulated in the Niepołomice Forest. Among these elements the highest input is achieved by lead and the lowest by cadmium. The outflow is relatively higher for Mn, Fe, Zn, and Ni, but it does not exceed 57.3%. The only exception is manganese (Table 6.10), with outflow exceeding input in the northern forest complex, indicating, that it is derived from the substrate there. The second lowest input is shown by nickel, with an outflow almost the same and rather high, 40% of input, in Traczówka and Śmigne Creeks.

6.5.6 Transport and Accumulation of Heavy Metals in the Niepołomice Forest

The effects of air pollution are clearly recognizable in the concentrations of some heavy metals in water and bottom sediments of Traczówka and Śmigne Creeks. The water contains increased total concentrations of Zn, Mn, and Fe, and at high water levels (thaws, rainy periods) also of Pb, Cu, and Ni. Much greater increases in the concentrations of those elements occur in the bottom sediment (see also Jop 1981). In fact, the concentration of zinc in bottom sediments may exceed 2,000 ppm following intense surface flow. Heavy metals are transported mostly during high water levels. Air flow is twice as strong above as inside the forest (Chap. 2) and hence air pollutants accumulate more at the top of the forest. Rains and melting snow wash air pollutants out of the tree canopies and undergrowth, and the subsequent surface flow transports them during these periods to the streams. When streams are fed chiefly by groundwater, the supply of these heavy metals is much smaller because they are accumulated to a large extent in the litter and soil. This is why the outflow of heavy metals from Traczówka watershed in 1977-1978 displayed a seasonal variation consistent in pattern with the variation in water flow in the streams. In 1977, the range of this variation was greater than in 1978. This may be due to the rainfall being more continuous and evenly distributed in 1978.

The forest causes a change in wind velocity, which affects the fall of pollutants which are greater at the eastern and western margins of the forest than in the center (Chap. 2). There is no clear relationship between the concentration of heavy metals and distance from the emission source, however the rain of heavy metals, especially Cu, Cd, and Ni, is decreased in the center of the forest. This study shows that the outflow of heavy metals per unit area is higher in the central part of the southern forest complex (Baczków Creek) than in the remainder of the Niepołomice Forest (Traczówka and Śmigne Creeks). The accumulation of heavy metals is lowest in the bottom sediments of Baczków Creek. Consequently, the central part of the southern forest complex is the part least polluted by industrial emissions. This is entirely consistent with bioindication tests (Chap. 1) on various plants and animals, which clearly show that deterioration of the ecosystem is only slight in the center of the southern forest complex, while it is moderate to severe in the remainder of the forest.

The existing data (stations 5, 7, and 8) suggest that the actual level of pollution does not significantly affect the rate of outflow of heavy metals from a watershed. This rate is correlated mainly with water acidity (pH from 4.2 to 5.6 for station 5, from 4.4 to 6.6 for station 7, and from 6.0 to 6.8 for station 9; section 6.4). Generally, the more acid the water, the higher the rate of outflow of heavy metals.

On the other hand, the rate of migration and accumulation of heavy metals is controlled not only by water, but also by soil properties. The soil is not only more acidic, but also poorer in clay minerals with high sorption ability (montmorillonite) in the southern forest complex. This is consistent with the lower rate of outflow and the higher accumulation of heavy metals in the northern forest complex.

The transport of heavy metals is also controlled by the relative significance of surface flow. The supply of groundwater to Śmigne Creek is of little importance, because the stream is maintained chiefly by surface flow (Sect. 6.1). The increased accumulation of heavy metals in the bottom sediments is partly due to the fact that the stream temporarily dries up in winter and in periods without rainfall. To the contrary, in the southern forest complex, streams are fed mostly by groundwater that is merely supplemented by surface flow (Sect. 6.1).

The high rate of accumulation of heavy metals is partly due to the flat terrain, the earthen embankments along the drainage ditches, and the dense coverage with herbaceous plants, because all of these factors counteract surface flow.

Deciduous forests are generally more resistant to industrial air pollution. However, the input of heavy metals here is almost the same in the deciduous and coniferous forests (Chap. 2). Moreover, the factors discussed above restrain the outflow of heavy metals from, and consequently reinforce their accumulation in, the deciduous forests. Total annual outflow of most heavy metals, except for manganese, is less than half in the northern forest complex. In 1977, it was 28.97 kg km^{-2} in Śmigne and 61.0 kg km^{-2} in Traczówka watershed. In 1978, the figures were 13.96 and 40.56 kg km^{-2}, respectively. The lack of any weakly deteriorated areas and the exclusive occurrence of moderately to severely deteriorated areas in the northern forest complex (Chap. 1) reflects this situation.

When compared to their concentrations in dust fall, the heavy metals that are the most accumulated in the Niepołomice Forest are lead, copper, and cadmium (Table 6.10) with accumulation rates of 93.7%, 89.3%, and 80.0%, respectively in Traczówka, and 94.3%, 91.4%, and 86.7%, respectively in Śmigne watershed. It should be kept in mind that these elements are being continuously introduced into the ecosystem. One should also be reminded of the increased concentration of

Table 6.10. Average proportion of heavy metals supplied to the watersheds by air pollution removed by Traczówka (a) and Smigne (b) Creeks in 1977

Element	Average annual input of heavy metals (see Chap. 2) ($kg\ km^{-2}$)	Outflow of heavy metals from the watershed ($kg\ km^{-2}$)		Outflow of heavy metals – % of input	
		a	b	a	b
Cu	18.37	2.20	1.6	10.7	8.6
Zn	123.1	53.2	22.1	43.2	18.0
Pb	31.5	2.3	1.8	7.3	5.7
Cd	1.5	0.3	0.2	20.0	13.3
Ni	6.6	2.4	2.6	36.4	39.4
Mn	72.1	41.3	80.6	57.3	111.8
Fe	3,030.0	1,215.0	515.8	40.1	17.0

mercury in the streams of the forest. The latter is among the most dangerous and toxic elements, with great capability of being accumulated in the trophic chain. Moreover, the toxic effects of microelements are often increased by synergistic relationships and the ability to accumulate in the trophic chain. Such are the reasons for the danger to ecosystems.

The variation in the accumulation of heavy metals in the environment discussed above is due to varying behavior in the soil. Lead, copper, and cadmium accumulate mostly in the litter (Chap. 5), the most important link in the forest ecosystem. Therefore, they are easily accessible and recycled. Having infiltrated the litter, water becomes considerably enriched in nickel and zinc (Chap. 5). This demonstrates that nickel and zinc undergo accumulation in deeper strata of the soil profile. However, their outflow accounts for a much higher proportion of their input to the ecosystem than lead, copper, and cadmium (Table 6.10). An interesting exception is provided by manganese in the northern forest complex (Table 6.10). This is the only heavy metal pollutant completely removed from the ecosystem; actually more is also derived from the substrate.

Micro- and macroelements differ in their concentration, function, and behavior in natural environments. In the Niepołomice Forest biogens are generally lost by the litter during decomposition, whereas metals undergo accumulation (Chap. 5). This behavior of heavy metals is caused by a chemical affinity with organic matter and by the sorption ability of the latter. Microelements are largely accumulated here, especially in Śmigne watershed. In turn, the main cations undergo an increased outflow. Furthermore, some macroelements are removed at a considerably higher rate from the deciduous forests than from the coniferous (Sect. 6.4).

The present study clearly indicates that only a few percent of the toxic heavy metals that are supplied to the Niepołomice Forest by industrial emissions are subsequently removed in streams. The majority are accumulated in the watershed. A certain proportion is temporarily accumulated in bottom sediments in the streams and ultimately removed from the watershed.

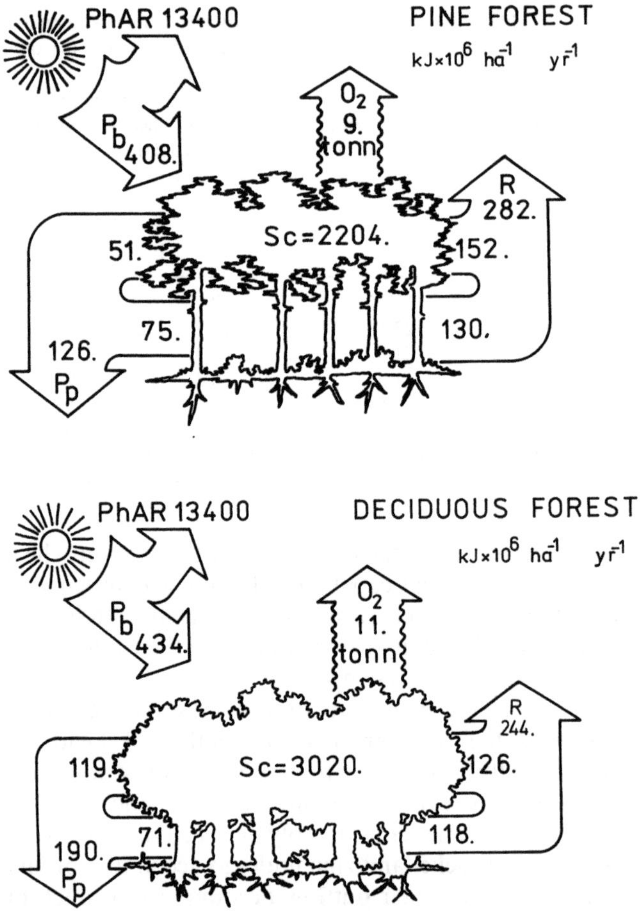

Fig. 7.1. Aboveground primary production of trees in the Niepołomice Forest. P_p net primary produc tion; *PhAR* photosynthetically active radiation; *R* respiration; P_b gross primary production (assimila tion); *Sc* aboveground biomass standing crop, in kJ 10^6 ha^{-1} yr^{-1}. O_2 production of oxyge (t ha^{-1} yr^{-1})

Bazilevich 1967, Molchanov 1971, Yurkievich and Yaroshevich 1974), taking in to consideration the age structure of stands. It was thus assumed that the bar constituted 10% of the dry mass of boles and roots and from 11% to 16% of th total tree biomass.

Caloric values of the dry mass of specific parts of plants have been taken fror various works, mostly from those concerning materials from the Niepołomic Forest (Molchanov 1971, Bobek et al. 1972, Bobek unpubl. data, Bandoła-Cioł czyk 1974, Sawicka-Kapusta et al. 1981).

Data were also available on the chemical composition of plant materials fror the Niepołomice Forest (Szulakowska 1974, Grodzińska 1980, Grodzińska et a 1983, Bobek unpubl., and others). Seed production in the deciduous section ha been taken from Bandoła-Ciołczyk (1974) and pine seed production was esti

Chapter 7

Energy, Nutrient, and Pollutant Budgets of the Forest Ecosystems

Chapter Authors: J. WEINER and W. GRODZIŃSKI

mated from values given by Molchanov (1971) supplemented by caloric values of seeds provided by Grodziński and Sawicka-Kapusta (1970).

The chemical composition of the forest herb layer and meadow vegetation was based on unpublished data of B. Bobek and on Rams (1978) after calculating, in both cases, weighted averages for more than ten dominant species. Heavy metal and sulfur content in tree leaves, small branches (twigs), bark and wood and also in forest herbaceous and meadow vegetation was taken from unpublished data of K. Grodzińska and B. Bobek, and from Grodzińska (1980) and Grodzińska et al. (1983) The chemical composition of roots has been assumed to be similar to that of wood.

7.2.2 Energy Budget of Primary Production

Figure 7.1 presents an outline of the energy budget for both parts of the forest. Gross primary production is slightly higher in the deciduous forest (434×10^6 kJ ha^{-1} yr^{-1}) than in the pine (408×10^6 kJ ha^{-1} yr^{-1}) which constitute 3.2 and 3.0% of PhAR, respectively. However, the net production of deciduous forest exceeds that of coniferous by as much as 50% (Fig. 7.1), while the efficiency of net primary production reaches 43.8 and 30.9% of gross primary production in deciduous and coniferous forests, respectively. These values constitute 1.42 and 0.94% of PhAR. In both forests about 10 t ha^{-1} of oxygen is released into the atmosphere.

7.2.3 Standing Crop of Biomass, Energy, and Chemical Elements

Tables 7.1 and 7.2 present values for the standing crop of dry mass, energy and specific nutrients in both types of forest stands. There is a noteworthy high

able 7.1. Average standing crop of dry mass, energy, nutrients, and pollutants in woody plants in pine forests, Niepołomice, Poland

em		Needles	Twigs	Wood		Bark	Total above-ground	Roots	Total
				Boughs	Boles				
ry mass g × 10^6 ha^{-1}		3.58	1.32	13.93	82.10	8.95	109.88	17.72	127.60
nergy kJ × 10^6 ha^{-1}		75.20	27.46	278.60	1,642.00	180.80	2,204.00	354.00	2,558.00
× 10^3 ha^{-1}	N	54.40	15.84	2.37	13.96	0.09	86.66	9.04	95.70
	P	4.33	1.32	4.18	24.63	–	34.46	5.32	39.78
	K	8.66	11.88	2.37	13.96	0.18	37.05	6.73	43.78
	Ca	7.84	5.15	5.29	31.20	24.52	74.00	7.09	81.09
	Mg	3.33	1.45	3.20	18.88	0.89	27.76	8.15	35.91
	S	6.34	0.80	2.93	17.24	11.81	37.12	3.72	40.84
	Fe	2.17	4.55	0.31	1.81	15.57	20.31	0.39	20.70
ha^{-1}	Zn	205.34	198.10	136.80	606.22	316.83	1,463.30	174.02	1,637.30
	Pb	82.73	70.00	36.22	213.46	489.92	892.30	46.07	938.37
	Cu	18.19	7.40	19.08	112.48	102.93	260.10	24.27	284.37
	Ni	33.47	1.90	39.84	234.81	49.49	354.50	50.68	410.18
	Cd	6.30	1.45	16.57	97.70	11.27	133.30	21.08	154.38

Table 7.2. Average standing crop of dry mass, energy, nutrients, and pollutants in woody plants in oak-hornbeam forest, Niepołomice, Poland

Item		Leaves	Twigs	Wood		Bark	Total above ground	Roots	Total
				Boughs	Boles				
Dry mass $g \times 10^6$ ha^{-1}		4.00	1.88	15.26	116.50	12.96	150.5	30.55	181.1
Energy $kJ \times 10^6$ ha^{-1}		82.40	36.38	305.20	2,330.0	265.2	3,019.0	611.0	3,630.0
$g \times 10^3$ ha^{-1}	N	81.60	19.93	15.56	118.88	87.33	323.3	93.48	416.7
	P	6.93	1.32	0.61	4.66	2.45	15.97	6.11	22.1
	K	12.88	2.93	1.53	11.65	5.03	34.02	3.06	37.1
	Ca	21.68	4.61	4.12	31.46	173.12	235.0	8.25	243.3
	Mg	4.72	2.82	2.44	18.64	4.39	33.01	9.78	42.8
	S	8.12	1.37	3.97	30.29	44.12	87.87	7.94	95.81
	Fe	13.19	0.54	0.23	1.75	24.81	40.52	0.46	40.97
g ha^{-1}	Zn	639.11	112.80	21.06	160.8	891.4	1,830.0	42.16	1,872.16
	Pb	199,61	21.81	67.45	514.9	1,057.8	1,860.0	135.03	1,995.03
	Cu	157.41	24.65	25.63	195.7	251.2	655.0	51.31	706.31
	Ni	72.02	6.77	27.93	213.2	110.0	430.0	55.91	485.91
	Cd	16.58	1.56	12.67	96.7	26.7	154.2	25.36	179.6

correlation between dry mass and energy content within the biomass categories. This is in sharp contrast to a lack of correlation between dry mass and content of chemical nutrients due to the variability in chemical composition between various parts of the plant.

In general, the standing crop of biomass maintained on a single hectare is higher in deciduous forest than in coniferous (151×10^6 and 111×10^6 g ha^{-1}, respectively). Similarly, the total content of nutrient elements (nitrogen, potassium, calcium, and magnesium) is higher in the deciduous part than in the coniferous, with the sole exception of phosphorus (Tables 7.1 and 7.2).

An uneven allocation of nutrient elements between categories of plant and animal biomass becomes more evident when presented as percentages (Table 7.3). During the vegetative season foliage retains most of the nutrients (nitrogen, phosphorus, and potassium) although leaves constitute a relatively small portion of total biomass. There is a characteristic difference in the site of calcium. In pine forest most of the calcium is stored in wood, whereas in deciduous it is in oak bark with large quantities of nitrogen (Table 7.3). The quantities of biomass and energy contained in the bodies of animals are minute, but it is worth noting the relative content of nutrients is larger by an order of magnitude than the content of energy or dry mass.

7.2.4 Nutrient Uptake by Forest Vegetation

The annual production of various categories of plant biomass (Chap. 3) combined with their chemical composition permits an estimate on the annual nutrient uptake from the soil. Data presented in Tables 7.4 and 7.5 indicate that deciduous

Table 7.3. Partitioning of dry mass, energy, nutrients, and pollutants in above-ground biomass of forests, Niepołomice, Poland. All values in per cent

Item	Coniferous					Deciduous				
	Leaves	Wood	Bark	Herbs	Consumers	Leaves	Wood	Bark	Herbs	Consumers[a]
Dry mass	3.3	88.0	8.1	0.6	0.008	2.6	88.5	8.5	0.3	0.007 (0.059)
Energy	3.4	87.9	8.2	0.5	0.004	2.7	88.2	8.8	0.3	0.009 (0.074)
N	56.1	33.2	0.1	9.5	1.11	24.2	5.9	26.0	3.6	0.56 (3.13)
P	12.1	83.9	0.0	2.6	1.25	38.2	36.3	13.5	8.8	3.25 (9.20)
K	21.9	71.5	0.5	5.9	0.15	25.7	32.1	10.0	31.9	0.19 (0.89)
Ca	10.5	55.8	32.8	0.9	0.04	9.0	16.7	71.9	2.1	0.26 (0.28)
Mg	11.7	82.6	3.1	2.6	0.0	14.2	71.8	13.2	3.9	0.0 (0.0)
Fe	9.5	11.3	68.3	1.9	9.1	30.1	5.8	56.6	0.2	7.28 (34.92)
Zn	13.5	61.8	20.8	3.7	0.1	33.3	15.4	46.5	4.5	0.09 (0.86)
Pb	9.2	35.4	54.3	1.2	0.0	10.7	32.5	56.9	0.02	0.004 (0.02)
Cu	6.9	52.5	38.9	1.8	0.1	23.9	37.3	38.1	0.7	0.003
Ni	9.1	75.6	13.5	1.7	0.08	16.6	57.2	25.4	0.74	0.004
Cd	4.7	85.9	8.4	0.9	0.1	10.7	71.5	17.2	0.51	0.11 (0.27)

[a] Values in parentheses calculated for an outbreak of *Tortrix viridana*

Table 7.4. Mean annual dry mass production, nutrient uptake and pollutant accumulation of trees in pine forests, Niepołomice, Poland

Item		Needles	Twigs	Wood		Bark	Total above-ground	Roots	Total
				Boughs	Boles				
Dry mass $g \times 10^6\ ha^{-1}\ yr^{-1}$		1.79	0.66	0.64	2.95	0.15	6.19	0.69	6.88
Energy $kJ \times 10^6\ ha^{-1}\ yr^{-1}$		37.59	13.73	12.74	59.0	3.03	126.09	13.8	139.9
$\times 10^3\ ha^{-1}\ yr^{-1}$	N	27.11	7.92	0.11	0.50	–	35.74	0.35	36.09
	P	2.16	0.66	0.19	0.86	–	3.90	0.21	4.11
	K	4.33	5.94	0.11	0.50	–	10.88	0.12	11.0
	Ca	3.92	2.57	0.24	1.12	0.41	8.26	0.28	8.54
	Mg	1.66	0.73	0.15	0.68	0.01	3.23	0.32	3.55
	S	3.17	0.40	0.13	0.62	0.20	4.52	0.14	4.60
	Fe	1.09	0.23	0.01	0.06	0.26	1.65	0.02	1.67
$ha^{-1}\ yr^{-1}$	Zn	102.67	99.1	6.26	28.97	5.31	242.31	6.78	249.1
	Pb	41.37	35.0	1.66	7.67	8.21	93.91	1.80	95.71
	Cu	9.09	3.7	0.87	4.04	1.73	19.43	0.95	20.38
	Ni	16.74	0.92	1.82	8.44	0.83	28.75	1.97	30.72
	Cd	3.15	0.73	0.76	3.51	0.19	8.34	0.82	9.16

forest takes up larger quantities of principal nutrients than does pine forest. This difference is striking in the case of nitrogen and phosphorus, although the standing crop of the latter is higher in pine than in deciduous forest.

When compared with relevant data obtained in other forest ecosystems which can be found in the literature, those for pine forests at Niepołomice fall within

Table 7.5. Mean annual dry mass production, nutrient uptake and pollutant accumulation of trees in deciduous forests, Niepołomice, Poland

Item		Leaves	Twigs	Wood: Boughs	Wood: Boles	Bark	Total above-ground	Roots	Total
Dry mass $g \times 10^6\ ha^{-1}\ yr^{-1}$		4.00	1.88	0.36	3.04	0.16	9.44	0.83	10.27
Energy $kJ \times 10^6\ ha^{-1}\ yr^{-1}$		82.40	36.38	7.20	60.80	3.29	190.07	16.60	206.70
$g \times 10^3\ ha^{-1}\ yr^{-1}$	N	81.60	19.93	0.37	3.10	1.08	106.10	2.54	108.60
	P	6.93	1.32	0.01	0.12	0.03	8.42	0.17	8.58
	K	12.88	2.93	0.04	0.30	0.06	16.21	0.08	16.30
	Ca	21.68	4.61	0.10	0.82	2.15	29.36	0.22	29.58
	Mg	4.72	2.82	0.06	0.49	0.05	8.14	0.27	8.40
	S	8.12	1.37	0.09	0.79	0.55	10.92	0.22	11.15
	Fe	13.19	0.54	0.01	0.05	0.31	14.09	0.01	14.10
$g\ ha^{-1}\ yr^{-1}$	Zn	639.11	112.80	0.50	4.19	11.06	767.70	1.19	768.90
	Pb	199.61	21.81	1.59	13.44	13.12	249.57	3.18	252.75
	Cu	157.41	24.65	0.60	5.10	3.11	190.87	1.43	192.30
	Ni	72.02	6.77	0.66	5.56	1.36	86.37	1.58	87.95
	Cd	16.58	1.56	0.30	2.52	0.33	21.29	0.71	22.00

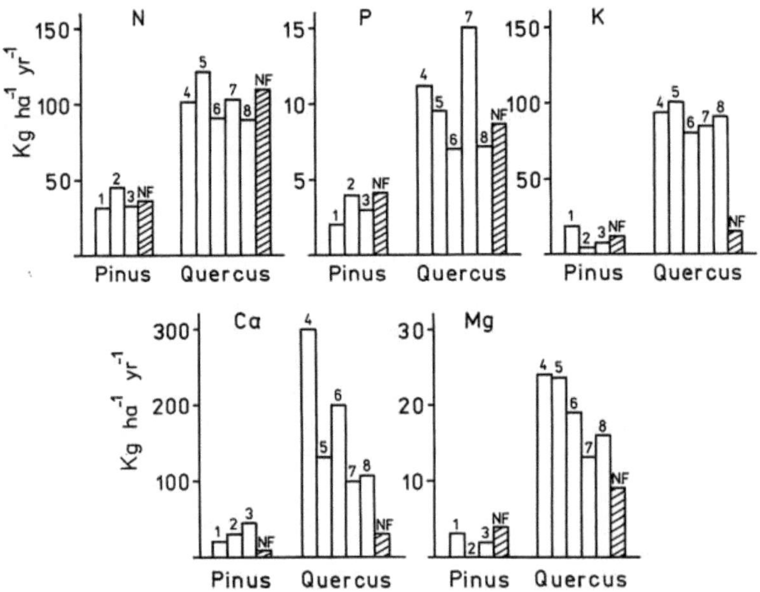

Fig. 7.2. A review of nutrient uptake estimates in various coniferous and deciduous forests of the temperate zone. *1 Pinus banksiana*, Foster and Morrison 1976; *2 Pinus silvestris*, Dengler 1930; *3 Pinus silvestris*, Ehwald 1957; *NF Pinus silvestris*, this study; *4 Quercus stellata/Q. marilandica*, Johnson and Risser 1974; *5, 6 Quercus robur/Fraxinus excelsior*, Duvigneaud and Denaeyer-de Smet 1970: *7, 8 Quercus aegopodialis*, Duvigneaud and Denaeyer-de Smet 1970; *NF Quercus robur*, this study

the range for such forests. Only calcium uptake appears to be slightly lower (Fig. 7.2; Rodin and Bazilevich 1967, Duvigneaud and Denaeyer-De Smet 1970). Oak forests at Niepołomice take up nitrogen and phosphorus in quantities similar to other known forests of the type, but significantly less calcium and potassium (Fig. 7.2).

The differences in the nutrient uptake by deciduous and coniferous trees at Niepołomice result from different levels of biomass production as well as from differences in average chemical composition of biomass produced. In deciduous forest the uptake of all nutrients, except potassium per unit of biomass produced always exceeds, sometimes even twofold, the corresponding values found in pine forest (Table 7.6).

A reverse relationship can be noted in biomass production and nutrient uptake in the herb layers of the two types of forest. Herbaceous vegetation in the

Table 7.6. Quantities of nutrients to synthetize 1 tonn of plant biomass in coniferous and deciduous forests, Niepołomice, Poland

Element	Nutrient uptake – kg t^{-1} biomass	
	Coniferous forest	Deciduous forest
N	5.15	10.5
P	0.59	0.83
K	1.57	1.58
Ca	1.21	2.90
Mg	0.51	0.82

Table 7.7. Average standing crop (=annual accumulation) of dry mass, energy, nutrients and pollutants in the herb layer of forests, Niepołomice, Poland

		Pine forests			Deciduous forests		
		Above-ground	Below-ground	Total	Above-ground	Below-ground	Total
Dry mass $g \times 10^6$ ha^{-1}		663.00	259.70	922.70	487.50	629.40	1,116.90
Energy $kJ \times 10^6$ ha^{-1}		11.90	4.66	16.56	8.75	11.30	20.10
$g \times 10^3$ ha^{-1}	N	9.22	3.32	12.54	12.09	14.35	26.44
	P	0.95	0.25	1.20	1.63	1.92	3.55
	K	2.31	0.44	2.75	16.01	9.94	25.95
	Ca	0.66	0.29	0.95	5.07	7.49	12.56
	Mg	0.73	0.14	0.87	1.27	0.82	2.09
	S	1.83	0.71	2.54	0.98	1.26	2.23
	Fe	0.43	0.17	0.59	0.08	0.10	0.18
g ha^{-1}	Zn	56.6	22.1	78.7	87.2	112.6	199.8
	Pb	10.8	4.2	15.0	0.44	0.56	1.00
	Cu	4.8	1.9	6.7	4.78	6.17	10.95
	Ni	6.3	2.5	8.8	3.22	4.15	7.37
	Cd	1.2	0.47	1.37	0.79	1.03	1.82

coniferous forest produces more biomass and retains more nutrients than the corresponding layer in deciduous forest (Table 7.7). Herbs constitute less than 1% of biomass and energy standing crop in both types of forest, but participation of this part of vegetation in the total nutrient standing crop and uptake is higher by one order of magnitude (Table 7.3). Unlike trees, the herb layer of pine forest plays a larger role in energy flow and matter cycling than does its counterpart in deciduous forest (Table 7.7).

7.2.5 Heavy Metal Accumulation

The concentrations of heavy metals and sulfur in specific parts of plants and in the bodies of animals, the standing crop of these elements and the annual biomass increment provide a basis for an estimate of the biotic pool and the annual rate of accumulation. Yet the data available do not permit distinction between that fraction of these elements which constitutes the physiological component of plant and animal bodies and that fraction which entered the organisms via mechanical and biomechanical processes caused by an enormous surplus of airborne compounds of these elements in the polluted environment. Moreover, there are no data pertaining to the normal physiological requirements of essential trace elements (micronutrients) in specific plant and animal species. However, since standing crop had been known, it was possible to follow their pathways and allocation in the forest ecosystem. Tables 7.1 and 7.2 give details of the distribution of elements in woody plants, and Table 7.7 provides corresponding information for herbaceous vegetation. Chapter 5 deals with heavy metal concentration in the litter layer of the pine forest. The partitioning of heavy metals among aboveground biomass components is given in Table 7.3.

In both sections of the Niepołomice Forest most of the heavy metal pool is concentrated in tree bark, less in leaves and wood. This indicates the mechanical deposition of elements originating from industrial pollution. The fact that a relatively large part of the pool is located in the bodies of consumers, in greater proportion than their share of total biomass, would indicate that some heavy metals are essential as micronutrients to many animals.

Estimates of the amounts of heavy metals contained in the aboveground vegetation indicate how small a fraction of these elements is actually stored there, in comparison to the pool stored in the litter (Chap. 5, Table 5.7). The litter pool of cadmium amounts to twice as much, and that of lead is as much as 35 times the corresponding values accumulated in the aboveground biomass in pine forest. Unfortunately, no relevant data from the deciduous forests are available. The amounts of nutrients in the litter there also exceed or at least equal those in the standing biomass.

7.2.6 Standing Crop, Production, and Turnover Rates of Biomass, Energy, Nutrients, and Pollutants

The ratio of annual production (annual accumulation of biomass, energy, and nutrients) to actual standing crop, i.e, turnover rate (Θ) characterizes the relative

Table 7.8. Standing crop, production and turnover rates of dry mass, energy and chemical elements in plants, Niepołomice Forest, Poland

		Coniferous			Deciduous		
		Production P	Standing crop Sc	Turnover rate Θ	Production P	Standing crop Sc	Turnover rate Θ
Dry mass $g \times 10^6\ ha^{-1}\ (yr^{-1})$		7.80	128.52	0.061	11.39	182.21	0.0625
Energy $kJ \times 10^6\ ha^{-1}\ (yr^{-1})$		142.65	2,574.56	0.055	226.75	3,650.05	0.0621
$g \times 10^3\ ha^{-1}\ yr^{-1}$ (P)	N	48.28	108.24	0.446	135.04	443.17	0.305
$g \times 10^3\ ha^{-1}$ (Sc)	P	5.31	40.78	0.130	12.13	25.63	0.473
	K	13.75	46.53	0.296	42.25	63.03	0.670
	Ca	9.49	82.04	0.116	42.14	255.81	0.165
	Mg	2.96	36.78	0.080	10.49	44.88	0.234
	S	7.21	43.38	0.166	3.38	98.04	0.136
	Fe	2.26	21.29	0.106	14.28	41.15	0.347
$g\ ha^{-1}\ yr^{-1}$ (P)	Zn	327.8	1,716.0	0.191	968.7	2,071.96	0.468
$g\ ha^{-1}$ (Sc)	Pb	110.71	953.37	0.116	253.75	1,996.03	0.127
	Cu	27.08	291.07	0.093	203.25	717.26	0.283
	Ni	39.52	418.98	0.094	95.32	493.28	0.193
	Cd	10.53	155.75	0.068	23.82	181.42	0.131

rate of processes in ecosystems. Turnover rates calculated for gross biomass and energy are slightly higher in the deciduous forest than the coniferous (0.0625 and 0.0621 as compared to 0.061 and 0.055). Similar, but more significant differences occur in the turnover rates of phosphorus, potassium, magnesium, and all heavy metals in deciduous forests, where the rates are twice as high as in the pine forests (Table 7.8).

7.3 Energy and Matter Flow Through Consumer Populations

7.3.1 Energy and Nutrient Intake by Consumers

Ecological studies on energy flow through terrestrial ecosystems thus far have led to the unequivocal conclusion that the participation of the consumer food chain is quite small, actually no more than a few percent of the total flow. The only exceptions noted have been special cases, intensively grazed grasslands, or outbreaks of some pests (Petrusewicz and Grodziński 1973, 1975).

Data on food intake in major groups of consumers obtained in both parts of the Niepołomice Forest was useful not only for the estimation of energy intake, but also for that of several principal nutrients. The total consumption removed from the producer level by all groups of consumers is provided in Table 7.9. In the course of the calculations several assumptions have been made; for example,

Table 7.9. Quantities of energy and chemical elements entering the consumers' trophic chain annually

Item		Coniferous			Deciduous		
		Annual production P	1st order consumption C	% C/p	Annual production P	1st order consumption C[a]	% C/p[a]
Energy kJ × 10^6 ha^{-1} yr^{-1}		142.65	3.16	2.2	226.75	9.7 (30.9)	4.3 (13.6)
g × 10^3 ha^{-1} yr^{-1}	N	48.28	10.18	21.1	135.04	18.6 (59.9)	13.7 (44.4)
	P	5.31	1.13	21.3	12.13	2.5 (8.9)	20.6 (73.4)
	K	13.75	3.88	28.2	42.25	2.3 (8.1)	5.4 (19.2)
	Ca	9.49	4.47	47.1	42.14	14.5 (53.6)	34.4 (127.2)
	S	7.21	0.234	3.2	13.38	0.70 (2.76)	5.2 (20.6)
	Fe	2.26	0.036	1.6	14.28	0.34 (1.33)	2.4 (9.3)
g ha^{-1} yr^{-1}	Zn	327.80	9.37	2.9	968.70	28.77 (75.36)	3.0 (7.8)
	Pb	110.71	1.12	1.0	253.75	5.46 (20.02)	2.2 (7.9)
	Cu	27.08	0.96	3.5	203.25	3.85 (15.33)	1.9 (7.54)
	Ni	39.52	1.14	2.9	95.32	1.83 (7.09)	1.9 (7.44)
	Cd	10.53	0.11	1.0	23.82	0.53 (1.74)	2.2 (7.30)

[a] Values in parentheses calculated for an outbreak of *Tortrix viridana*

plant material makes up about 25% of the diet of birds, 78% of the diet of rodents from the deciduous forest, whereas in rodents from pine forest it constitutes over 90% (Chap. 4).

In both forest types the energy intake by consumers has not exceeded 4.5% of annual production, the only exception being during oubreaks of the oak leafroller (*Tortrix viridana*) when as much as 13% of the energy of annual production may be consumed (Table 7.9).

There is quite the reverse situation when it comes to a consideration of matter flux in consumer populations. The combined effects of preferences for foods with higher nutrient content and an uneven distribution of nutrients within various parts of plants (Tables 7.1–7.3) contributes to a wide range of values of nutrient flow through the consumer food chain, the variation is from 4.5% to as much as 45% of all nutrients retained annually in plant biomass production.

In coniferous forests consumers take in more than 21% of the nitrogen and phosphorus, 28% of the the potassium and nearly half of the calcium accumulated in plant production. Though consumers have the larger share of total energy flow in the deciduous forest, they participate to a lesser degree in the nutrient flux (more than 13% of nitrogen, 21% of phosphorus, only 5% potassium, and 34% of calcium) (Table 7.9).

During leafroller outbreaks annual consumption may even exceed annual accumulation in total primary production. This paradox may be attributed to an acceleration of nutrient turnover rates due to intensive grazing and subsequent regrowth of leaves, although possible errors cannot be excluded.

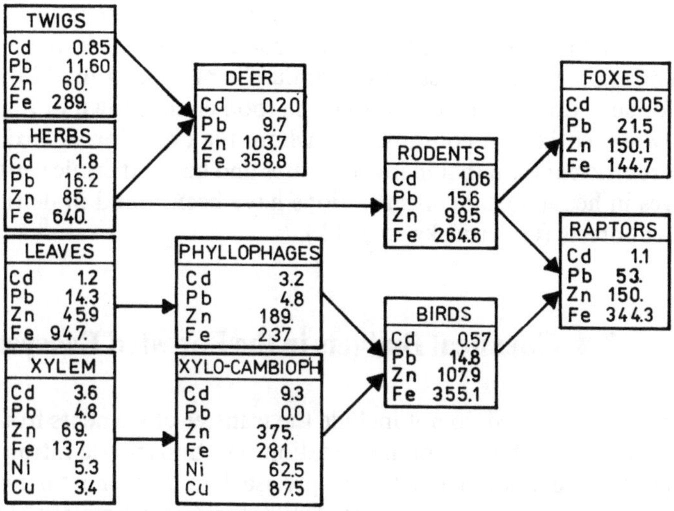

Fig. 7.3. Concentration of selected heavy metals (ppm) in various links of the trophic web of the Niepołomice Forest

The nutrients taken in by herbivores are returned almost completely to the system in excreta and to some extent as dead bodies. The amount of biomass transported out of the ecosystems studied by emigration of insects, birds, and deer, is considered negligible. Thus in a relatively short period nutrient elements are re-cycled in the form of compounds that are easily decomposable and suitable for uptake by plants (see also nutrient consumption as a percent of primary production, Table 7.9). The quantities of nutrients made available in this way correspond roughly to those applied in aerial fertilization (Grodziński and Lesiński 1978).

7.3.2 Intake and Accumulation of Pollutants by Consumers

Intake of heavy metals has been calculated from the chemical composition of biomass consumed (Chap. 4), then converted to percent of total heavy metal burden in plants consumed by animals. These percentages do not differ from the percentage of consumed total biomass and from the percentage of energy consumed (Table 7.9). This suggests that the intake of heavy metals in food is indiscriminate.

Pathways of pollutants (heavy metals, sulfur, fluorides, pesticides, etc.) in ecosystems have traditionally attracted the interest of ecologists, since these elements and compounds were shown to accumulate in trophic chains and finally reach levels dangerous to organisms. Such accumulation has been reported in respect to mercury and other heavy metals in aquatic ecosystems but has not yet been confirmed for forest ecosystems, this study included.

Figure 7.3 depicts part of the food web of forest ecosystems and the concentrations of heavy metals in specific organisms. A marked increase in heavy metal concentration can be noted in the case of herbivorous insects, particularly for zinc, nickel, and copper concentrations which may be attributed to normal physiological factors. In other groups of consumers even a decrease in heavy metal concentration has been observed (cadmium in mammals and birds). The explanation for such diferences may be that the food chain shown in Fig. 7.3 consists of mainly homeothermic consumers in whom the excretion of toxic elements may be more efficient than that in invertebrates and heterothermic vertebrates. Similar changes in heavy metal concentrations have been found in detritus food chains in these forests (Grodzińska unpubl.).

7.4 Chemical Budgets in the Forested Ecosystems

The data collected do not include the transfer of elements from parental rock via weathering and also lack information on atmospheric nitrogen fixation and nitrogen volatilization. Nevertheless precise data on elemental input in dusts and gases (Chap. 2, see also Manecki et al. 1981) and an estimation of nutrient export in watershed outflow (Chap. 6) allow determination of an approximate overall budget of nutrients and pollutants.

The data on total elemental influx (Chap. 2) do not include nutrients. The atmospheric inflow in precipitation was, however, studied by Zieliński in coniferous forests to facilitate the study of elemental budgets in forest litter (Chap. 5). Because some methods applied to pollutants have given results similar to those obtained by Manecki and others (Chap. 2), it appeared acceptable to apply the results of Zieliński on nutrients only to overall budgets.

The net difference between the amount of an element going into the system and that leaving it in watershed outflow or retention may be compared with the accumulation of the element in organic matter or the biotic pool, calculated as the sum of levels of this element in the annual biomass increment of plant and animals. Obviously, no quantitative conclusions can be drawn from such a comparison, due to differences in calculation procedures and errors involved. It may, however, suffice for the formulation of quantitative hypotheses on pathways and accumulation sites of chemical elements.

The outflow was measured from two woodland watersheds considered the most representative for the two respective forest types. They were used for an estimate of total outflow from the two forest complexes (Chap. 6).

Overall budgets, retention, and accumulation of nutrients as well as pollutants in both sections of the Niepołomice Forest are in Table 7.10. Two nutrients, calcium and magnesium, demonstrate marked negative net differences in both deciduous and coniferous forests. This is undoubtedly associated with the levels of these elements in the soil where they are leached by water. The same holds for manganese, which is present in high levels in the soils of the northern section of the forest (Chap. 1) so that more is exported from the ecosystem than has entered by industrial dusts (Table 7.10). All other elements studied have positive net differences. For example there is approximately 20 kg of nitrogen retained annually, 16–30 kg of sulfur and 23-26 kg of iron ha^{-1} in both forest types (Table 7.10). Amounts of most nutrients, as well as sulfur and iron retained do not show any

Table 7.10. Budgets of chemical elements in forest ecosystems, Niepołomice, Poland

Element		Input[a]	Coniferous			Deciduous		
			Output[b]	Retention	Accumulation[d]	Output[a]	Retention[c]	Accumulation[d]
$g \times 10^3$ $ha^{-1} yr^{-1}$	N	20.08	3.64	17.16	48.28	1.76	19.04	135.04
	P	1.83	0.165	1.665	5.31	0.46	1.37	12.13
	K	8.59	2.70	5.89	13.75	2.14	6.45	42.25
	Ca	26.27	40.9	−14.63	9.49	75.48	−49.21	42.14
	Mg	2.54	9.5	− 6.96	4.42	15.74	−13.20	10.49
	S	60.3	33.3	27.0	7.21	44.36	15.94	13.38
	Fe	30.3	12.15	18.15	2.26	5.16	25.14	1.59
	Zn	1.231	0.532	0.699	0.328	0.221	1.010	0.969
	Pb	0.315	0.023	0.292	0.111	0.018	0.297	0.254
	Cu	0.187	0.020	0.167	0.027	0.016	0.171	0.203
	Ni	0.066	0.024	0.042	0.040	0.026	0.040	0.095
	Cd	0.015	0.003	0.012	0.011	0.002	0.013	0.024
	Mn	0.721	0.413	0.308		0.806	− 0.085	

[a] N, P, K, Ca, Mg after Zieliński (Chap. 5); others after Manecki et al. (Chap. 2)
[b] After Reczyńska-Dutka et al. (Chap. 6)
[c] Retention = input − output
[d] Accumulation in biotic pool (after Chap. 3 and 4; cf. Table 7.4, 7.5, 7.7, and 7.9)

significant differences in the two ecosystems. Of the heavy metals deposited in dust only zinc displays a tendency for higher retention in deciduous areas, while other metals are retained in almost equal quantities in deciduous and coniferous sections.

Accumulation in the biotic pool is somewhat higher in deciduous forests. This holds true for all metals, particularly copper, but with the exception of iron. All nutrients are taken in by producers in considerably larger quantities than actually retained. This implies that the intrasystem cycling is much more intensive than transfers in and out.

In coniferous forest cadmium and nickel accumulation actually equal retention. This indicates that these two metals are concentrated in living organisms and are later transferred to litter (Chap. 5). In the deciduous forests cadmium and nickel as well as copper and iron accumulate in the biotic pool at even higher rates than they enter the system (Table 7.10). This may indicate that they cycle within the system much more intensively than in the coniferous forests.

Lead, cadmium, and copper retention are equal and almost complete in both types of forest. Nickel is similarly accumulated in both forests but only to 60% (Fig. 7.4). Iron and zinc are retained in larger quantities in deciduous forests, while sulfur is retained in greater amounts in coniferous forest.

Relative accumulation in the biotic pool is always higher in deciduous forest, except for iron (Fig. 7.4). Accumulation of copper, nickel, and particularly cadmium exceeds the current input, which indicates that remarkable quantities of these metals are taken up from deposits previously stored in litter and soil.

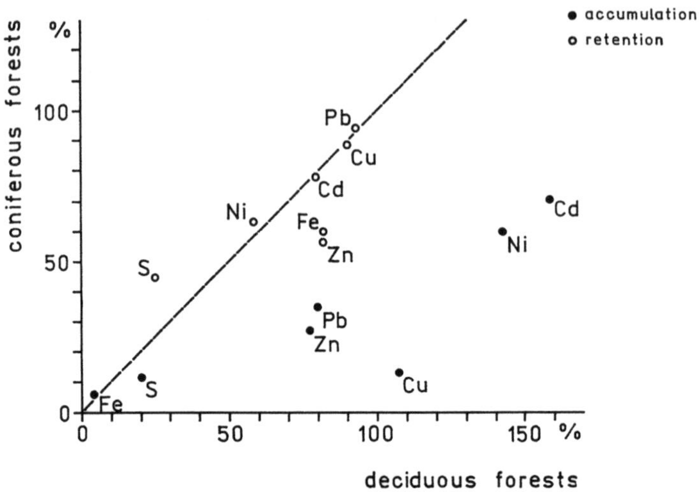

Fig. 7.4. Corresponding values of retention in the whole ecosystem (*open circles*) and accumulation in the biotic pool (*solid circles*) of heavy metals in coniferous and deciduous forests at Niepołomice, Poland

7.5 Discussion

7.5.1 Primary Production

Net primary production in the oak-hornbeam forest at Niepołomice has been studied by Medwecka-Kornaś et al. (1974), Bandoła-Ciołczyk (1974) and also by Banasik (1978a). All these studies were undertaken on the same 1-ha plot of 70–100-year-old Tilio-Carpinetum forest. The total aboveground net primary production recalculated from data given in these papers reaches 197×10^6 kJ ha^{-1}, yr^{-1}, which is equivalent to 9.77 metric tons dry weight ha^{-1} yr^{-1}. In this chapter net primary production of deciduous forest has been estimated to be 190×10^6 kJ yr^{-1}, i.e. 9.44 t ha^{-1} yr^{-1} (Fig. 7.1, 7.9), which agrees well (3.5% difference) with the previous estimates (Medwecka-Kornaś et al. 1974). In addition, the estimates presented here involved more precise determinations of the biomass standing crop of trees and growth rates for various age classes and habitats throughout the whole deciduous complex. The stand chosen by Medwecka-Kornaś and others was apparently quite representative of the entire deciduous section.

The primary production for pine forest given here (see also Chap. 3) constitutes the first estimate for this forest. Average primary production for the entire pine forest is 126×10^6 kJ ha^{-1} yr^{-1}, which is equivalent to 6.19 metric tons of dry weight of annual increment (Fig. 7.1). Estimates of production and standing crop for both types of forest fall within the range of values determined for European temperate lowland forests (for comparison see Rodin and Bazilevich 1967, Cannell 1982, Reichle 1981). Actually the Niepołomice Forest has lower than average net production of deciduous stands and definitely a lower value than average production of pine forests. Long-term affects of air-borne pollution and changes in water regime (lowering of the water table) may explain this lower productivity.

7.5.2 Nutrient Budgets

A better understanding of biogeochemical cycling in forest ecosystems has been achieved by determining input–output balances in woodland watersheds (e.g., Bormann and Likens 1967, Likens et al. 1977, Swank and Waide 1980). Such an experimental approach to the whole ecosystem enabled ecologists to discover many of the functions of forest ecosystems including elemental budgets. Nutrient and water budget estimates can be determined more precisely in mountainous watersheds (Swank 1981) than in those of the lowlands. This technique has been employed in both sections of the Niepołomice Forest despite many inaccuracies of measurement in water budget of this lowland watershed (Chap. 6). Thirteen nutrients and pollutants whose pathways have been studied quantitatively (Fig. 7.10–7.20 in Appendix) may be compared with those in four long-studied forest ecosystems in the USA. These represent various geographical and climatological conditions but have been studied using methods similar enough to enable comparisons (Swank and Waide 1980).

The four watershed ecosystems include oak-hickory forest in the Coweeta basin, Southern Appalachian Mts, NC (Johnson and Swank 1973, Swank and Douglass 1977), oak-hickory forest in Walker Branch, Eastern Tennessee, TN (Van Hook et al. 1977, Henderson et al. 1978), northern hardwoods in the Hubbard Brook Experimental Forest, White Mts, NH (Likens et al. 1977, Bormann and Likens 1979), and Douglas-fir forest in the Western Cascade Mts, OR (Grier et al. 1974, Sollins et al. 1980). In European forests early elemental budgets were determined by Duvigneaud and Denaeyer-De Smet (1970) in mixed oak forest in Belgium. This study, however, did not follow the input–output design.

Each watershed study is unique, representing specific features of vegetation with a local history, hence no comparisons of elemental pools in biomass of various ecosystem components can easily be made. It is possible to attempt to compare input–output budgets for several nutrients (Ca, K, and N) which are common to those studies.

Precipitation is considered the way in, while streamflow from the watershed is the way out of the system. In the North-American watersheds the calcium budgets (= retentions) are always negative. Output is higher than input by 3–136 kg ha^{-1} yr^{-1}, whereas in the Niepołomice watersheds in the pine and deciduous sections of the forest this value is correspondingly -15 and -49 kg ha^{-1} yr^{-1}. Such an agreement of results (Table 7.11) indicates a universal phenomenon in that calcium is almost always surplus in the soil from which large quantities are leached (Likens et al. 1967, 1977).

It is more difficult to compare nitrogen budgets because they involve both organic and inorganic nitrogen. Inorganic nitrogen budgets were always positive in the North-American watersheds (0.5–7.8 kg of N ha^{-1} yr^{-1}), while at Niepołomice nitrogen retention is even higher, reaching 17–19 kg of N ha^{-1} yr^{-1}. This seems to be a reliable result because of known high input of atmospherically derived nitrogen compounds (Table 7.11).

Table 7.11. Nutrient budgets kg ha^{-1} yr^{-1} – calcium, potassium and nitrogen in the Niepołomice Forest and a comparison with selected North American forest ecosystems. Budgets computed for various forest watersheds. Inputs in precipitation, outputs in streamflow, Niepołomice Forest, Southern Poland from Table 7.10 and Figs. 7.10, 7.12, 7.13 (see Appendix). Coweeta, NC, and Walker Branch, TN, from an outline in Swank and Waide (1980), Hubbard Brook, NH, from Bormann and Likens (1979), whereas for H. J. Andrews, OR from Sollins et al. (1980)

Nutrient Site and forest type	N (inorganic)			K			Ca		
	Input	Output	Retention	Input	Output	Retention	Input	Output	Retention
Niepołomice S. Poland oak-hornbeam	20.8	1.8	+19.0	8.6	2.1	+6.5	26.3	75.5	−49.2
Niepołomice S. Poland pine	20.8	3.6	+17.2	8.6	2.7	+5.9	26.3	40.9	−14.6
Coweeta, NC oak-hickory	4.5	0.1	+ 4.4	2.1	5.6	−3.5	4.8	7.7	− 2.9
Walker Branch, TN, oak-hickory	9.3	1.5	+ 7.8	3.0	7.0	−4.0	12.0	148.0	−136.0
Hubbard Brook, NH, Northern hardwoods	6.5	4.0	+ 2.5	0.9	1.9	−1.0	2.2	13.7	− 11.5
H.J. Andrews, OR, Douglas-fir	2.0	1.5	+ 0.5	0.9	9.5	−8.6	3.6	123.1	−119.5

Budgets for potassium are usually negative (Table 22 in Likens et al. 1977). In the four North-American watersheds discussed here the loss of potassium ranged from 1 to 9 kg ha^{-1} yr^{-1} (Table 7.11). In both deciduous and coniferous sections of the Niepołomice Forest the net difference in potassium budgets are conclusively positive (5.9 and 6.5 kg). This may be due to unusually heavy deposition in precipitation (8.6 kg ha^{-1} yr^{-1}) and is in sharp contrast to values summarized by Likens et al. (1977) from as low as 0.8 to 3.2 kg of K ha^{-1} yr^{-1}.

Only three nutrient elements (N, K, Ca) have been discussed here. Data for three others (P, Mg, and S) are presented in Table 7.10 and Figs 7.10, 7.12. It is worth noting that net budgets for all nutrients except potassium conform well to the common pattern in various forest ecosystems summarized by Likens and others (1977, their Table 22).

These comparisons could have been extended to intrasystem cycles of these nutrients including standing crop and transfers but, for the sake of brevity, the reader should refer to comprehensive works by Likens et al. (1967), Sollins et al. (1980), and Swank and Waide (1980).

7.5.3 Pollutant Budgets

Of the North-American watersheds discussed, the one most similar to the Niepołomice Forest is undoubtedly the Walker Branch forest. Located near coal-

Table 7.12. Pollutant budgets of zinc, lead and cadmium in forest ecosysystems, Niepołomice, Southern Poland, and in Walker Branch, TE. Atmospheric inputs and stream outputs from those watersheds in g ha^{-1} yr^{-1}

Pollutant Site and forest type	Zn			Pb			Cd		
	Input	Output	Retention	Input	Output	Retention	Input	Output	Retention
Niepołomice[a] S. Poland oak-hornbeam	1,231	221	+1,010	315	18	+297	15	3	+13
Niepołomice[a] S. Poland pine	1,231	532	+ 699	315	23	+292	15	3	+12
Walker Branch, TN[b] oak-hickory	538	140	+ 398	286	6	+280	21	7	+14

[a] This Chapter, cf. Table 7.10 and Figs. 7.18–7.20 in Appendix
[b] Van Hook et al. (1977), for similar atmospheric deposition see Lindberg and Harriss (1981)

fuelled power plants, this site is characterized by heavy atmospheric deposition including 286 g of Pb, 21 g of Cd, and 538 g of Zn (Van Hook et al. 1977) and 6.3 kg of S ha^{-1} yr^{-1} (Lindberg and Harris 1981). Lead and calcium depositions are quite similar to those at Niepołomice but there is only half the zinc and one tenth the sulfur.

Budgets summarized by Van Hook and others (1977) closely resemble those obtained for the same pollutants at Niepołomice (Figs. 7.18–7.20, Table 7.10). This relates not only to accumulations in the litter layer but also to the retention of large amounts of toxic heavy metals and sulfur in the forest. Unfortunately the budgets for heavy metals are ominously positive. The accumulation of the most toxic element, cadmium, in the pine and deciduous forests at Niepołomice is 80% and 87%, respectively, whereas in Walker Branch the relevant figure was 67%. Similar comparisons of lead and zinc accumulations show that the Niepołomice Forest retains 93% and 94% of lead and 57% and 82% of zinc in the deciduous and coniferous sections respectively while in Walker Branch 98% of lead and 74% of the zinc is retained (Table 7.12). Sulfur has very high net retention in both parts (27 and 16 kg ha^{-1} yr^{-1}, Table 7.10, Fig. 7.10), whereas in slightly polluted watersheds of Hubbard Brook its input in bulk precipitation was always smaller than hydrologic output (Eaton et al. 1980). Fluorine compounds which are extremely harmful to forests (Amundson and Weinstein 1980) have not been included in these studies. Tentative estimates of their fall are 4.4 kg ha^{-1} yr^{-1} for 1978 (Bik and Zajac 1980). Fluorine accumulation in the bones of large herbivorous mammals such as roe deer (*Capreolus capreolus L.*) reached over 2,000 ppm (Grodzińska et al. 1983, Grodziński unpubl.). The fluorine danger has gradually decreased since 1981 when the electrolytic production of aluminum in an aluminum plant near Cracow was discontinued.

7.5.4 The General Pattern of Energy, Nutrient, and Pollutant Behavior in Forest Ecosystems

The patterns of matter, nutrient, and pollutant cycling discussed for the Niepołomice Forest and other forest watersheds permit one to attempt to describe the general behavior of these elements in forest ecosystems (Fig. 7.5). The simplest diagram of a forest ecosystem, including only producers, consumers, and decomposers shown with the soil and a pool of plant biomass has been used. Energy entirely from outside flows mainly via producers and decomposers and leaves the system in the form of dissipated heat. In forests, a large portion of this energy is temporarily stored in wood and is often exported from the system as timber (Fig. 7.5 a).

Quantities of nutrients transferred in and out are minute when compared with internal streams, particularly those between producers and decomposers. It can be concluded from our studies that consumers have a much larger share in nutrient cycling than can be inferred from their share in the energy flow. Nutrient accumulation in wood is negligible (Fig. 7.5 b).

The heavy metal cycles represent quite reverse characteristics. Heavy metals in ecosystems are mainly allochtonous. They do not leave the systems, but accumulate mainly in the litter and soil. They also cycle in the grazing food chain but only to a slight degree (Fig. 7.5 c).

Sulfur, which is actually a nutrient and becomes a pollutant ($SO_4^{=}$) only when present in excessive amounts, does not conform to the general pattern outlined here (Fig. 7.10 in Appendix).

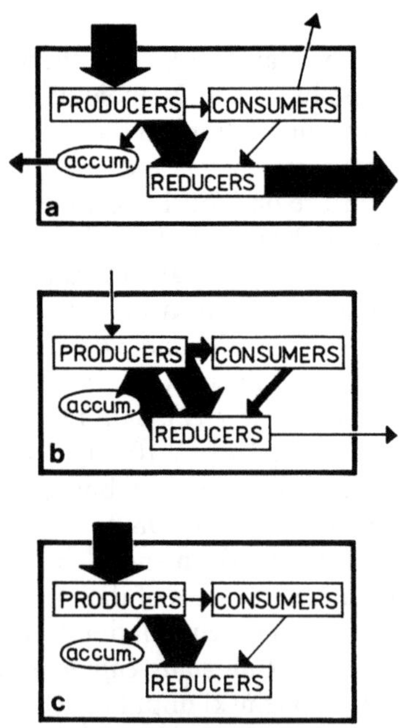

Fig. 7.5 a–c. A general scheme for the dynamics of energy (*a*), nutrients (*b*), and pollutants (*c*) in the forest ecosystems. *Thickness of arrows* indicates the intensity of flux

7.5.5 Disruptions in Elemental Cycles and the Future of the Niepołomice Forest

The current input of pollutants into the forest system at Niepołomice can be characterized as chronic exposure to moderate levels of airborne pollution (Grodziński and Yorks 1981). In the late 1970's this input represented overall 60 kg of sulfur and 33 kg of heavy metals ha^{-1} yr^{-1}, with a high water-soluble fraction (Chap. 3). At such levels of pollutant deposition the ecosystem ceases to act as a patient sink. Initial subtle changes can be traced in ecosystem functions particularly in forest nutrient cycling and forest metabolism (Smith 1974, 1981). At the system level two important functions of the forest may be impaired by disruptions in the mineral cycling (Zinke 1980), a decline in net tree production, and inhibition in litter decomposition (Grodziński and Yorks 1981). The first is mainly associated with high SO_2 concentrations and wet and dry sulfur deposition, the second with the accumulation of toxic heavy metals.

Scotch pine makes up 71% of the total area and 72% of biomass of the entire Niepołomice Forest complex (Chap. 3, Table 3.1; Grodziński and Lesiński 1978). The pine is the most susceptible to the effects of airborne SO_2. At the existing levels of SO_2 prevalent in the area (Table 2.3), about 60% of the stands demonstrate foliage damage of severe to moderate intensity (Fig. 1.16). The life span of pine needles has been reduced and potential photosynthetic activity has declined by 13–18% (Maczek 1977, also Chaps. 1 and 3). A 6–18% reduction in pine diameter growth related to these effects has been documented for the last 25 years

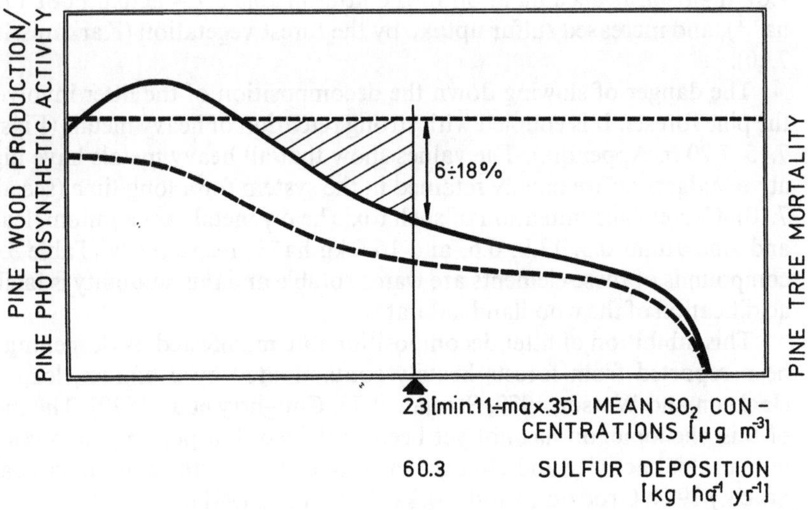

Fig. 7.6. Projections of pine forest (*Pinus silvestris*) production as a function of sulfur dioxide concentration and airborne sulfur deposition. Annual mean concentration of SO_2 as well as annual means of minimal and maximal concentrations indicated as presented in Table 2.3, whereas total sulfur deposition is taken from Table 2.5. *Solid line* pine wood production and/or pine photosynthetic activity arbitrarily adopted from Chapters 3 and 8, and from Greszta (1975) and Wolak (1977); *broken line* pine tree mortality in terms of deadwood volume; *horizontal broken line* level of wood production in an unpolluted pine forest

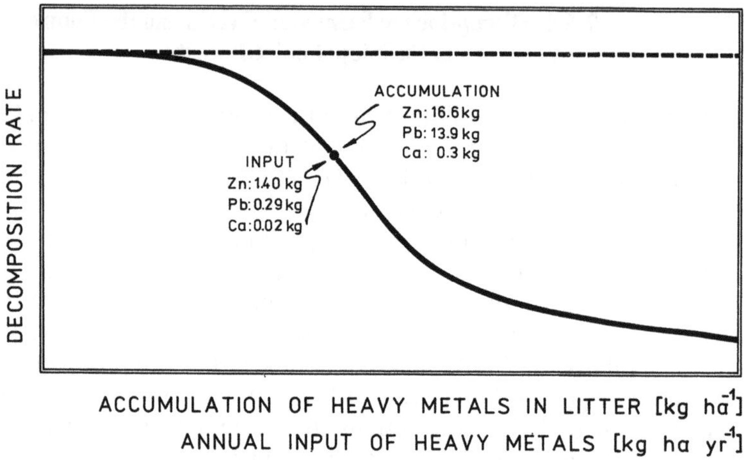

Fig. 7.7. Inhibition of litter decomposition rates as a function of heavy metal accumulation in the litter layer. Projections for pine forest at Niepołomice, Poland. *Horizontal broken line* rate of litter decomposition in an unpolluted pine forest. Values represent the actual retention of lead, cadmium, and zinc in the litter and annual input of these heavy metals to the litter (Tables 5.6 and 5.7)

(Białobok 1978, also Chap. 8). The real decline in pine production is actually greater and reaches 20–30%. This includes tree mortality and an increase in deadwood volume, as shown in Fig. 7.6. These processes are connected with the arresting of airborne sulfur by the whole system, particularly the pine forest (Fig. 7.6), high sulfur accumulation in the litter (Table 5.7 – actual pool 172 kg of S ha^{-1}), and increased sulfur uptake by the forest vegetation (Karkanis 1976, Fig. 7.10).

The danger of slowing down the decomposition of the litter involves mainly the pine forests. It is coupled with strong retention of heavy metals (Figs. 7.7, and 7.15–7.20 in Appendix). The values show that all heavy metals have highly negative budgets or are mostly retained in the system for a long time (see also Table 7.10). Current accumulation of such toxic heavy metals as cadmium, lead, nickel, and zinc attain 0.3, 13.9, 0.6, and 16.6 kg ha^{-1}, respectively (Table 5.7). Some compounds of these elements are water-soluble and this solubility is facilitated by acidification of the woodland habitat.

This inhibition of litter decomposition rate manifested by deepening litter has been reported from forests heavily contaminated by cadmium, lead, and zinc (Jackson and Watson 1977, Strojan 1978, Coughtry et al. 1979). The mechanism of this phenomenon has not yet been explained, but possibly involves a decline in microbial activity and changes in composition of the soil fauna (Babich and Stotzky 1974, Grodziński and Yorks 1981, Smith 1981).

This process has probably just begun in the Niepołomice Forest (Chap. 5). There are also indications that with further loading with heavy metals it may intensify and last for long periods (Fig. 7.7). Disruption of elemental budgets will increase if the area is further exposed to airborne pollution. The forecasts projected beyond the year 2000 differ in respect to industrial gases and dusts (Fig. 7.8). Regarded on the scale of Poland, estimates of sulfur dioxide emission and

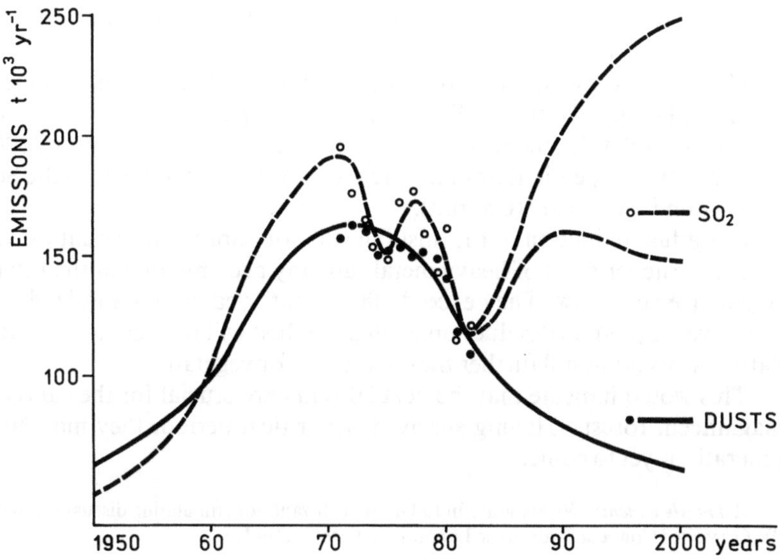

Fig. 7.8. A prognosis of air pollution in the Cracow region for the period 1950–2000. Emission of industrial dusts and SO_2 to 1982 taken from Statistical Yearbooks (GUS 1971–1983). Future projections based on surveys by the Cracow Council of Environmental Protection.

deposition stand at $2,500 \times 10^3$ t yr^{-1} and $2,825 \times 10^3$ t yr^{-1}, respectively, while predicted values for emission and deposition for the year 2002 are 3,070 and $3,097 \times 10^3$ t yr^{-1} respectively (Highton and Chadwick 1982).

Observed in the context of the Cracow region the present emission of SO_2 of 120×10^3 t yr^{-1} is among the highest per unit area, close to that for Silesia (Statistical Yearbook GUS 1983). During the last few years the emissions have dropped considerably, by as much as 26–32% (Statistical Yearbooks GUS 1976–1983). A forecast for the year 2000 projected by the Council of Environment Protection in Cracow envisages possible scenarios anticipating emissions within the range of 150 to 250×10^3 t yr^{-1}.

Emissions of industrial dusts in Poland total $1,901 \times 10^3$ t yr^{-1} of which 109 t fall on the Cracow region (Statistical Yearbook GUS 1983). Dust emissions in the region have decreased by 30%, starting in the early 1970's and realistic projections indicate that in the year 2000 they will reach the low level of the early 1950's (Fig. 7.8). Knowing these local forecasts one can anticipate future developments in the Niepołomice Forest. Such a simulation model was described for the Ponderosa pine forest (Kercher and Axelrod 1980) but the input parameters and variables now available are insufficient to provide the basis for a realistic model here.

The general effects of various air pollutants upon forest vegetation are well known (Mudd and Kozłowski 1975, Hughes et al. 1980, Overrein et al. 1981, Smith 1981, also Chap. 8). Such effects have been the topics of recent international symposia (e.g., Drablos and Tollan 1980, Hutchinson and Havas 1980, Papers presented ... 1980, Proceedings of symposium ... 1980, International conference ... 1981). There are, however, only a few quantitative functions coupling such processes as forest growth rates and SO_2 concentration, decomposition rates and heavy metal accumulation, etc.

Using the best available data on the influence of emissions on pine forests in Southern Poland (Greszta 1975, Wolak 1977, Sporek 1983) a forecast has been attempted of developments in the Niepołomice Forest to the year 2000. Any increases in the SO_2 emissions in the area will cause grave danger to the Scotch pine forests dominating here (Chap. 8).

A gradual reduction of the dust fall will not improve the situation very quickly. The turnover rates of heavy metals already accumulated in the litter-soil subsystem are very slow. They exceed 100 years for cadmium and lead (Table 5.8). Moreover, a predicted reduction in alkaline dust will produce a more acid precipitation, worsening still further the conditions for vegetation.

This would indicate that the next 20 years are crucial for the survival of these magnificent forests. Having survived the critical period, they may be saved for generations yet to come.

Acknowledgments. We are grateful to Dr. W. T. Swank for stimulating discussions when one of us (W. G.) visited Coweeta Hydrologic Laboratory, Otto, NC in 1982.

7.6 Appendix

Diagrams off Energy, Nutrient, and Pollutant Budgets in the Niepołomic Forest Ecosystems

The diagrams in Figs. 7.9–7.20 represent energy, nutrient, and pollutant transfers within the deciduous and coniferous forest ecosystems. Because of the lack of complete data only four major components of the ecosystems are distinguished: producers (all plants), consumers (animals), forest litter (including decaying organic matter and the detritus food chains), and inorganic soil. The elemental budgets are more complete for pine than for deciduous forests.

The arrow thickness is roughly proportional to the order of magnitude of energy or matter transfers. The loops denote the intensity of accumulation of elements in particular compartments or retention within the whole ecosystem. All values in kJ or $g\,h^{-1}\,yr^{-1}$.

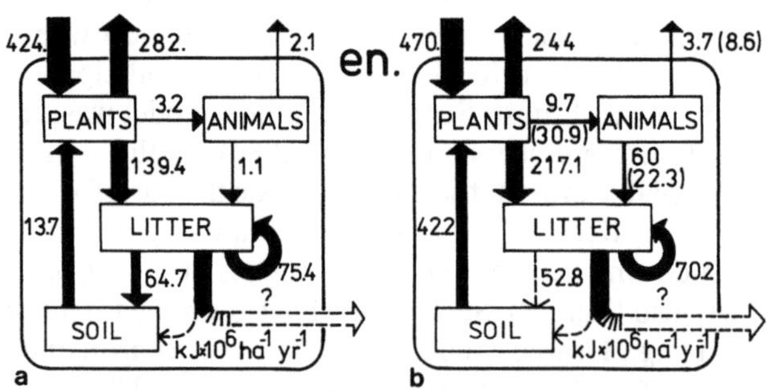

Fig. 7.9

Figs. 7.9–7.20. explained in Appendix

Diagrams 7.9 a, b, show the networks of energy flow in coniferous and deciduous forests. The energy entering ecosystem (gross primary production) is represented by incoming arrows whereas the energy leaving ecosystems in form of heat (respiration) is denoted by outcoming arrows.

The following figures represent nutrients (N, P, K, Ca; Figs. 7.11–7.14), sulfur (Fig. 7.10) and heavy metals (Fe, Ni, Cu, Zn, Cd, Pb; Figs. 7.15–7.20). Here the incoming arrows represent atmospheric deposition, whereas outcoming arrows show export with water.

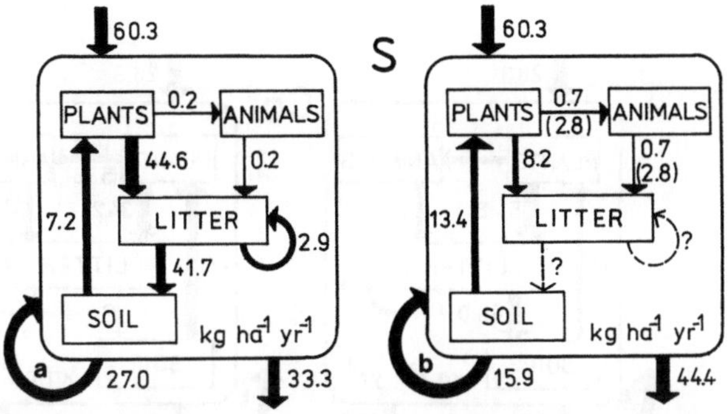

Fig. 7.10

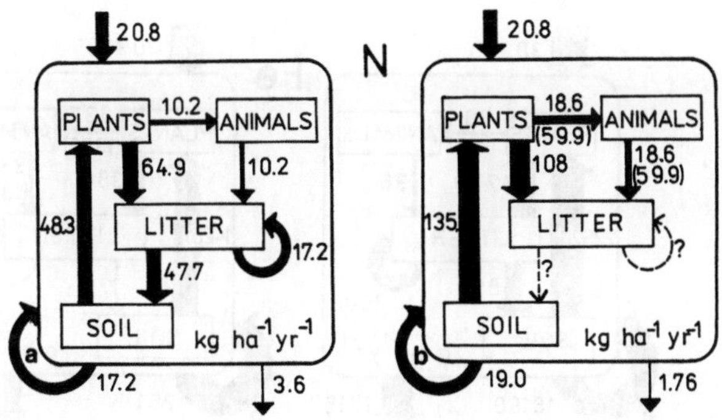

Fig. 7.11

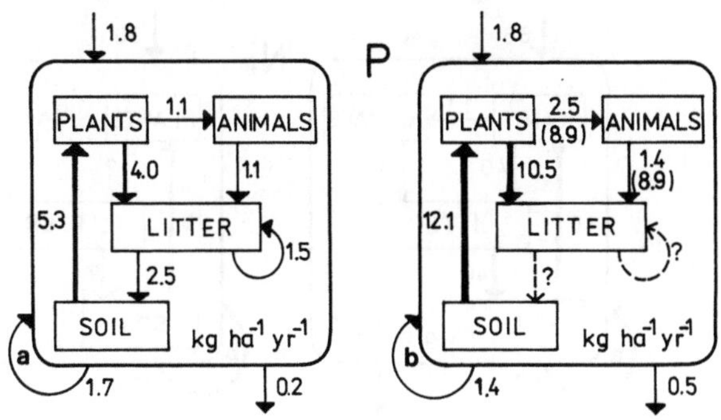

Fig. 7.12

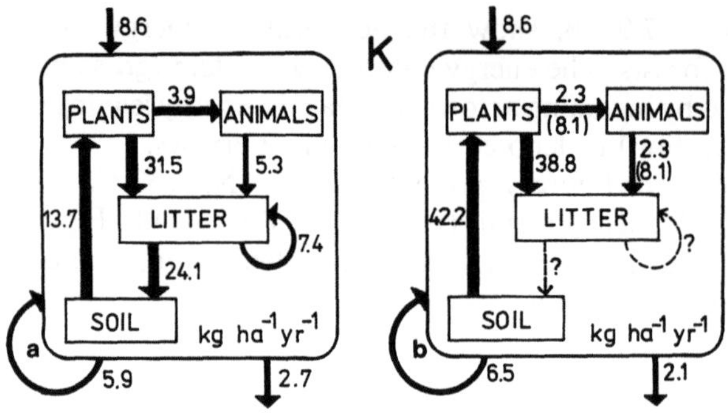

Fig. 7.13

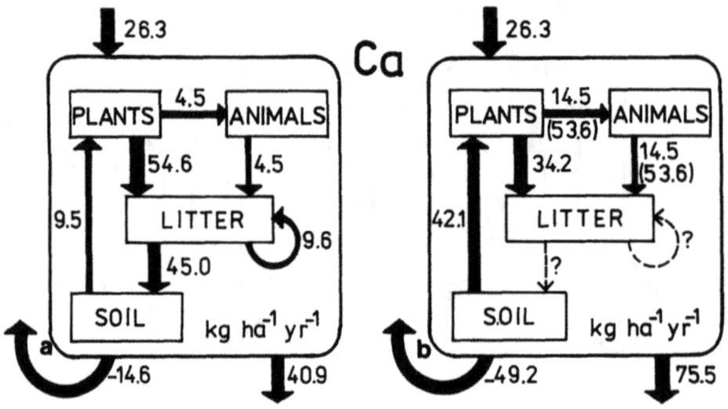

Fig. 7.14

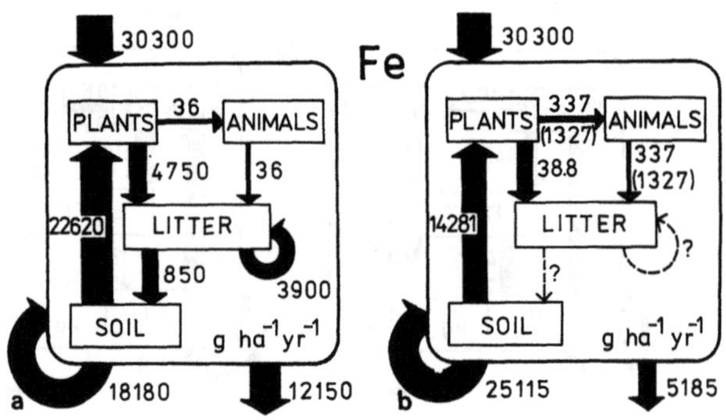

Fig. 7.15

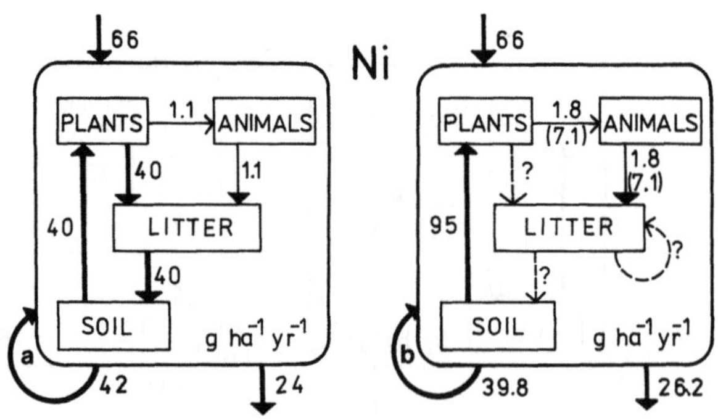

Fig. 7.16

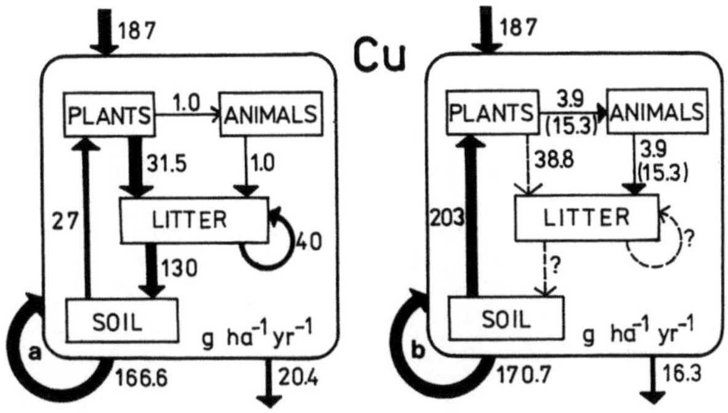

Fig. 7.17

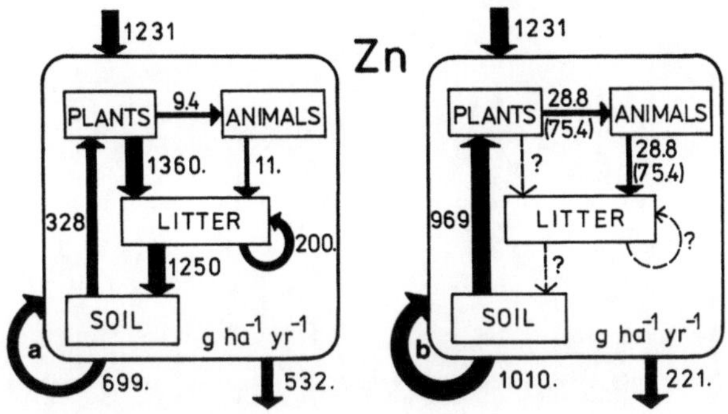

Fig. 7.18

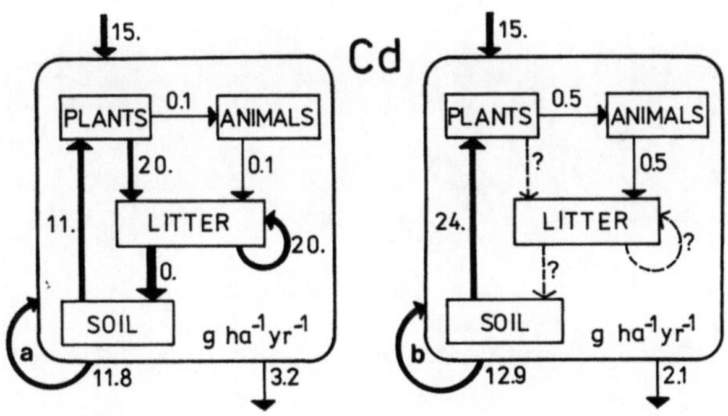

Fig. 7.19

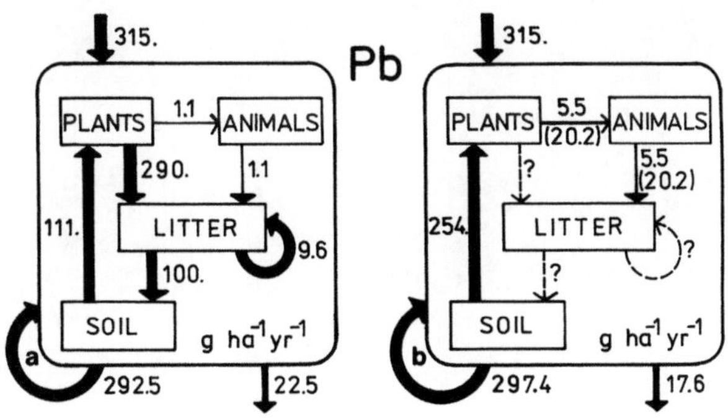

Fig. 7.20

Chapter 8

Concepts of Forest Management in Industrial Regions

Chapter Editors: S. BIAŁOBOK and J. FABIJANOWSKI

Contributors:
B. ADAMCZYK, S. BIAŁOBOK, J. FABIJANOWSKI, J. GRESZTA, J. A. LESIŃSKI, J. OLEKSYN, and L. RACHWAŁ

8.1 Forest Functions in Industrial Regions

J. Fabijanowski and J. A. Lesiński

Owing to the variety of exploitation and management programs, forests should be adapted to serve many functions. However, it often becomes necessary to determine which function has priority in a given area (Anderson 1970, Papanek 1974). Relative to the situation in Polish forestry this is to a certain extent expressed by the designation of production forests and conservation forests of various categories (Lesiński 1974, 1983). Socially suitable hierarchies relative to the importance of various forest functions change with time and depend to a large extent on the state of social consciousness as determined by the degree of cultural development. However, the final selection of major function is determined primarily by location of the forest in question, on its suitability to serve a particular function, which in turn depends primarily on site conditions, specific composition, and state of health. Thus one may speak of forests that are adapted or nonadapted to specific social functions. The latter category would include artificial pine and spruce plantations growing on unsuitable sites and endangered by various calamities.

The majority of the functions served by a forest depend upon the very nature of forest ecosystems and hence can be obtained simultaneously and without reduction in efficiency (Myczkowski and Lesiński 1982). It is only the intensification of one of these functions, for example, a densely populated game preserve, that may cause reduction in the efficiencies of other functions (Clawson 1975). In many instances, however, productive and nonproductive functions of a forest ecosystem can be reconciled, and obviously, changes in management and more specific qualifications of the forest service are needed to this end (Fabijanowski and Jaworski 1978). The only forest function that conflicts consistently with all others, is that of a "biological filter of polluted air" (Lesiński 1982). This function is required of forests affected by industrial air pollution, which in Poland occupy almost 500,000 ha (Lesiński 1974, 1983). Because of their location near large industrial centers, these forests are expected also to serve touristic and recreational functions. This is so with the Forest Protection Belt surrounding the Upper Silesian Industrial Region. At Wałbrzych as at Jelenia Góra, forests affected by industry are required to protect watersheds and climatic health resorts. However, a substantial decline in their effectiveness in protecting water reserves, soils, and wildlife has been signalled by Materna (1980).

The polluted air-filtering mechanisms of forest have not as yet been sufficiently recognized (Wentzel 1975, Bernatzky 1978, Knabe 1977, 1979). Some studies in this field have recently been conducted by Iwanek (1980). The filtering capacity of forest ecosystems is estimated to be of the order of several scores of tons of dust and a much smaller volume of gaseous pollutants ha^{-1} yr^{-1} (Knabe 1977, Madders and Lawrence 1982). According to Materna (1980) the highest limit of spruce forest filtering capacity is 0.3 t of sulfur ha^{-1} yr^{-1}. However, filtering functions of forests frustate the aims and purposes of forestry (Piekara 1975). As noted by Lesiński (1974), the rate of increase in area of forests affected by industrial air pollution provides an objective measure of the failure to provide adequate protection of air purity by industry. According to Smith (1974, 1981) forest ecosystems can function as a sink or dump for air pollution only so long as the pollutant concentrations are low and do not lead to visible disturbances in the ecosystem. As the load of contaminants of the ecosystem increases, it enters a phase of reduced biomass productivity due to disturbances in mineral cycling, and to reduction in photosynthesis, respiration, and transpiration due to changes in species composition by the establishment of plants less sensitive to pollutants, and the development of diseases and mass outbreaks of pests, which leads finally to a general breakdown in biological balance (cf. Chap. 7). The consequent serious disturbances (the third phase of Smith 1974, 1981) result in major changes in water relations, microclimate, soil processes etc. (Greszta 1975, Lesiński and Suliński 1982).

Forests in industrial regions are frequently not adapted to perform a filtering function, because they are generally even-aged, artificial pine or spruce monocultures, and very sensitive to air pollution. An increase in the efficiency of forests for the air-purifying function may be achieved in time by restructuring the forest and effecting changes in species composition (Fabijanowski and Lesiński 1980). Two types of forest restructuring are possible. On richer sites, the return to an initially more diversified ecosystem may be biologically justified. On poorer soils, however, such a procedure enforced upon the forest is in disagreement with the natural biological potential, and frequently additional treatments such as fertilization, the introduction of appropriate species, even of foreign origin, individual selection for decreased sensitivity, etc. are required.

As may be judged from the above discussions the "filtering" or "fume" absorbing functions of a forest in industrial regions (passive protection of air purity) are in conflict both with the natural character of the forest as an ecosystem and with the social aspects of a forest, both in terms of production and protection or other nonproductive functions (Lesiński 1981 a, b). As a result there is an inverse relationship between function and the other values, and the greater the absorption of emissions by a forest, the less suitable it will be for the physical and psychic regeneration of man, for the protection of water resources, for the production of wood etc. A successful detoxification of gaseous pollutants by forest vegetation is unattainable.

Perhaps the outcome of ecological studies conducted in the Niepołomice Forest may be applied to the entire Polish lowlands, as well as to the lowland regions of eastern and western Europe, in order to reduce the effects of industrial emissions. The forest ecosystems of this complex resemble, in fact, not only other Polish lowland forests, but also those of East and West Germany, as well as those in the mid-western parts of the USSR.

The similarities are related primarily to the following features: (1) the species composition of the forest ecosystems, (2) the dominance of pinewoods over forests with deciduous species, (3) a considerable proportion of pine stands from seeds of foreign provenance, (4) many generations of negative selection of tree ecotypes with the economically valuable being first utilized and, (5) the genetically determined high sensitivity of Scotch pine and to a lesser degree some deciduous species (pedunculate oak, sycamore, poplar), to the action of industrial emissions (Davis and Wilhour 1976, Białobok 1980). The high share of Scotch pine relative to broadleaved trees in various regions and forest ecosystems results in their high sensitivity to air pollution. The proportion of Scotch pine in forests of central and northern Poland ranges from 63.0% to 91,0% (Białobok 1970). On the other hand, the proportion of deciduous trees such as oak, ash, sycamore, maple, elm, and lime in the same regions ranges from 2.7% to 6.2%.

In the lowland sections of East Germany (Potsdam and Magdeburg regions), the share of Scotch pine in forests is very high, similar to that in western Poland. On the other hand, the proportion of deciduous trees (oak, birch, aspen, alder) is insignificant, although slightly higher in Poland (Dengler 1944). According to this author, the strong dominance of Scotch pine relative to deciduous species is typical also in some regions in East and West Germany. One can expect that these forests are thus very sensitive to emissions by expanding industry.

Dengler (1944) also emphasizes that all the forests in West Germany north of the Main and west of the Harz and Thüringer Wald are composed of introduced pine and spruce of foreign provenance. The proportion of pine stands of foreign provenance ranges to 80% in northern and central East Germany, according to Busse and Troeger (1938). In the Central European Lowland, between Berlin and Sovetsk, Tilsit in the USSR, and between Stralsund in the north and Opole in the south, the proportion of Scotch pine of foreign origin in 70-year-old stands averages 30–50%. It is even higher in the Niepołomice Forest, often exceeding 50%. Foreign pine populations usually have much lower viability and productivity than the native local populations.

8.2 Stand Conformity with Site

B. Adamczyk, J. A. Lesiński, and S. Białobok

Pedological and phytosociological studies indicate that, given a set of climatic conditions, soil, and plants form mutually interdependent entities (Braun-Blanquet et al. 1954, Adamczyk 1962, Adamczyk and Zarzycki 1963, Fabijanowski and Zarzycki 1965). In such harmonious entities individual plant species and their associations not only show greater viability and productivity (Schmidt-Vogt 1968, Puchalski and Prusinkiewicz 1975), but also possess greater resistance to injurious agents such as industrial or urban dust and gases (Buchwald and Engelhardt 1968–69). One can therefore expect that the sensitivity of forests, particularly those of artificial origin, to air pollutants decreases with increasing conformity of plants to site conditions. This assumption forms the basis for the restructuring of tree stands in regions affected by urban and industrial pollution.

The concept of sensitivity of the whole community to the action of air pollutants cannot be equated with sensitivities of single trees, because even in the

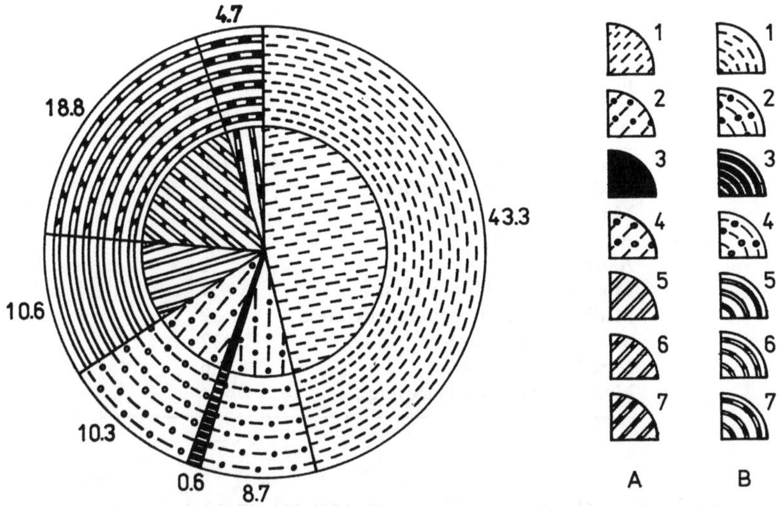

Fig. 8.1 A, B. Soil–site relations in the Niepołomice Forest expressed as a percentage of area occupied: **A** Soil subtypes; **B** Forest site types after Gruszczyk et al. (1976). *1* peaty and mucky podzolic soils – site of moist oak-pine forest; *2* iron podzols and podzolic soils – site of medium moist oak-pine forest; *3* oligotrophic peat and peat-gley soils – site of swamp pine forest; *4* podzolized lessive soils, podzolized brown and cryptopodzolic soils – site of medium moist deciduous forest; *5* mesotrophic muck-gley soils and oligotrophic-mesotrophic variants of mucky podzolic soils – site of moist deciduous forest; *6* "old warp soils" (gleyed humic brown soils, gleyed typical and leached brown soils, pseudogley soils, occasional humic gley soils) – site of moist hornbeam-oak forest; *7* mesotrophic-eutrophic peat soils, peat-gley, mucky-gley and humic-cley soils, shallowly gleyed humic brown soils – sites of alder swamp forest and carr

most injured forests individual trees can be found without visible symptoms of injury. The question of the individual resistance of some species and populations considered to be generally vulnerable to air pollution will be discussed in Section 8.4. At this time, only the discrepancy between the specific composition of plant communities of the Niepołomice Forest and site conditions will be evaluated.

Under relatively uniform climatic conditions throughout the area, observed site variation depends primarily on differences in soil conditions, especially water availability and trophic value (cf. Chap. 1.2). In the southern part labile sites predominate, associated with different variants of exclusively sandy oils or layered soils with more compact sandy-clayey material (Chap. 1.3.2). This category of soils and sites is very sensitive to inappropriate species composition.

The soil–site relationships in the forest are illustrated in Fig. 8.1. The soil (A) and site (B) units 1–3 pertain primarily to the pine forest complex occupying the southeastern part, units 4–5 to the mixed pine-oak forest complex in the southwestern part, and units 6–7 primarily to the deciduous forest complexes in the northern areas. Small patches of moist deciduous forest and alder forest (item B. 6–7) occur also in the mixed pine-oak complex and along the northern and southern borders of the pine complex. Abundant moisture is typical not only of the mesotrophic-eutrophic alder and hornbeam sites and the moist deciduous forests, but also of the oligotrophic-mesotrophic sites of mixed pine and mixed deciduous

Table 8.1. Contribution of Scotch pine on various forest sites in the southern part of the Niepołomice Forest

Forest site type	Adequately cultivated area (in %)
Medium moist pine forest	100
Medium moist deciduous forest	93
Moist hornbeam-oak forest	92
Ash-alder forest	90
Alder forest	70
Moist mixed deciduous forest	59
Medium moist mixed deciduous forest	19
Mixed bog pine forest	12
Mixed moist pine forest	6
Mixed medium moist pine forest	4

forests of the southern area. The high proportion of moist sites in mixed forests, and particularly in mixed pine forests, permits the introduction of more hydrophyllous species of trees and shrubs. The structure of the soil profiles and the most important soil properties of the soil–site complexes are presented in a simplified range map (Chap. 1.2).

A comparison of the existing species composition of tree stands with the recommendations provided by the *Silvicultural norms* (1969) for sites represented in the Forest indicates an excess of Scotch pine on most sites (Table 8.1). This inconsistency in the specific composition of tree stands relative to site conditions concerns primarily pine on mixed pine-oak forest sites of high and medium humidity and on mixed deciduous forest sites of medium humidity and to a lesser extent holds true as well for other sites.

Based on phytosociological criteria (Ćwikowa and Lesiński 1981; Chap. 1.1), one can consider that approximately 37% of the area of the southern pine zone (complexes of pine- and mixed pine-oak forests) represents distinctly deviant forest communities. This concerns primarily the transitional associations between Tilio-Carpinetum and Pino-Quercetum, and between Carici elongatae-Alnetum and Pino-Quercetum, which developed following clear cutting and the planting of pine on alder and hornbeam sites. These abnormal associations are characterized by extensive communities, heavily populated with *Carex brizoides* or *Molinia coerulea* and other species, which are typical of secondary communities (Ćwikowa and Lesiński 1981). Alexandrowicz and Łaszczyński (1979) describe these modified communities as sedge or moor grass forest types, respectively. According to these authors, they characterize an aging phase of forest communities on various sites. The sedge forest type is thought to indicate modification of forest communities on mixed deciduous forest sites of medium and high humidity, whereas the moor grass type one on moist mixed pine-oak forest sites, and possibly also those of alder forest.

In the northern parts of the forest, where naturally stable loamy soils of alluvial origin prevail, nonconformities of tree composition with site conditions seem related to imbalance in various components of the stands. An excess of oaks and black alder is observed, while other deciduous species such as lime, hornbeam, ash, etc., are under represented. There are transitional associations between the flooded forest and the moist hornbeam forest (Ćwikowa and Lesiński 1981, Chap. 1.1), which indicate a succession induced by the regulation of the Vistula. In fact, they represent a harmonious entity with moist deciduous forest sites developed subsequently in these areas.

One should add that the restructuring of the species composition and structure of these tree stands, which has been undertaken by the Forest Service in the Niepołomice Forest, is purposeful.

8.3 Site Formation

B. Adamczyk

Studies carried out by Keller (1964) and IUFRO (FAO/IUFRO Int. Symp. Forest. Fertiliz. 1973) show unequivocally that it is only under optimal site conditions that tree species and their ecotypes may exhibit increased resistance to the action of urban and industrial pollution. The question arises as to whether and to what extent the potential site conditions of a specific forest complex are suitable to permit restructuring without additional ameliorative treatments.

It is well known that an admixture of deciduous trees in pure pine stands results generally in an improvement of site microclimate, as well as in an increase in biological activity in the soil, greater absorption of mineral nutrients, and a more efficient turnover of biogenic elements (Chodzicki 1933, 1934, Wittich 1951, Puchalski and Prusinkiewicz 1975, Obmiński 1977). However, it has been correctly emphasized by Chodzicki (1934) that vegetation alone is incapable of transforming naturally poor soils and sites into fertile units. Soil and vegetation proceed toward the formation of unique ecological climaxes and paraclimaxes which, at least in temperate climates, are predetermined by substrate or occasionally by landscape configuration. Site conditions and the specific composition of plant communities depends primarily upon trophism and the water relations of soil (Adamczyk 1966). This predetermination has been underestimated, for instance, in the restructuring program of deciduous plantations in the Upper Silesian and Kraków industrial regions. The deciduous trees introduced to poor pine sites are relatively easily attacked by fungi and insects (Schneider 1971, Dominik 1977, Domański 1978) and the considerable losses observed subsequently in such stands do not foster optimism. Starzyk and Starzyk (1981) also point out the significance of some members of the family Cerambycidae that to date have been underestimated and may be of importance. Species such as *Saperda scalaris* and *Leiopus nebulosus* occur abundantly on oaks (*Quercus robur*, and *Q. sessilis*) and black alder (*Alnus glutinosa*) in the Niepołomice Forest, and are the main tree destroyers, particularly in younger age classes. The establishment of deciduous stands solely by the restructuring of pine stands is thus possible only on richer sites, whereas additional amelioration treatments are indispensable on poorer sites.

Due to the moister conditions in the southern forest complex, the labile and naturally poor sandy soils (Chap. 1.2) would respond more positively to the implantation of deciduous trees (Sect. 8.4). In general, however, such soils are unable to satisfy the higher edaphic requirements of deciduous species. They require deacidification, enrichment in biogenic components to dilute the increased concentration of Zn, Pb, Cd, and S, and replacement of the pine forest type of nitrogen turnover ($N-NH_4$) with the deciduous type ($N-NO_3$). An increase in the biological activity of the soil is also required, and maintenance of a relatively high ground water level because the ability of the resultant tree stands to raise capillary water to subsurface soil horizons is low. Treatments that would hasten mineralization of organic matter in the soil are not recommended in spite of the fact that low rates of litter decomposition are generally considered to be a negative factor in silviculture (Assmann 1961). Here, an increased accumulation of organic matter in the soil will cause a greater accumulation of toxic pollutants in it (Adamczyk and Maciaszek 1977, Adamczyk and Irzabek 1978). In labile sandy soils the organic matter is in fact the only filter and buffer able to counteract unfavorable dilutions (Adamczyk 1977), and this holds true for insufficiently adjusted mineral nutrition (Baule and Fricker 1967), as well as for any unexpected increase in the emission of toxic pollutants.

On excessively degraded or polluted soils complete reforestation can be facilitated by deep plowing and parallel treatment with bentonite, peat, and mineral fertilizers (Bolewski et al. 1968, Królikowski and Kowalkowski 1977, Mucha and Ratajszczak 1977, Mucha and Sienkiewicz 1977). However, the extent of soil pollution in the Niepołomice Forest, the absence of a distinct iron-pan horizon, the generally thin 2–3 cm (horizon with raw humus (A_0F+H) and favorable moisture indicate that such radical actions are not required. The following measures are recommended for those soils and sites of the southern complex: (1) Fertilization with crushed dolomite ($CaCO_3 \cdot MgCO_3$), some 2 t ha^{-1} to deacidify and to supplement deficiencies of Ca and Mg, (2) Full NPK fertilization with 90 kg N ha^{-1} (30 kg ha^{-1} annually) in the form of ammonium nitrate (NH_4NO_3), 80–90 kg P ha^{-1} as borinated or triple superphosphate [$Ca(H_2PO_4)_2 \cdot H_2O$], 80–90 kg ha^{-1}K as potassium chloride (KCl 56%), and supplementary fertilization with molybdenum as 1.5 kg ha^{-1} of $(NH_4)_2MoO_4$. Ammonium and potassium sulfates should not be used because of the high concentration of sulfur in the soils here (Adamczyk and Irzabek 1978). (3) Introduction, in situations where a sparse canopy permits light penetration, of leguminous plants to achieve phytomelioration, e.g., lupine (*Lupinus polyphyllus*), white sweet clover (*Melilotus albus*), birds-foot-trefoil (*Lotus corniculatus*). Prusinkiewicz recommends (Puchalski and Prusinkiewicz 1975) *Sorbaria sorbifolia* as a very promising understory-nutrient-enriching agent, (4) Maintenance of the groundwater level at a depth of 60–80 cm on sites of moist mixed pine-oak forest and moist mixed deciduous forest, and of 40–50 cm on alder forest sites. In fact, there are favorable conditions to accomplish this in the area. As judged from a preliminary survey (Adamczyk et al. 1978), the groundwater level is either very shallow (0–49 cm) or shallow (50–99 cm) throughout 60% of the total area.

In strongly acidic soils with raw humus (A_0F+H) and a mucky Am in podzolic soils of the mixed pine-oak forests, some 58% of the total amounts of Zn are found in the soluble, available form. On the other hand only 6% of the Zn occurs in the soluble form in less acidified varieties of mull humus A_1 in the mesotrophic-eutrophic soils of deciduous forests (Adamczyk and Maciaszek 1977, Adamczyk et al. 1978). Mineral fertilization with NPK contributes to a gradual change of pine forest sites into moderately rich deciduous forest sites. These assumptions were the basis for recommending phytomelioration and the raising of groundwater levels. Puchalski and Prusinkiewicz (1975) claim that European forest trees achieve maximum growth rates on sites with groundwater levels ranging from 0.8 to 1.5 m in depth. Uggla and Uggla (1979) also considered the influence of structural soil features and concluded that the optimum groundwater depth in sandy soils falls between 0.8 and 1.2 m. Recent studies in the Niepołomice Forest (Adamczyk and Maciaszek 1977, Adamczyk et al. 1978) suggest that higher levels in sandy soils counteract chemical denudation of the soil, contribute to increased trophism of the site, and to some extent assist in lowering the concentration of some pollutants, particulary soluble forms.

Mineral fertilization should be repeated in similar or adjusted concentrations every 5 to 10 years. This corresponds roughly to the time needed to achieve discernible effects of fertilization. Much attention must be paid to nitrogen. It shows the most dramatic effects (Baule and Fricker 1967), but overdosing extends the vegetative season and thereby reduces resistance to frost. Ammonium nitrate is recommended instead of less soluble urea $(CO/NH_2)_2$ because of the unfavorable ratio of $N–NH_4 : N–NO_3$. Fertilization should be extended for 3 years because nitrate nitrogen is absorbed only biologically and hence attempts to use a single heavy dose would be unreasonable.

The proposed amelioration should improve the general condition of the soils and sites. However, these are merely ad hoc prophylactic measures. The possibility of improving the sanitary state of the soils, sites, and vegetation cannot be identified with the formation of a forest ecosystem that would be entirely resistant to injurious pollutants. The fate of existing patches of natural forest will depend primarily on cleansing measures to be undertaken by existing and potential emission sources.

8.4 Selection of Trees and Shrubs for Forest Restructuring in Industrial Regions

S. Białobok, J. Oleksyn, and L. Rachwał

Air pollution containing injurious industrial gases severely affects forest ecosystems. This is well exemplified in the Niepołomice Forest, where severely injured pine needles occur over 47% of the forested area and moderately injured over the remaining 53% (Grodzińska et al. 1978).

A chronic action of highly concentrated emissions causes the formation of plant associations typical of an industro-climax zonation, in which trees are often dwarfed and frequently also with one-sided and sparse crowns (Bohne 1972, Wentzel 1963, 1971, 1978, Wolak et al. 1977). Under such conditions Mamajew and Shkarlet (1972) in the Urals and Mrkva (1969) in Czechoslovakia have discovered a reduction in the potential for generative reproduction in Scotch pine.

Single old tree individuals, which persist in a polluted environment may be selected for the breeding of varieties that are more tolerant to injurious emissions (Rohmeder and Schönborn 1965, Knabe 1967, Guderian 1977, Białobok and Karolewski 1978). Genetic variation may also exist in populations that occur under the influence of industrial pollution (Huttunen 1978, Szmidt 1978, Białobok 1980, Mejnartowicz 1981, Karnosky and Steiner 1981, Huttunen and Törmälehto 1982). Also some progeny of trees tolerant to SO_2, HF, and O_3 have retained this resistance (Rohmeder and Schönborn 1965, Białobok et al. 1980, Scholz et al. 1980).

8.4.1 Tolerance of Trees and Shrubs to Industrial Fumes

As an example of the reconstruction of spruce and pine stands in the Erzgebirge the procedure described by Lampadius (1969) can be cited. New species were gradually introduced, such as red oak, indigenous oak, beech, larch, and white pine, to replace spruce and pine.

In Central Europe the native species most sensitive to industrial emissions are Scotch pine, Norway spruce, and silver fir. Antipov (1979) has compiled data on the sensitivity of 19 pine species to industrial emissions and shown that of 61 authors studying the problem, 75% have described Scotch pine as sensitive.

Data on the sensitivities of some forest species of trees and shrubs to industrial emissions are given in Table 8.2. Some species show great variation in these responses, which may be due to various conditions of the experiment and methods of observation, as well as to phenotypic and genotypic variability. One can, nevertheless, single out some species that are distinctly sensitive, as well as those distinctly tolerant of, various gases. Relatively few tree species are sufficiently tolerant of sulfur dioxide and ozone pollution. Tree sensitivity to nitrogen oxides has been little studied thus far.

Concerning the action of the gases mentioned in Table 8.2, especially sensitive are: *Pinus silvestris, Picea abies, Abies alba*, and *Pinus strobus*.

Relatively low sensitivity to SO_2 is exhibited by *Acer platanoides, Alnus glutinosa, Quercus robur, Q. rubra*, and *Robinia pseudoacacia. Acer platanoides, Betula verrucosa, Quercus robur, Q. rubra,* and some other species are also tolerant to ozone.

Unfortunately, there are relatively few confirmed reports concerning the maximum tolerance, to various gases or mixtures for woody plants cultivated in industrial regions. The maximum concentration of SO_2 in air in the Niepołomice Forest (Table 2.3) is of such magnitude that the future growth of Scotch pine there is endangered. Guderian and Stratmann (1968) determined levels of SO_2 injurious to Scotch pine under field conditions. During ($\bar{c}_i$) exposure time t_i these were 0.58–0.69 mg m^{-3}, while for the whole experiment during $\bar{c}_m$ monitoring time t_m the levels were 0.026–0.053 mg m^{-3}. Knabe (1972) has established that in the Ruhr Region, where the mean concentration of SO_2 in the growing period exceeded a value of 0.08 mg m^{-3}, the cultivation of Scotch pine was risky.

Table 8.2. Degree of resistance of trees and shrubs to toxic gases according to various authors

cies	Toxic gases														
	SO_2			O_3			NO_x			Fluorine and its compounds			Mixture of various gases		
ciduous															
r platanoides	$T^4T^{15}T^{21}$	$I^2I^{10}I^{17}I^{18}$	S^{11}	$T^1T^6T^{10}T^{12}$			T^{17}	I^9	S^1	T^5					
r pseudoplatanus	T^{16}	$I^2I^{11}I^{18}$			I^{16}									I^3	
us glutinosa	T^{18}	I^2I^{19}									I^5		T^3		
us incana		I^2I^{19}	S^{20}		I^{16}							S^5	T^3		
ula verrucosa	T^{18}	$I^2I^4I^{11}I^{16}I^{19}$	$S^{15}S^{17}S^{20}$	$T^1T^6T^{12}$	$I^{10}I^{16}$			I^{17}	S^9		I^5			I^3	
ula pubescens		I^2	$D^{11}S^{20}$					I^{17}							
pinus betulus	T^1	I^2	S^{20}				T^9		S^1			S^5			
us silvatica	T^1T^{14}	$I^2I^{16}I^{19}$	$S^{13}S^{18}S^{20}$	$T^{12}T^{16}$			T^9		S^1		I^5				
xinus excelsior		$I^2I^{11}I^{19}$	$S^{13}S^{16}S^{18}S^{20}$		I^{16}						I^5			I^3	
ulus tremula		I^2I^{18}	$S^{16}S^{17}$	T^{16}				I^{17}						I^3	
ercus robur	$T^1T^{10}T^{14}$	$I^2I^{11}I^{16}$	S^{20}	$T^1T^6T^{10}T^{12}$	I^{16}		T^9T^{17}			T^5				I^3	
ercus rubra	$T^1T^2T^{10}T^{14}T^{19}T^{21}$	I^{18}	$S^{11}S^{20}$	$T^1T^6T^{10}T^{12}T^{16}$							I^5		T^3		
binia pseudoacacia	$T^1T^2T^{10}T^{11}T^{18}$	I^{19}		T^{12}		S^1S^{10}	T^9		S^1		I^5		T^3		
nbucus nigra	$T^2T^4T^{16}T^{19}$	$I^{11}I^{19}$					T^9		S^1				T^3		
bus aucuparia		$I^2I^{16}I^{19}$	$S^1S^{11}S^{17}S^{18}$	T^1T^{16}		S^6S^{12}		I^{17}						I^3	
a cordata	T^{15}	I^2	$S^{11}S^{13}S^{17}S^{19}S^{20}S^{21}$	$T^6T^{10}T^{12}$		S^1		I^9	S^1S^{17}			S^5			S^3
niferous															
es alba		I^8	S^2S^{20}	T^6											
ix decidua		I^2I^{13}	$S^8S^{19}S^{20}$		I^{12}	$S^1S^7S^{10}$		I^1	S^9			S^5		I^3	
ix leptolepis	T^{14}	$I^2I^4I^8I^{20}$	S^{19}		$I^{12}I^{16}$	$S^1S^7S^{10}$			S^9			S^5		I^3	
ea abies	T^4	I^{16}	$S^2S^8S^{11}S^{13}S^{14}S^{18}S^{19}S^{20}S^{21}$	$T^1T^7T^{10}T^{12}T^{16}$				I^9				S^5			S^3
ea pungens	$T^2T^{13}T^{15}T^{20}$	$I^8I^{11}I^{18}I^{19}$		$T^1T^7T^{10}T^{12}$				I^9	S^1			S^5	T^3		
us strobus	$T^{13}T^{14}$	$I^2I^{18}I^{20}$	$S^1S^4S^8S^{10}S^{12}$	T^{16}		$S^1S^7S^{10}S^{12}$			S^1			S^5	T^3		
us contorta		$I^1I^2I^8$			I^{12}						I^5				
us nigra	$T^1T^8T^{10}T^{13}T^{18}T^{20}$	$I^2I^4I^{16}$	$S^{12}S^{19}$		I^{16}	$S^1S^7S^{10}S^{12}$	T^9		S^1			S^5		I^3	
us silvestris		$I^4I^{10}I^{16}$	$S^2S^8S^{14}S^{18}S^{19}S^{20}S^{21}$		I^{12}	$S^1S^7S^{10}$						S^5			S^3
ja plicata	$T^1T^2T^8T^{19}$														
udotsuga menziesii	T^8	$I^1I^2I^4I^{10}I^{16}$	S^{13}	$T^1T^7T^{10}$	$I^{12}I^{16}$										

tolerant, S – sensitive, I – intermediate sensitivity. Numbers 1,2,3, etc. indicate the literature quotations referred to below and quoted in detail in the bibliography. Anonim 1973, 2 – Antipow 1979, 3 – Białobok and Rachwał 1975, 4 – de Cormis 1970, 5 – Dässler et al. 1972, 6 – Davis and Wood 1968, 7– Davis and Wood 1972, 8 – Enderlein l Vogl 1966, 9– Haut van 1975, 10 – Jensen et al. 1976, 11 – Kondratiuk et al. 1980, 12 – Lacasse and Treshow 1978, 13 – Lampadius 1965, 14 – Lampadius 1969, 15 – Linzon l. 1973, 16 – Mooi 1976, 17 – Nikołajewskij 1979, 18 – Pelz et al. 1963, 19 – Ranft and Dässler 1970, 20 – Schönbach et al. 1968, 21 – Spahova and Riazanceva 1978

Table 8.3. Proposal of IUFRO of maximum permissible levels of SO_2 and HF to protect forests (Resolution adopted at the Xth meeting of the S 2.09 IUFRO Working Party)

	Mean concentration μg m^{-3}		
	Annual	per 24 hr exceeded maximally 12 times per 6 months	97.5 percentile 30 min in vegetation period
SO_2			
Highest permissible levels of gas for full plant productivity	50	100	150
To maintain protection of forests serving a social function in extreme environments (high elevation forests in the boreal zone)	25	50	75
HF			
Highest permissible levels of gas for full plant productivity	0.3	–	0.9

To ensure protection of forests in Europe IUFRO Section S2.09.00 proposed mean maximal concentrations of SO_2 and HF for the safety of spruce forests. One can assume these levels hold also for pine and they are presented in Table 8.3. For forests serving a social function and growing under critical site conditions and for montane forests and those in the boreal zone, the mean maximum concentrations of SO_2 must be no more than half the values shown in Table 8.3.

However, Wentzel (1981) is of the opinion that the maximum values of SO_2 concentration proposed by IUFRO should be adjusted for the various biotic and abiotic conditions of different forest regions in Europe. Wentzel (1978) has also proposed that mean critical values of SO_2 in industrial regions over long periods of time should be adjusted to a lower level of 0.05 mg m^{-3}, in order to ensure forests sufficient protection. Low concentrations of SO_2, with an annual average of 0.07 mg m^{-3} lower wood production by 20%. However, an annual average concentration of 0.095 mg m^{-3} results in a lowering of volume production by 40% (Jeffree 1980).

In the Niepołomice Forest 10-year measurements of mean annual concentrations of SO_2 primarily in the western part do not have significant fluctuations and attain a level of approximately 0.020 mg m^{-3}. However, during the colder 6-month period, concentration increases almost threefold and under certain weather conditions may attain even higher values. Mean concentrations of sulfate aerosol in the eastern part are lower than in the western, being 0.011 mg m^{-3} (Grodzińska et al. 1978). The magnitude of these concentrations during a year are similar to the values, $\bar{c}_m$, during monitoring time t_m, given by Guderian and Stratmann (1968). As a consequence about 40% of the Scotch pines have injured needles (Grabowski 1981 a, b).

Despite the fact that the danger of injury by fluoride compounds in the Niepołomice Forest has declined substantially since the closing of the aluminum smelter, the danger still exists from mixtures of SO_2 and fluorides emitted by the steel complex 10–15 km to the west. The injurious affects of mixtures of SO_2 and fluorides on pine forests has been observed previously in close proximity to aluminum smelters and phosphate fertilizer plants.

However, this phenomenon has not been critically investigated and biologically is little understood. One might suspect there are similar relationships in the degree of Scotch pine sensitivity of fluoride compounds as has been found to exist for Norway spruce (Scholz et al. 1978). In that study the genetic variability of various populations was related to levels of sensitivity to hydrogen fluoride. There is the probability of selecting provenances less susceptible to air pollution, because in the case in question additive variance and dominance variance amounted to 60% of phenotypic variance (broad sense hereditability = 0.60).

The nearby steel complex is emitting NO_x mixed with SO_2. The toxicity of this mixture has not yet been sufficiently studied in respect to Scotch pine. Białobok et al. (1980) have found injury caused by NO_2 similar to that by SO_2. Hicks (1978) considers the injurious effects of the mixture of these two gases to be additive in respect to aspen and more than additive for such annual plants as tobacco, soya, radish, and beans. Such a mixture undoubtedly may also prove dangerous to the health of Scotch pine.

In Table 8.2 species of trees which may be cultivated in the Niepołomice Forest following reconstruction are listed. Information is given on sensitivity to emissions of SO_2, O_3, NO_x, fluorides, and mixtures of gases.

In 1976–1977, the Institute of Dendrology of the Polish Academy of Sciences established an experimental provenance area with black pine, Norway spruce, and oaks in the forest. It is hoped that observations on survival, degree of injury, viability, and growth intensity will permit a more appropriate selection of provenances of these trees suitable for use under the conditions existing in the Niepołomice Forest. On the basis of analyses of the rate of growth and health of some populations of native and introduced tree species (*Pinus strobus, P. nigra, Quercus robur, Larix leptolepis, L. decidua, L. eurolepis*) that grow in young plantations in the experimental area, their utility for these sites has been confirmed. Particularly intense growth has been observed in larches.

8.4.2 Principles of Seed and Nursery Work

The principles of selecting seed stands, and trees, the establishment of a seed orchard for Scotch pine, and Norway spruce and preparation of planting material, as well as the methods of propagating trees and shrubs, have been discussed by Białobok (1971), Hejmanowski et al. (1975), Przybylski et al. (1976). For such tree species as European larch, Polish larch, Japanese larch and silver birch it would be possible to establish seed orchards from seedlings.

In forests of the Silesian-Kraków region single individuals or populations of Scotch pine trees may be found that survive in the polluted environment rather well with a sufficient biomass production and after vegetative propagation can be used for the establishment of seed orchards from grafts. For Scotch pine, the establishment of 1.6 ha for such a seed orchard is proposed. These orchards will supply seeds for pine plantations in the Niepołomice Forest. An assortment of several forest tree species prepared for the reconstruction of the Niepołomice Forest stands will be based on local populations, on trees from the Cracow-Silesian region, as well as on seed orchards including provenances beyond the industrial regions but tolerant to industrial emissions. This should counteract at least a part of the dangers mentioned by Scholz (1983).

For *Betula verrucosa, B. pubescens, Larix polonica, L. decidua* and *L. leptolepis,* seed orchards may be established from seedlings. However, these species should be subjected to an even longer period of testing in order to determine the resistance of pollutants.

Seeds of pedunculate oak, red oak, black alder, ash, maple, black locust, and lime could be obtained from the forest by standard selection of seed stands (Hejmanowski et al. 1975). This holds true also for birch but later, seed orchards from selected birch seedlings should be established.

To date, only pedunculate oak seeds with good silvicultural characteristics are produced in the Niepołomice Forest in a seed stand. The available seed stands of pine have not been considered for restructuring the Niepołomice Forest with the objective of greater resistance to pollution. They are too small and inadequate for this purpose.

No protective action for plant cover will produce the desired effects if technical approaches to holding emissions below levels of maximum plant tolerance are not fully utilized. It is for this reason that the IUFRO working group on Industrial Pollution passed a resolution to protect European forests (Table 8.3).

Lower emissions of SO_2 and fluorides may save the existing pine stands from the direct action of these gases, however cultivation may suffer from acid rain and the injurious affects of industrial emissions on the soil.

The latter report emphasizes that techniques to reduce levels of environmental pollution do exist. The same approach should be followed in other countries, because cooperation between technicians and biologists concerned with environmental protection provides the only real hope for protecting our natural environment.

8.5 The Niepołomice Forest as an Example of a Forest Affected by Industrial Stress

S. Białobok, J. Fabijanowski, J. Greszta, and J. A. Lesiński

The studies presented in this chapter have important regional significance but will also apply to similar ecosystems throughout Europe where the necessity will undoubtedly arise to reduce the scale and rate of changes caused by the development of heavy industry. It is significant that Niepołomice is probably the only location in Central Europe where such diversified ecosystems with so many different functions have been subjected to detailed ecological studies immediately at the outset of industrial development nearby.

These studies have not only permitted the recognition of zones of severe injury to pine forests, but also types of general zonation of pollution influences. These studies have also permitted a recognition of pollutant pathways based on an analysis of pollutant input to the forest communities, outflow from the forest complex, and the direct penetration and effects on the plant cover.

Associated with these studies of injuries to the vegetation are concomitant investigations on the function of industrial forests and on the selection of trees and shrubs suitable for restructuring.

A diverse forest ecosystem is less sensitive to environmental stresses than a monoculture. Tree species better adapted to site conditions are also more resistant to the influences of industry. Consequently, mixed pine-oak and mixed deciduous forests are seen to be less sensitive to industrial emissions than pure pine plantations.

Much importance has been attached to these studies, because they permit a proper course of stand restructuring to engender plant communities that will be more tolerant to industrial emissions. Soils here are quite rich, and consequently it is possible to achieve a substantial increase in the number of forests containing deciduous trees (pedunculate and sessile oak, beech, and others) as well as conifers such as larch, and a reduction in the proportion of pine to about 45% (Fabijanowski and Lesiński 1978). The reintroduction of less sensitive Scotch pine into industrial regions will depend on the development of inexpensive methods for the mass propagation of improved resistant populations. Orchards for the production of seed from ramets of Scotch pine ortets selected for resistance to polluting gases may permit the introduction of less sensitive populations into industrial regions. Unfortunately, this is a lengthy procedure.

These studies of the forest ecosystems of the Niepołomice Forest have also formed the basis for recommendations of the State Forest Administration (Grodziński and Lesiński 1978), which are now being put into practice. The desired results of protecting the vegetation cannot be achieved, however, if industrial emissions are not drastically reduced, below the maximum levels tolerated by tree species.

References

Abaturov BD, Zubkova LV (1969) Influence of small sousliks on hydro-physical properties of solonetz soils of Zavolžije semi-desert. Pochvovedenie 10:59–69

Abaturov BD, Rakova MV, Seredneva TA (1975) Influence of little sousliks on production of semidesert vegetation of North Caucasus. In: Zlotin RJ, Isakov JA, Hodašova KS (eds) The role of animals in ecosystem functioning. Izd Nauka MOIP, Moskva: 15–18 pp

Adamczyk B (1962) Studies on the relations of pedology and phytosociology in the Mała Łąka Valley, Tatry Mts, Poland, Acta Agrar Silvestria 2:45–116

Adamczyk B (1966) Studies on the development of relation between parent rock and soil. Part II Forest soils formed on Flysch rocks of the Magura nappe in the Gorce Mts. Acta Agrar Silvestria Ser Silvestris 6:3–48

Adamczyk B (1977) The question of absolute and relative fertility of sandy forest soils in relation to mineral fertilization (in Polish). Postępy techn w leśn 28:14–20

Adamczyk B, Irzabek T (1978) Fertilization of the Niepołomice Forest (in Polish). In: Grodziński W, Lesiński JA (eds) Ecological principles for rebuilding the region of Niepołomice Forest, Jagiellon Univ, Kraków; 37–43 pp

Adamczyk B, Maciaszek W (1977) General characteristics of the Niepołomice Forest soils and their suitability for secondary alterations (in Polish). Ann techn Report Project 10.2.10, Inst Ecol Pol Acad Sci Warszawa; 1–113 pp

Adamczyk B, Zarzycki K (1963) Soils of the forest plant communities in the Western Bieszczady Mts (Polish Eastern Carpathians). Acta Agrar Silvestria Ser Silvestris 3:133–175

Adamczyk B, Niemyska-Łukaszuk J, Maciaszek W, Hara-Drożdż M, Suliński J (1978) Some properties of the Niepołomice Forest soils and their resistance to degradation (in Polish). Ann techn Report Project 10.2.10, Inst Ecol Pol Acad Sci Warszawa; 77 pp

Adamczyk B, Niemyska-Łukaszuk J, Maciaszek W, Hara-Drożdż M, Suliński J (1978) The fractional composition of humic bonds, content of heavy metals and various forms of sulphur and an introductory characteristic of groundwater in the soils of Niepołomice Forest (in Polish). Ann techn Report Project 10.2.10, Inst Ecol Pol Acad Sci Warszawa; 78–92 pp

Alexandrowicz BW, Łaszczyński A (1979) Initial discerning of the seat types and the forest habitat of the Niepołomice great forest. Acta Agrar Silvestria Ser Silvestris 18:3–18

Ambroziak K (1979) Body composition in the blackcap (*Sylvia antricapilla* L.) during breeding season (in Polish). MSc thesis – Jagiellon Univ, Kraków; 16 pp

Amundson RG, Weinstein LH (1980) Effects of airborn F on forest ecosystems. Proc Symp Effects air pollutants on Mediterranean and temperate forest ecosystems, Pacific Southwest Forest Range Exp Station Berkeley CA; 63–78 pp

Andersen J (1954) The food of the Danish badger (*Meles meles danicus* Degerbd) with special reference to the summer months. Dan Rev Game Biol:1–75

Anderson RC (1970) An Outline for multiple use planning. J For 68:558–560

Andersson M, Erlinge S (1977) Influence of predation on rodent populations. Oikos 29:591–597

Anonim (1973) Trees for polluted air. USDA Forest Service Publ 1230:1–12

Antipov WG (1979) Resistance of woody plants to industrial fumes (in Russian). Nauka i Technika, Minsk; 214 pp

Assmann E (1961) Waldertragskunde. BLV München Bonn Wien; 490 pp

Auerbach SI (1969) Radionuclide cycling in terrestrial ecosystems. Oak Ridge Nat Lab Publ 320:76–112

Babich H, Stotzky G (1974) Air pollution and microbial ecology. CRC Crit Rev Environ Control 4:353–409

Bac S (1968) Role of the forest in water balance of Poland. Folia For Pol Ser A Lesn 14:5–61

Bališ M (1980) Red deer (in Czech). Prir, Bratislava: 335 pp

Banasik J (1973) Ground flora production in stand of *Tilio-Carpinetum*. Bull Acad Pol Sci Cl II 21:593–599

Banasik J (1978 a) Seasonal development and net production of the ground flora of two oak-hornbeam stands in the Niepołomice Forest. Stud Nat Sci A 14:67–134

Banasik J (1978 b) A bibliography of the natural history of the Niepołomice Forest with special consideration of the period 1946–1977. Stud Nat Sci A 14:205–223

Banasik J, Jankowska K (1978) Ground flora net production of the oak-hornbeam stands in the Lipówka nature reserve in the Niepołomice Forest. Stud Nat Sci A 17:135–147

Bandoła-Ciołczyk E (1974) Production of tree leaves and energy flow through the litter in *Tilio-Carpinetum* association (International Biological Programme area). Stud Nat Sci A 9:29–91

Bandoła-Ciołczyk E, Witkowski Z (1976) Energy flow through oak leaves and caterpillars feeding on them in an oak-hornbeam ecosystem of the Niepołomice Forest. Bull Acad Pol Sci Cl II 24:385–392

Baradziej E (1974) Net primary production of two marsh communities near Ispina in the Niepołomice Forest (southern Poland). Ekol pol 22:145–172

Barett GW, Van Dyne GM, Odum EP (1976) Stress ecology. Bioscience 26:192–194

Baszyński T, Izdebski K, Malicki S, Udziak S (1968) Preliminary studies of the yield of selected meadow ecosystems in the environs of Sobieszyn on the Wieprz. Ann Univ Mariae-Curie-Skłodowska 23 C:67–97

Batzli GO (1978) The role of herbivores in mineral cycling. In: Adriane DC, Brisbin LL (eds) Environmental chemistry and cycling processes, USDE Springfield; 95–112 pp

Baule H, Fricker C (1967) Die Düngung von Waldbäumen. BLV, München–Basel–Wien: 259 pp

Baumgartner A (1968 a) In: Schnelle F (ed) Frostschutz im Pflanzenbau. BLV, München–Basel–Wien, PWRiL, Warszawa; 47–118 pp

Baumgartner A (1968 b) In: Schnelle F (ed). Frostschutz im Pflanzenbau. BLV, München–Basel–Wien, PWRiL, Warszawa; 119–168 pp

Belskij NV (1962) Kestrel and long-eared owl in Botanical Garden (in Russian). Ornit Izd Moskovsk Univ, Moskva; 2:316–324

Benecke P, van der Ploeg RR (1978) Wald und Wasser II. Quantifizierung des Wasserumsatzes am Beispiel eines Buchen- und eines Fichtenanstaltbestandes im Solling. Forstarch 2:26–32

Berdau F (1859) Flora cracoviensis (in Latin). Typis CR Jagiellon Univ, Kraków; 448 pp

Berg B, Wessen B, Ekbohm G (1982) Nitrogen level and decomposition in Scots pine needle litter. Oikos 38:291–296

Bernatzky A (1978) Tree ecology and preservation. Elsevier Amsterdam Oxford New York; 135–140 pp

Białecka Z (1979) Changes in contents of fat and mineral nutrients (N, P, K, Ca) in the body of great tits during breeding season (in Polish). MSc thesis – Jagiellon Univ, Kraków; 21 pp

Białobok S (1970) The pine in Polish forests since the late glacial period. In: Białobok S (ed) Popular Sci Monogr Our Forest Trees (Tanslation) USDA and NSF, Washington DC; 16–51 pp

Białobok S (1971) Establishment of forest seed orchards. PWRiL, Warszawa; 143 pp

Białobok S (1978) Dendrological measurement of selected plots in Niepołomice Forest (in Polish). Ann techn Report Project MR-II/15, Inst Dendrol Pol Acad Sci, Kórnik near Poznań; 1–6 pp

Białobok S (1980) Identification of resistant or tolerant strains and artificial selection or production of such strains in order to protect vegetation from air pollution. Papers presented to Symp Effects of Air-Borne Pollution on Vegetation, Warsaw; 253–264 pp

Białobok S (1980) Forest Genetics and air pollutant stress. Proc Symposium Effects of Air Pollutants on Mediterranean and Temperate Forest Ecosystems. Riverside; 100–102 pp

Białobok S, Karolewski P (1978) Evaluation of the degree of resistance of maternal Scots pine trees and their progeny to the action of SO_2, O_3 and a mixture of these gases. Arbor Kórnickie 23:299–310

Białobok S, Rachwał L (1975) Tentative list of ornamental trees and shrubs selected for rapidly developing and industrial towns (in Polish). Ministry of Administration, Regional Management and Environment Protection, Warszawa; 17 pp

Białobok S, Karolewski P, Oleksyn J (1980) Sensitivity of Scots pine needles from mother trees and their progenies to the action of SO_2, O_3, a mixture of these gases, NO_3 and HF. Arbor Kórnickie 25:289–303

Bieszczad-Kosch M (1979) Changes in mineral nutrient contents (Ca, K, N, P, Mg) during the postnatal development of the great tit, *Parus major* L. (in Polish). MSc Thesis – Jagiellon Univ, Kraków; 15 pp

Bik A, Zając KP (1980) Studies on fluoride content in the rainfall of Niepołomice Forest area (in Polish). Ochrona Powietrza 14:57–59

Bobek B (1973) Net production of small rodents in a deciduous forest. Acta theriol 18:403–434

Bobek B (1974) Production and standing crop of deer population in a deciduous forest. 1st Int Theriol Congr, Moskva; p 76

Bobek B (1980) Model for optimization of roe deer management in Central Europe, J Wildl Manage 44:837–848

Bobek B, Borowski S, Dzięciołowski R (1975) Browse supply in various forest ecosystems. Pol Ecol Stud 1:17–32

Bobek B, Weiner J, Zieliński J (1972) Food supply and its consumption by deer in a deciduous forest of Southern Poland. Acta theriol 17:187–202

Bobek B, Perzanowski K, Siwanowicz J, Zieliński J (1979) Deer pressure on forage in a deciduous forest. Oikos 32:373–380

Bocock KL, Gilbert O, Copstick CK, Twinn DC, Waid JS, Woodman MJ (1960) Changes in leaf litter when placed on the surface of soils with contrasting humus types I. Losses in dry weight of oak and ash leaf litter. J Soil Sci 11:1–9

Bohne H (1972) Klärung eines Rauchschadensfalles bei Kiefernbeständen im Ruhrgebiet. Mitt Forstl Bundes-Versuchsanst 97:141–150

Bolewski A, Michałek Z, Skawina T (1968) Application of morillonite clays as sorbents to improve the effectiveness of mineral fertilizers. Zesz probl Nauki Pol 84:209–216

Bolewski A, Fijał J, Kłapyta Z, Manecki A, Ptasińska M, Żabiński W (1976) Geochemical changes in natural environment within zones of industrial emission of fluorine compounds. Pr Miner 50:7–57

Bombówna M, Bucka H, Huk W (1978) The influence of towns in Southern Poland on the nutrient content in waters of the Carpathian rivers. Acta Hydrobiol 20:245–261

Bormann FH, Likens GE (1967) Nutrient cycling. Science 155:424–429

Bormann FH, Likens GE (1979) Pattern and process in a forested ecosystem. Springer, New York Heidelberg Berlin; 253 pp

Borowiec S, Skrzypczyński T, Kucharska T (1978) Migration of mineral compounds from the soils of the Szczecińska Lowland. Soc Sci Szczecin 47:1–68

Borowski M (1974) Increment of trees and tree stands (in Polish). PWRiL, Warszawa; 353 pp

Bowen HJM (1966) Trace elements in biochemistry. Academic Press, London, New York; 235 pp

Bratton SP (1974) The effect of the European wild boar (*Sus scrofa*) on the high-elevation vernal flora in Great Smoky Mountains Natl Park. Bull Torrey Bot Club 101:198–206

Braun-Blanquet J, Pallman H, Bach R (1954) Pflanzensoziologische und bodenkundliche Untersuchungen im Schweizerischen Nationalpark und seinen Nachbargebieten. II. Vegetation und Böden der Wald- und Zwergstrauchgesellschaften (*Vaccinio-Piceion*). Ergebn wiss Unters Schweiz Natn Parks 4:1–200

Bray JR (1963) Root production and the estimation of net productivity. Can J Bot 41:65–72

Brosset C (1978) Water-soluble sulphur compounds in aerosols. Atmos Environ 12:25–38

Brożek S (1979) Nitrogen mineralization in the soils of the Niepołomice Forest (in Polish). Ann techn Report Project MR-II/15; Inst Ecol Pol Acad Sci, Warszawa; 30–42 pp

Buchwald K, Engelhardt W (ed) (1968–69) Handbuch für Landschaftspflege und Naturschutz (Schutz, Pflege und Entwicklung unserer Wirtschafts- und Erholungslandschaften auf ökologischer Grundlage). Bayerischer Landwirtschafts Verlag, München; Bd. I–IV: 245, 502, 271, 252 pp

Budyko MI (1974) Changes in climate (in Russian). Gidrometeoizdat, Leningrad; 280 pp

Busse T, Treger R (1938) Fremdrassigkeit und Schlechtrassigkeit im ostdeutschen Kieferwalde. Z Forst Jagdw 4:177–194

Bzowski M (1973) The relief and hydrology of Vistula valley bottom in the region of Niepołomice Forest. Stud Nat Sci 7:7–37

Cabejszek J (1959) Summer seston of the Niemen river. Pol Arch Hydrobiol 5:8–28

Cajs L (1963) Studies on food composition in several species of owls (in Polish). Zesz nauk Uniw Pozn Biol 4:22–40

Cannell MGR (1982) World forest biomass and primary production data. Academic Press, New York; 448 pp

Chamberlain AC (1975) The movement of particles in plant communities. In: Monteith JL (ed) Vegetation and the Atmosphere. Academic Press, London; 155–203 pp

Chew RM (1974) Consumers as regulators of ecosystems: an alternative to energetics. Ohio J Sci 74:359–370

Chodzicki E (1933) Mikrobiologische Untersuchungen über den Einfluß der Holzarten Zusammensetzungs-Veränderung im Bestande auf den Zustand des Bodens. Kasa Mianowskiego, Warszawa; 298 pp

Chodzicki E (1934) Buchen-Beimischung in Kiefernbeständen, als Edaphischer Faktor auf Diluvialen, Sandigen Podsol- und Braunerden. Kasa Mianowskiego, Warszawa; 254 pp

Chodzicki E (1947) Forest physiographic regions in SW Poland (in Polish). Sylwan 1–4:32–82

Ciuła M (1979) Food habits of red fox (*Vulpes vulpes* L.) in the Niepołomice Forest (in Polish). MSc thesis – Jagiellon Univ, Kraków; 29 pp

Claeson A, Forsberg Å (1978) Algal assay procedure with one or five species, minitest. Mitt Int Ver Theor Angew Limnol 21:21–30

Clawson M (1975) Conflicts and strategies in forest land management. J Soil Water Conserv 63–67

Cormis de L (1970) Atmospheric pollution and plants (in Polish). Ochr Powietrza 5:10–13

Coughtrey PJ, Jones CH, Martin MH, Shales SW (1979) Litter accumulation in woodlands contaminated by Pb, Zn, Cd and Cu. Oecologia (Berl) 39:51–60

Crocker RL, Major J (1955) Soil development in relation to vegetation and surface age at Glacier Bay, Alaska. J Ecol 43:427–448

Cromack K Jr, Monk CD (1975) Litter production, decomposition, and nutrient cycling in a mixed hardwood watershed and white pine watershed. In: Howell FG, Gentry IB, Smith MH (eds) Mineral Cycling in Southeastern Ecosystems, ERDA Symp Seri 609–624 pp

Ćwikowa A (1974) Floristic and geographic characteristics of the distribution of the native elms (*Ulmus* spp.) in the state forest of Cracow (in Polish). PhD thesis – Acad Agr, Kraków; 121 pp

Ćwikowa A, Lesiński JA (1981) Floristic differentiation of communities of present forest vegetation of Niepołomice Forest. Studia Ośrodka Dokumentacji Fizjograficznej 9:159–196

Czarnowski M (1972) Ecophysiological methods for studies of plant photosynthetic production. Biul Warzywniczy 13 Suppl: 99–127

Czarnowski M (1973) Ecophysiological studies on the rating method for evaluation the photosynthetic production of leaves of selected plant species (in Polish). Pol Acad Sci Dep Plant Physiol Lab Photos, Kraków; 112 pp

Czarnowski M (1974) Photosynthetic production of tree leaves in a city habitat. Bull Acad Pol Sci Cl II 22:603–608

Czarnowski M (1975a) Studies on ecophysiological method for estimation of photosynthetic production in leaves of chosen plant species. Bull Acad Pol Sci Cl II 23:413–420

Czarnowski M (1975b) Photosynthesis and solar energy conversion. In: Petrusewicz K (ed) Polish participation in the International Biological Programme 1964–1973. Pol Acad Sci, Pol Nat Comm IBP, Warszawa; 175–181 pp

Czarnowski M (1975c) Potential photosynthesis of tree leaves polluted by industrial emissions. Bull Acad Pol Sci Cl II 27:605–612

Czarnowski M (1977a) Seasonal changes of photosynthetic rate in deciduous tree growing within the region of industrial emissions. Bull Acad Pol Sci Cl II 25:443–450

Czarnowski M (1977b) Ecophysiological estimation of photosynthetic production in tree leaves. In: Huzulak J, Masarovičova E (eds) Photosynthesis and water regime of woody plants. Proc Symp on Dendrophysiology, Modra-Piesky; 239–250

Czarnowski M (1978) Plant photosynthetic production in oak-hornbeam forest ecosystem. Stud Nat Sci A 14:165–190

Czarnowski M (1981) Ecology of photosynthesis Niepołomice Forest trees. Studia Ośrodka Dokumentacji Fizjograficznej 9:313–334

Czuraj M, Radwański B, Strzemeski S (1966) Tables of standing tree volume (in Polish). PWRiL, Warszawa

Dässler HG, Ranft H, Rehm KH (1972) Widerstandsfähigkeit von Gehölzen gegenüber Fluorverbindungen und Schwefeldioxid. Flora 161:289–302

Davis DD, Wood FA (1968) Relative sensitivity of twenty-two tree species to ozone. Phytopatology 58:p 399

Davis DD, Wood FA (1972) The relative susceptibility of eighteen coniferous species to ozone. Phytopatology 62:14–19
Davis DD, Wilhour RG (1976) Susceptibility of woody plants to sulfur dioxide and photochemical oxidants. Ecol Res S EPA-600, 3:76–102
Dengler A (1930) Waldbau auf ökologischer Grundlage. Springer Berlin
Dengler A (1944) Waldbau auf ökologischer Grundlage, 3rd edn. Springer Berlin; 596 pp
Denisiuk Z (1976 a) Meadows of the northern part of the Niepołomice Forest, Stud Nat Sci A 13:7–100
Denisiuk Z (1976 b) Les caricaies humides et inondables de la Grande-Pologne, leur écologie et leur valeur économique. Coll phytos 5:227–244
Denisiuk Z (1980) Sedge meadows of the Wielkopolska Region (*Phragmitetea* class). Stud Nat Sci A: 1–140
Denisiuk Z, Medwecka-Kornaś A (1976) Distribution of the plant communities and the potential natural vegetation in the northern part of the Niepołomice Forest. Stud Nat Sci A 13:171–195
Denisiuk Z, Dziewolski J, Ferchmin M, Medwecka-Kornaś A, Michalik S (1977) Vegetation map of the northern part of the Niepołomice Forest. Stud Nat Sci A 13:1–195
Derlaga A (1981) Effect of industrial pollution upon the rate of decomposition of organic matter (in Polish). MSc thesis – Jagiellon Univ, Kraków; 24 pp
Doktorowicz-Hrebnicki S, Różycki S (1949) General geological map of Poland (in Polish). Sect E 4, PIG, Kraków-Warszawa
Domański S (1978) Fungi occurring in forests injured by air pollutants in the Upper Silesia and Cracow Industrial Regions of Poland. VI Higher fungi colonizing the roots of trees in converted forest stand. Acta Soc Bot Pol 67:285–295
Dominik J (ed) (1977) Forest protection. PWRil, Warszawa; 531 pp
Douglas LA, Tedrow JCF (1959) Organic matter decomposition rates in arctic soils. Soil Sci 88:305–312
Dovland H, Joranger E, Semb A (1976) Deposition of air pollutants in Norway. In: Braeke FH (ed) Impact of acid precipitation on forest and freshwater ecosystems in Norway. Res Rep 6/76 of SNSF Project, NISK, Oslo-Ås; 15–35 pp
Drablos D, Tollan A (eds) (1980) Ecological impact of acid precipitation. Proc Int Conf Ecol impact acid precip, Sandefjord, Norway, SNSF project, Oslo-Ås; 383 pp
Drożdż A (1966) Food habits and food supply of rodents in the beech forest. Acta theriol 11:363–384
Drożdż A (1968) Digestibility and assimilation of natural foods in small rodents. Acta Theriol 13:367–389
Dubiel E (1971) Vascular plants of the northern part of the Niepołomice Forest. Stud Nat Sci A 6:13–52
Dubiel E (1973) Plant associations of the old Vistula beds in the region of Niepołomice Forest. Stud Nat Sci A 7:67–124
Dulka J, Risby TH (1976) Ultratrace metals in some environmental and biological systems. Anal Chem 48:640A–653A
Duvigneaud P (1967) L'ecologie science moderne de synthese. II Ecosystemes et biosphere. Min Educat Nation Cult, Bruxelles; 27–64 pp
Duvigneaud P, Denaeyer-De Smet S (1970) Biological cycling of minerals in temperate deciduous forests. In: Reichle DE (ed) Analysis of temperate forest ecosystems. Springer, Berlin Heidelberg New York, pp 199–225
Dworak TZ (1978) Investigations of atmospheric aerosol pollutions by the method of multi-spectral pictures. In: Teledetekcja Atmosferyczna – mat sesji naukowej, IMGW, Warszawa; 104–118 pp
Dynowska I (1971) Type of the rivers regimes of Poland. Zesz nauk Jagiellon Univ Pr Geogr 50:1–150
Dziewolski J (1974) Standing crop, increment and structure of the stand in oak-lime-hornbeam forest, *Tilio-Carpinetum* (International Biological Programme area). Stud Nat Sci A 9:7–28
Dziewolski J (1976) General outline of forest management history in the Niepołomice Forest. Stud Nat Sci A 13:101–106
Eaton JS (1977) The input of gaseous and particulate sulphur to forest ecosystem. In: Institute of Terrestrial Ecology workshop on methods in acid precipitation in forest ecosystems, ITE, Edinburgh; 1–12 pp
Eaton JS, Likens GE, Bormann FH (1980) Wet and dry deposition of sulfur at Hubbard Brook. In: Hutchinson TC, Havas M (eds) Effects of acid precipitation on terrestrial ecosystems, Plenum, New York London; 69–75 pp
Ehwald E (1957) Über den Nährstoffkreislauf des Waldes. Deut Akad Landw Wiss Sutz 6:1–56

Ellenberg H (1964) Stickstoff als Standortfaktor. Ber Dtsch Bot Ges 77:82–92

Ellenberg H (ed) (1971) Integrated experimental ecology. Ecological Studies 2. Springer, Berlin Heidelberg New York; 214 pp

Elton ChS (1966) The pattern of animal communities. Methuen, London, Wiley, New York; 432 pp

Enderlein H, Vogl M (1966) Emperimentelle Untersuchungen über die SO_2 Empfindlichkeit der Nadeln verschiedener Koniferen. Arch Forstwesen 15:1207–1224

Escherich K (1923) Die Forstinsekten Mitteleuropas, II Band. Parey, Berlin; 664 pp

Fabijanowski J, Jaworski A (1978) Forest and their protection. In: Michajłow W, Zabierowski K (eds) Protection and management of the natural environment. PWN, Warszawa-Kraków; 187–235 pp

Fabijanowski J, Lesiński JA (1978) Suggestions concerning reconstructing of forest stands (in Polish). In: Grodziński W, Lesiński JA (eds) Ecological principles for rebuilding the region of Niepołomice Forest. Jagiellon Univ, Kraków; 63–71 pp

Fabijanowski J, Lesińska JA (1980) Reconstruction of species composition of stands as a necessary condition for the survival of the Niepołomice Forest. In: Drablos D, Tollan A (eds) Proc Int conf ecol impact acid precip SNSF project, Oslo-Ås Norway; 164–165 pp

Fabijanowski J, Zarzycki K (1965) The Vegetation of the Świnia Góra forest reserve, Holy Cross Mountains, Central Poland. Acta Agrar Silvestria Ser Silvestris 5:61–103

FAO/IUFRO International Symposium on Forest Fertilization (1973) Ministre l'Agriculture, Service des Forets, Paris; 395 pp

Faulkner R (ed) (1975) Seed orchards. A joint production by specialist members of the IUFRO Working Party on Seed Orchards S2.03.3, Forestry Commission Bull 54:1–149, London; Her Majesty's Stationary Office

Ferchmin M (1976) Wet alder wood *Carici Elongatae-Alnetum* and *Alno-Padion* alliances in the northern part of the Niepołomice Forest. Stud Nat Sci A 13:107–142

Ferchmin M, Medwecka-Kornaś A (1976) Oak-hornbeam woods in northern part of the Niepołomice Forest. Stud Nat Sci A 13:143–169

Förstner U (1976) Schwermetallverunreinigung in Gewässern und ihre Erfassung durch Sedimentuntersuchungen. Polizei Technik Verkehr 1:25–32

Foster NW, Gessel SP (1972) The natural addition of nitrogen, potassium and calcium to a *Pinus banksiana* Lamb. forest floor. Can J For Res 2:448–455

Foster NW, Morrison IK (1976) Distribution and cycling of nutrients in a natural *Pinus banksiana* ecosystem. Ecology 57:110–120

Fruziński B, Łabudzki L, Wlazełko M (1975) Debarking impact upon silvicultural value of deciduous thickets. Pol Ecol Stud 1:51–60

Galloway JN, Likens GE, Edgerton ES (1976) Hydrogen ion in the acid precipitation of the North-Eastern United States. Water Air Soil Pollut 6:423–433

Garland JA, Atkins DHF, Readings CJ, Coughey SJ (1974) Deposition of gaseous sulphur dioxide to the ground. Atmos Environ 8:75–79

Gębczyński M, Górecki A, Drożdż A (1972) Metabolism, food assimilation and bioenergetics of three species of dormice (*Gliridae*). Acta theriol 17:271–294

Geiger R (1961) Das Klima der bodennahen Luftschicht. Viewig, Braunschweig; 327–331 pp

Glutz V., Blotzheim UN (1966) Handbuch der Vögel Mitteleuropas. Akad Verlagsges, Frankfurt; 1:483 pp

Glutz V, Blotzheim UN (1971) Handbuch der Vögel Mitteleuropas. Akad Verlagsges, Wiesbaden; 4:947 pp

Głowaciński Z (1975) Birds of the Niepołomice Forest (A faunistic ecological study). Acta Zool Cracov 20:1–87

Głowaciński Z (1978) Birds of the Lipówka nature reserve in the Niepołomice Forest. Stud Nat Sci A 17:169–189

Głowaciński Z (1981) Breeding avifauna of forests creating of an allogenic successional series in the Niepołomice Forest. Studia Ośrodka Dokumentacji Fizjograficznej 9:229–253

Głowaciński Z, Weiner J (1977) Energetics of bird communities in successional series of a deciduous forest. Pol Ecol Stud 3:147–175

Głowaciński Z, Weiner J (1980) Energetics of bird fauna in consecutive stages of a semi-natural pine forest. Ecol pol 28:71–94

Głowaciński Z, Weiner J (1983) Successional trends in energetics of the forest bird communities in Southern Poland. Hol Ecol 6:305–314

Golley FB, Ryszkowski L, Sokwi JT (1975) The role of small mammals in temperate forests, grasslands and cultivated fields. In: Golley FB, Petrusewicz K, Ryszkowski L (eds) Small mammals their productivity and population dynamics. Cambridge Univ Press, Cambridge; 223–243 pp

Górecki A (1965) Energy values of body in small mammals. Acta theriol 10:333–352

Górecki A (1968) Metabolic rate and energy budget in the bank vole. Acta theriol 13:341–365

Górecki A, Gębczyńska Z (1962) Food conditions for small rodents in a deciduous forest. Acta theriol 6:275–295

Górecki A, Szafranek I (1981) Biogen contents in the bodies of two species of rodents. Acta theriol 26:341–347

Gosz JR (1978) Nitrogen inputs to stream water from forests along an elevational gradient in New Mexico. Water Resour 12:725–734

Gosz JR, Likens GE, Bormann FH (1976) Organic matter and nutrient dynamics of the forest and forest floor in the Hubbard Brook Forest. Oecologia (Berl.) 22:305–320

Götz R (1978) Die Mobilisierung von Schwermetallen (Pb, Zn, Cu und Ni) im Wasser durch Chuminsäuren aus schwerlöslichen Bodenkörpern. Forum Staedte-Hyg 29:33–37

Götz R (1979) Die Mobilisierung von den Schwermetallen Blei, Zink, Nickel und Kupfer aus schwerlöslichen Bodenkörpern im Wasser durch Ligninsäure. Vom Wasser 52:145–151

Goszczyński J (1977) Connections between predatory birds and mammals and their prey. Acta theriol 22:399–430

Gotzmann J, Jabłoński B (1972) Our bird's nests (in Polish). PZWS Warszawa; 282 pp

Grabowski A (1978) Method for determination of the degree of pine stands damages. In: Grodzińska K (ed) Acidification of forest environment (Niepołomice Forest) caused by SO_2 emissions from steel mills, Project PL-FS-75 FG-PO-355 JB 22, Inst Bot Pol Acad Sci, Kraków; 9–13 pp

Grabowski A (1981 a) Generalized estimation on degree of endangering of Niepołomice Forest ecosystems by air pollution. Studia Ośrodka Dokumentacji Fizjograficznej 9:369–375

Grabowski A (1981 b) Malformations of crowns of Scots pine in Niepołomice Forest. Studia Ośrodka Dokumentacji Fizjograficznej 9:357–367

Graczyk R (1974) The nourishment composition of hollow birds nestlings in the gradation region of pine lappest moth (*Dendrolimus pini* L.) in Wyszyna forest district. Roczn Akad Roln Poznań 65:29–42

Greszta J (1975) The influence of emissions on forest habitats and pine stands in the Cracow-Silesian industrial region. Stud Nat Sci A 10:1–183

Greszta J, Braniewski S, Nosek A (1982 a) The sensitivity of the seedlings of *Pinus silvestris, Pinus nigra, Picea abies, Larix decidua* and *Alnus glutinosa* to dusts from 10 industrial works introduced into the soil. Fragm florist Geobot 28:53–65

Greszta J, Braniewski S, Nosek A (1982 b) The effects of dusts from different emittors on the height increment of the seedlings of selected tree species. Fragm florist Geobot 28:67–75

Greszta J, Olszowski J, Godzik S (1969) The effect of air pollution on wood volume increment in the common pine (*Pinus silvestris*). Ekol pol A 17:827–846

Grier CC, Cole DW, Dyrness CT, Frederiksen RL (1974) Nutrient cycling in 37 and 450-year-old Douglas-fir ecosystems. In: Waring RH, Edmonds RL (eds) Integrated Research in the Coniferous Forest Biome, Conif Forest Biome Bull Univ Washington, Seattle; 1:21–34

Grochowski J (1973) Forest Mensuration (in Polish). PWRiL, Warszawa; 593 pp

Grodzińska K (1971) Acidification of tree bark as a measure of air pollution in Southern Poland. Bull Acad Pol Sci Cl II 19:189–195

Grodzińska K (1976) Acidity of tree bark as a bioindicator of forest pollution in Southern Poland. In: Dochinger LS, Seliga TA (eds) Proc 1st Int Symp Acid Precipitation. Columbus, Ohio; 905–911 pp

Grodzińska K (1977) Changes in the forest environment in Southern Poland as a result of steel mill emissions. In: Miyawaki A, Tüxen R (eds) Vegetation Science and Environmental Protection. Proc Int Symp Tokyo, Protec Environm Exc Veget Sci Japan 1974, Maruzen Co, Tokyo; 207–215 pp

Grodzińska K (1978 a) Acidity of tree bark as a bioindicator of forest pollution in Southern Poland. Water Air Soil Pollut 7:3–7

Grodzińska K (1978 b) Mosses as bioindicators of heavy metal pollution in Polish National Parks. Water Air Soil Pollut 9:83–97

Grodzińska K (1978c) Acidification of the Niepołomice Forest ecosystem under exposure to SO_2 emissions (in Polish). Ann techn Report Project 10.2.10: Inst Ecol Pol Acad Sci, Warszawa; 35 pp

Grodzińska K (1979) Tree bark – sensitive biotest for environment acidification. Environm Internat 2:173–176

Grodzińska K (1980) Acidification of forest environment (Niepołomice Forest) caused by SO_2 emissions from steel mills. Final Report, Project PL-FS-75 FG-PO-355 JB 22 Inst Botany Pol Acad Sci, Cracow; 143 pp

Grodzińska K (1982a) Plant contamination caused by urban and industrial emissions in the region of Cracow city (Southern Poland). In: Bornkamm R, Lee JA, Seaward MRS (eds) Urban Ecology The Second European Ecological Symposium (1980). Blackwell, Oxford–London–Edinburgh–Boston–Melbourne; 149–160 pp

Grodzińska K (1982b) Monitoring of air pollutants by mosses and tree bark. In: Steubing L, Jäger HJ (eds) Monitoring of air pollutants by plants – Methods and Problems. Proc Intern Workshop (1981). Junk, The Hague Boston London; 33–42 pp

Grodzińska K, Lesiński JA, Zając KP (1978) Actual state and air pollution stress on the Niepołomice Forest (in Polish). In: Grodziński W, Lesiński JA (eds) Ecological principles for rebuilding the region of Niepołomice Forest. Jagiellon Univ Kraków; 9–27 pp

Grodzińska K, Grodziński W, Zeveloff SI (1983) Contamination of roe deer forage in a polluted forest of Southern Poland. Environ Pollut A 30:257–276

Grodziński W (1968) Energy flow through a vertebrate population. In: Grodziński W, Klekowski RZ (eds) Methods for ecological bioenergetics. Pol Acad Sci, Warszawa-Kraków; 239–252 pp

Grodziński W (1975) Energy flow through a vertebrate population. In: Grodziński W, Klekowski RZ, Duncan A (eds) Methods for ecological bioenergetics. Blackwell, Oxford, London, Edinburgh, Melburne; 65–94 pp

Grodziński W (1978) Functioning and management of forest ecosystems in an industrial region of Southern Poland. Abstracts Lectures 2nd Int Congr Ecology, Jerusalem; 3:p 6

Grodziński W, French NR (1983) Production efficiency in small mammal populations. Oecologia (Berl) 56:41–49

Grodziński W, Górecki A (1967) Daily energy budgets of small rodents. In: Petrusewicz K (ed) Secondary productivity of terrestrial ecosystems. PWN, Warszawa; 295–314 pp

Grodziński W, Lesiński JA (1978) Ecological principles for rebuilding the region of the Niepołomice Forest (in Polish). Jagiellon Univ Kraków; 80 pp

Grodziński W, Bobek B, Drożdż A, Górecki A (1970) Energy flow through small rodent populations in a beech forest. In: Petrusewicz K, Ryszkowski L (eds) Energy flow through small mammal populations. PWN, Warszawa; 291–298 pp

Grodziński W, Sawicka-Kapusta K (1970) Energy values of tree-seeds eaten by small mammals. Oikos 21:52–58

Grodziński W, Wunder BA (1975) Ecological energetics of small mammals. In: Golley FB, Petrusewicz K, Ryszkowski L (eds) Small mammals – their productivity and population dynamics. Cambridge Univ Press, Cambridge; 173–204 pp

Grodziński W, Yorks TP (1981) Species and ecosystem level bioindicators of airborne pollution: an analysis of two major studies. Water Air Soil Pollut 16:33–53

Grodziński W, Pucek Z, Ryszkowski L (1966) Estimation of rodent numbers by means of prebaiting and intensive removals. Acta theriol 1:297–314

Gruszczyk A (1981) Soils of Niepołomice. Studia Ośrodka Dokumentacji Fizjograficznej 9:71–88

Gruszczyk A, Janowicz C, Krynicki R (1976) Soil and habitat survey of the Niepołomice Forest (MS in Polish). Biuro Urządzeń Lasu OZLP. Kraków; 122 pp

Guderian R (1977) Air pollution, phytotoxity of acid gases and its significance in air pollution control. Springer Berlin, Heidelberg, New York; 127 pp

Guderian R, Stratmann H (1968) Freilandversuche zur Ermittlung von Schwefeldioxidwirkungen auf die Vegetation. III. Grenzwerte schädlicher SO_2-Immissionen für Obst- und Forstkulturen sowie für landwirtschaftliche und gärtnerische Pflanzenarten. Forsch Ber Landes Nordrhein-Westfalen Nr 1920, Westdeutscher Verlag, Köln Opladen

Gumiński R (1951) Meteorology and climatology for agriculturists (in Polish). PWRiL, Warszawa; 250 pp

Haber A (1969) Wild boar – biology and management (in Polish). PWRiL, Warszawa; 216 pp

Haber A, Pasławski T, Zaborowski S (1975) Wildlife management (in Polish). PWN, Warszawa; p 369

Hanley DP (1976) Tree biomass and productivity estimated for three habitat types of Northern Idaho. Univ Idaho Forest Wildl and Range Exp Stat Bull 14:1–15

Harper HA (1969) Principles of physiological chemistry (in Polish). PZWL, Warszawa; 663 pp
Hájková A, Květ J (1970) Analysis of primary productivity in two types of inundated meadows. In: Dykyjova D (ed) Productivity of terrestrial ecosystems. Production processes. Czechosl. IBP/PT-PP Rep No 1, Praha: 47–50
Haut van H (1975) Kurzzeitversuche zur Ermittlung der relativen Phytotoxität von Stickstoffdioxid. Staub Reinhalt Luft 35:187–193
Hawksworth DL, Rose F (1970) Quantitative scale for estimating sulphur dioxide air pollution in England and Wales using epiphytic lichens. Nature 227:145–148
Hayward GF, Phillipson J (1979) Community structure and functional role of small mammals in ecosystems. In: Stoddart DM (ed) Ecology of small mammals. Chapman and Hall, London; 136–211 pp
Heinroth O, Heinroth M (1927) Die Vögel Mitteleuropas in allen Lebens- und Entwicklungsstufen photographisch aufgenommen und in ihrem Seeleben bei der Aufzucht vom Ei ab beobachtet. Hugo Bermühler, Berlin; 160 pp
Hejmanowski S, Antosiewicz Z, Czapiewski S, Kocięcki S, Wolski R (1975) Program of seed management and introduction of forest genetics achievements into practice as employed in the state forests over the years 1975–1980 (in Polish). Naczelny Zarząd Lasów Państwowych, Warszawa
Henderson GS, Swank WT, Waide JB, Grier CC (1978) Nutrient budgets of Appalachian and Cascade region watersheds: a comparison. For Sci 24:385–397
Herbichowa M (1969 a) Primary production of a potato field. Ekol pol A 17:73–86
Herbichowa M (1969 b) Primary production of a ryefield. Ekol pol A 17:343–350
Hess M (1965) Vertical climatic zones in the Polish Western Carpathians. Zesz nauk Jagiellon Univ Prace Inst Geogr 33:1–276
Hess M (1969) The climate of the Cracow subregion. Folia geogr S geogr phys 3:5–63
Hicks DR (1978) Diagnosis vegetation injury caused by air pollution. Environmental Protection, Washington DC; 182 pp
Highton NH, Chadwick MJ (1982) The effects of changing patterns of energy use on sulfur emissions and depositions in Europe. Ambio 11:324–329
Hornbeck JW, Likens GE, Eaton JS (1976) Seasonal patterns in acidity of precipitation and their implications for forest stream ecosystems. In: Dochinger LS, Seliga TA (eds) Proc 1st Int Symp Acid Precipitation. Columbus, Ohio; 597–609 pp
Hryniewicz R (1973) Determination of ammonia and ammonium ions in the atmosphere. Ochrona Powietrza 4:106–109
Huber-Pestalozzi G (1955) Das Phytoplankton des Süßwassers. 4. Euglenophyceen. In: Thienemann A (ed) Die Binnengewässer. Schweizerbarth, Stuttgart; p. 114, p. 605
Hughes MK, Lepp NW, Phipps DA (1980) Aerial heavy metal pollution and terrestrial ecosystems. In: Macfadyen A (ed) Advances Ecol Research, Academic Press London New York; 11:217–314 pp
Huhtala K, Sulakava S (1976) Breeding biology of goshawk (in Finnish). Suomen Luouto 35, 6:299–303, 374
Hutchinson TC, Havas M (eds) (1980) Effects of acid precipitation on terrestrial ecosystems. NATO Conf Series I Ecology Plenum, New York London; 654 pp
Huttunen S (1978) The effects of air pollution on provenances of Scots pine and Norway spruce in Norway spruce in northern Finland. Silva Fenn 12:1–16
Huttunen S, Törmälenhto H (1982) Air pollution resistance of some Finnish *Pinus silvestris* L. provenances. Aquilo Ser Bot 18:1–9
Inman JC, Parker GR (1978) Decomposition and heavy metals dynamics of forest litter in northwestern Indiana. Environ Pollut A 17:39–51
International conference Heavy metals in the environment – Amsterdam – September 1981 (1981) Comm Europ Communities, WHO, CEP Consultants, Edinburgh; 732 pp
International Symposium Sulphur in the Atmosphere (1978) Atmos Environ 12:1–16
Iwanek J (1980) Numerical model of the effect of plants on air pollution conditions. In: Papers presented Symp Effects of Air-Borne Pollution on Vegetation, Warszawa; p 295
Izdebski K, Popiołek Z (1976) Productivity of undergrowth in selected forest communities of Central Roztocze. Ann Univ Mariae Curie-Skłodowska C 31:55–67
Izmaiłow B (1975) Origin and age of the sand dunes in the Niepołomice Forest. Folia geogr S geogr phys 9:43–61

Jackson DR, Watson AP (1977) Disruption of nutrient pools and transport of heavy metals in a forested watershed near a lead smelter. J Environ Qual 6:331–338

Janert B (1961) Lehrbuch der Bodenmelioration. VEB Verlag für Bauwesen, Berlin; 31–97 pp

Jankowska K (1967) Seasonal changes of vegetation and net primary production of the fresh meadow *Arrhenatheretum elatoris*. Stud Nat Sci A 1:153–173

Jankowska K (1971) Net primary production during a three-year succession on an unmowed meadow of the *Arrhenatheretum elatioris* plant association. Bull Acad Pol Sci Cl II 19:709–794

Jankowska K (1975) The ecology and primary production of the fresh meadow in the Ojców National Park and the xerothermic grassland in the Skowronno steppe reserve near Pińczów. Stud Nat Sci A 11:1–79

Jankowska K (1978) Ground flora net production in an oak-hornbeam stand in the Niepołomice Forest, estimated by means of a simplified method. Stud Nat Sci A 14:135–144

Janota-Bassalik L, Kermen J (1978) Changes in the microflora of forest soils irrigated some kind of pollution with organic compounds. Zesz probl Postęp Nauk roln 204:121–131

Jaskowski J (1956) Mammals of old valleys of Barycz and Orla (in Polish). Pr Kom Biol PTPN 18:151–168

Jeffree CE (1980) Plant damage caused by SO_2. Papers presented to the Symposium on the Effects of Air-Borne Pollution on Vegetation, Warszawa; 328–354

Jelonek M (1977) Population structure and dynamics of the wild boar in Niepołomice Forest and their influence upon agricultural crops (in Polish). MSc thesis – Jagiellon Univ, Kraków; 24 pp

Jensen K, Dochinger LS, Roberts BR, Townsend AM (1976) Pollution Responses. In: Miksche JP (ed) Modern Methods in Forest Genetics. Springer, Berlin Heidelberg New York; 189–216 pp

Jensen B, Sequeira DM (1978) The diet of the red fox (*Vulpes vulpes* L.) in Denmark. Dan Rev Game Biol 10:1–16

Jezierski W, Myrcha A (1975) Food requirements of a wild boar population. Pol Ecol Stud 1:61–84

Johnson FL, Risser PG (1974) Biomass, annual net primary production, and dynamics of six mineral elements in a post oak-blackjack oak forest. Ecology 55:1246–1258

Johnson PL, Swank WT (1973) Studies on cation budgets in the Southern Appalachians on four experimental watersheds with contrasting vegetation. Ecology 54:70–80

Jop K (1979) Quality evaluation of roe deer antlers from an industrial region in Southern Poland. Acta theriol 24:23–34

Jop K (1981) Ecology of the forest stream Lane Błoto in the Niepołomice Forest. 1. Chemism of water and bottom sediments and its changes under the influence of industrial pollution. Acta Hydrobiol 23:107–123

Jordan HJ, Lechevalier MP (1975) Effect of zinc smelter emission on forest soil microflora. Can J Microbiol 21:1855–1866

Kabawa RR, Kelley WD, Curl EA (1978) Proteolytic activity of *Trichoderma virida* in mixed culture with *Sclerotium folfsii* in soil. Can J Microbiol 24:486–490

Karkanis M (1973) Woodland and meadow soils in the northern part of Niepołomice Forest and its conservation II. Stud Natur Sci A 7:39–65

Karkanis M (1974) Seasonal changes of soil components in relation to nutrient circulation in oak-hornbeam forest. Fragm Florist Geobot 20:477–496

Karkanis M (1975) Decomposition of litter of various species of deciduous trees and its effect on soil environment. Fragm Florist Geobot 21:71–97

Karkanis M (1976) The circulation of sulphur in the forest ecosystem *Tilio-Carpinetum* in the northern part of Niepołomice Forest. Fragm Florist Geobot 22:351–363

Karnosky DF, Steiner KC (1981) Provenance and family variation in response of *Fraxinus americana* and F. *pennsylvanica* to ozone and sulfur dioxide. Phytopathology 71:804–807

Kasina S (1971) Distribution of atmospheric pollutants in the region of the Niepołomice Forest. Ochrona Powietrza 5:5–8

Kasina S (1977 a) Methods in wet deposition research of sulphur compounds in the region of Cracow. In: Institute of Terrestrial Ecology workshop on methods in acid precipitation in forest ecosystems, ITE, Edinburgh; 1–7 pp

Kasina S (1977 b) Preliminary results of the investigations of aerosols, atmospheric sediments and precipitation acidity in the region of Cracow. In: Int symp on meteorological aspects of atmospheric pollution, Voeikov Main Geophys Obs, Leningrad; 1–9 pp

Kasina S (1978 a) Sulphur budget in the region of Cracow. Időjárás 6:320–324

Kasina S (1978 b) Precipitation acidity in the Cracow region. In: Central Electric Res Lab workshop on ecological effects of acid precipitation. CERL, Leatherhead; CP2:1–4

Kasina S, Kwiek J, Lewińska J, Bzowski M, Rapacz J, Walat W, Zgud K (1978) Air pollution in the Niepołomice Forest. In: Grodzińska K (ed) Acidification of forest environment (Niepołomice Forest) caused by SO_2 emissions from steel mills, Project PL FS75-FG PO 355 JB 22 Inst Botany Pol Acad Sci Cracow; 9–13 pp

Katz M (1967) Guide to the selection of methods for measuring air pollutants. WHO, Geneva; 49–46 pp

Kawecka B (1964) Communities of algae in the lower part of the Rogoźnik stream. Acta Hydrobiol 6:119–128

Kaźmierczak T (1978) The *Schneumonidea* (*Hymenoptera, Ichneumonidae*) of the Lipówka Nature Reserve in the Niepołomice Forest. Stud Nat Sci A 17:207–221

Keller Th (1964) Industriale Immissionen und Forstwirtschaft. Schweiz Z Forstwes 4:228–255

Keller T (1973) Enge Wechselbeziehungen zwischen Wald und Luft – Gewässerschutz. Lufthygiene 97:1–4

Kemp ALW, Thomas RL (1976) Impact of man's activities on the chemical composition in the sediments of Lake Ontario, Erie and Huron. Water Air Soil Pollut 5:469–490

Kendeigh MS, Dolnik VR, Gavrilov VM (1977) Avian energetics. In: Kendeigh MS, Pinowski J (eds) Granivorous birds in ecosystems. Cambridge Univ Press, Cambridge London New York Melbourne; 127–204 pp

Kercher JR, Axelrod MC (1980) Modelling the impact of SO_2 pollution on the community structure and succession in a coniferous forest of the western. U.S.A. In: Drablos D, Tollan A (eds) Proc Int Conf Ecol impact acid precip, Sandefjord, Norway; 174–155 pp

Khangarot BS (1981) Effect of complexing agents on the acute toxicity of copper to common guppy *Lebistes reticulatus* (Peters). Acta Hydrochim-Hydrobiol 5:505–509

Khangarot BS, Durve VS, Raibamshi VK (1981) Toxicity interactions of zinc-nickel, copper-nickel and zinc-nickel-copper to a freshwater teleost, *Lebistes reticulatus* (Peters). Acta Hydrochim-Hydrobiol 5:495–503

Kiciński T (1964) Share of the groundwaters in the total outflow of rivers (in Polish). Gospod wodna 5:173–185

Kiszka J (1964) The lichens of the Sandomierz Lowland, I Lichens of Niepołomice Forest district. Fragm Florist Geobot 10:527–564

Kiszka J (1977) The effect of town and industry emissions on the lichen flora of Cracow and the Niepołomice Forest. Pr monogr WSP, Kraków; 19:1–133

Kiszka J (1981) Lichens flora of Niepołomice Forest. Studia Ośrodka Dokumentacji Fizjograficznej 9:335–356

Kleiber M (1961) The fire of life – an introduction to animal energetics. Wiley and Sons, New York; 454 pp

Klein J (1977) Licht changes in the interior of an oak-hornbeam stand and their relation to the development of leaves and trees and shrubs in the northern part of the Niepołomice Forest. Fragm Florist Geobot 23:201–214

Klein J (1978) The local climate of Vistula valley in the northern part of Niepołomice Forest. Stud Nat Sci A 14:9–66

Klotz RL, Matson EA (1978) Dissolved organic carbon fluxes in the Shetucket River of eastern Connecticut USA. Freshwater Biol 8:347–355

Knabe W (1967) Methoden der Auslese und Züchtung immissionsresistenter Gehölze. XIV IUFRO – Kongreß München 1967, Sektion 24:1–19

Knabe W (1972) Immissionsbelastung und Immissionsgefährdung der Wälder im Ruhrgebiet. Mitt Forstl Bundesversuchsanst 97:53–87

Knabe W (1977) The role of tree stands for reducing air pollution. Proc 4th Int Clean Air Congr, Tokyo; 952–957 pp

Knabe W (1979) Capacity and efficiency of vegetation in reducing air-borne pollution in urban and industrial areas. Papers presented Symp Effects of Air-Borne Pollution on Vegetation; Warsaw; 278–294 pp

Koehler W (1978) An introduction to hylopathology (in Polish). PWN, Warszawa; 406 pp

Kondratiuk EN, Tarabin VP, Baklanov VI, Burda RI, Kharkhota AI (1980) Industrial Botany (in Russian). Naukova Dumka, Kiev; 260 pp

Kononowa MM (1963) Organic matter of soils, structure, propriety and methods of research. Izd Akad Nauk SSSR, Moskva; 100–107 pp

Kosinskaja V (1960) Cryptogamic Flora of Soviet Union 5. *Conjugatae* 2. *Desmidiales* 1. Izd Akad Nauk SSSR, Moskva-Leningrad; 706 pp

Kotańska M (1967) Biomass dynamics of underground plant organs in some grassland communities of the Ojców National Park. Bull Acad Po Sci Cl II 15:625–631

Kotańska M (1970) Morphology and biomass of the underground organs of plants in grassland communities of the Ojców National Park. Stud Nat Sci A 4:1–109

Kotańska M (1973) Productivity of underground plant organs on some meadow and marsh communities of the Niepołomice Forest (Southern Poland). Bull Acad Pol Sci Cl II 21:555–560

Kotańska M (1975) Seasonal changes of biomass of underground plant organs in some meadow communities. Zesz Nauk Uniw Jagiellonsk Pr bot 3:23–47

Kowalczewski A, Wasilewski L (1966) Differentiation of biomass of *Phragmites communis* Trin. and its production in Mikołajskie Lake. Bull Acad Pol Sci Cl II 14:219–223

Kozłowski J, Górecki A, Bobek B (1981) Evaluation of computer models in the estimation of energy flow through rodent populations. Bull Acad Pol Sci Cl II 28:497-503

Królikowski L, Kowalkowski A (1977) Effect of hydro-phytomeliorations on the content of readily soluble components in podsolic soils and on the chemical composition of Scots pine (*Pinus silvestris* L.) needles (in Polish). Postepy techniki w leśn 29:49–58

Krupa J (1876) List of plant species collected in Cracow Grand Duchy and in the Niepołomice Forest in the year 1876 (in Polish). Spraw Komis Fizjograf PAU 11:84–128

Kruuk H (1978) Foraging and spatial organisation of the European badger, *Meles meles* L. Behav Ecol Sociobiol 4:75–89

Krzysik F (1974) Knowledge of timber (in Polish). PWRiL, Warszawa; 403–427 pp

Krzywański D, Samosiej L (1976) Differentiation of biomass in aquatic communities in the old-river beds of Pilica river. Acta Universit Lodziensis, Fol bot 2:189–206

Kulczycki A (1964) Studies on food composition of owls from Beskid Niski (in Polish). Acta zool Cracov 9:529–559

Květ J, Ondok JP, Nečas K, Jarvis PG (1971) Methods of growth analysis. In: Šestak Z, Čatsky J, Jarvis PG (eds) Plant photosynthetic production. Manual of methods. Junk The Hague; 343–391 pp

Lacasse NL, Treshow M (eds) (1978) Diagnosing vegetation. Injury caused by air pollution. US Environ Prot Agency, Research Triangle Park NC; 1–4, 11 pp, Appendix C, 1-C, 28

Lamb D, Florence RG (1975) Influence of soil type on the nitrogen and phosphorus content of radiata pine litter. NZ J For 5:143–151

Lambor J (1962) Methods of hydrological forecasts (in Polish). Wyd Komunikacji i Łączności, Warszawa; 260–270 pp

Lambor J (1971) Engineering Hydrology (in Polish). Arkady, Warszawa; 324 pp

Lampadius F (1965) Die Gefährdung der Industriebegrünung durch phytotoxische Emissionen. Tech Univ Dresden Fakult f Forstw, Tharandt; 17–19 pp

Lampadius F (1969) Der Burkhardtwald bei Aue als klassisches Beispiel waldbaulicher Rauchschadenabwehr. Akademia Ost Berlin 50,3. From review in Arch Forstw (1970) 19:204–206

Lear van DH, Goebel NB (1976) Leaf fall and floor characteristics in loblolly pine plantations in the South Carolina, Piedmont. Soil Sci Soc Am J 40:116–119

Lee JJ, Inman DL (1975) The ecological role of consumers: an aggregated systems view. Ecology 56:1455–1458

Leipnitz W (1970) Die Wirkung der Klarwasserberegnung bei der Steigerung der Ertragsfähigkeit und Ertragssicherheit von Dauerweiden in der Altmärkischen Wische bei Anwendung von Stickstoffdünger. Z Landeskult 11:29–44

Lemke J (1973) Quantitative characterization of needles and branches bearing needles in younger pine stands. Folia for Pol Ser A Lesn 21:173–197

Lemke J (1978 a) Quantitative characteristics of foliage and twigs with needles in old pine stands. Folia for Pol Ser A Lesn 23:53–66

Lemke J (1978 b) Variabilität des Nadelgewichtes in Astquirlen der Kiefern verschiedener Altersklassen. Sylwan 2:13–21

Lesiński JA (1974) Protective forests (in Polish). Aura 1:6–8

Lesiński JA (1981 a) Losses on timber production in the Niepołomice Forest. Studia Ośrodka Dokumentacji Fizjograficznej 9:377–383

Lesiński JA (1981 b) Diminution of protective and recreation values of the Niepołomice Forest. Studia Ośrodka Dokumentacji Fizjograficznej 9:393–400

Lesiński JA (1982) Functions of forests in Poland. Studia Ośrodka Dokumentacji Fizjograficznej 10:201–213

Lesiński JA (1983) Effects of acid precipitation on forests in Poland: selected topics. Water Quality Bull 8:156–159 and 169–170

Lesiński JA; Suliński J (1982) Water- and climate-protective forests in the Niepołomice Forest. Studia Ośrodka Dokumentacji Fizjograficznej 10:214–222

Lewińska J (1977) Climatic problems of town-industrial agglomerations. Człowiek i Środowisko 1:5–33

Lewińska J, Rutkowski D (1977) The influence of air pollution on the loss of solar energy. In: Lashihar AB (ed) Proc of the Conf on the phys of solar energy. Arab Dev Inst, Tripoli; 73–80 pp

Lewis WM, Canfield D (1977) Dissolved organic carbon in some dark Venezuelan waters and a revised equation for spectrophotometric determination of dissolved organic carbon. Arch hydrobiol 79:441–445

Liberti A, Passanzini M, Vicedomini M (1972) The determination of the non-volatile acidity of rain water by a coulometric procedure. Analyst 97:352–356

Lihačev GN (1956) Some remarks on the ecology of badger in broad-leaved forests of the Tula region. In: Jurgenson PB (ed) Sbor Mat Rez Izuč Mlekop Gos Zap Moskva; 14:123-141

Lihnell D (1968) Sulphate contents of tree leaves as an indicator of SO_2 air pollution in industrial areas. Air Pollution, Proc I Eur Congress on the Influence of Air Pollution on Plants and Animals, Wageningen; 341–352 pp

Likens GE, Bormann FH, Johnson NM, Pierce RS (1967) The calcium, magnesium, potassium, and sodium budgets in a small forested ecosystems. Ecology 48:772–785

Likens GE, Bormann FH, Pierce RS, Eaton JS, Johnson NM (1977) Biogeochemistry of a forested ecosystem. Springer, New York Heidelberg Berlin; 146 pp

Lindberg SE, Harriss RC (1981) The role of atmospheric deposition in an eastern United States deciduous forest. Water Air Soil Pollut 16:13–31

Linzon SN, McIlveen WD, Temple PJ (1973) Sulphur dioxide injury to vegetation in the vicinity of a sulphite pulp and paper mill. Water Air Soil Pollut 2:129–134

Lipa JJ, Madziara-Borusiewicz K (1976) Microsporidians parasiting the green tortrix (*Tortrix viridiana* L.) in Poland and their role in the collapse of the tortrix outbreak in Niepołomice Forest during 1970–1974. Acta protozool 15:529–536

Łoziński W (1935) The Niepołomice Forest soils. Pr rol-leśne PAU 17:1–21

Luckey TD, Venugopal B (1977) Metal toxicity in mammals. 1 – Physiologic and chemical basis for metal toxicity. Plenum, New York London; 238 pp

Mackiewicz M (1954) The space distribution of insolation in Poland (in Polish). Przegl met 7:5–15

Maczek W (1977) Photosynthesis production of *Pinus silvestris* L. in the Niepołomice Forest within the range of industrial emission. Bull Acad Pol Sci Cl II 25:685–693

Maddars M, Lawrance MIG (1982) The role of woodland in air pollution control. Q J For 32:257–261

Madziara-Borusiewicz K (1984) Heavy metals in the food (*Pinus silvestris* L.) and in the organism and excrements of its consumers *Blastophagus piniperda* L. in the Niepołomice Forest near Cracow (South Poland). Z Angew Entom (in press)

Madziara-Borusiewicz K, Strzelecka H (1977) Conditions of spruce (*Picea excelsa* L.) infestation by the engraver beetle (*Ips typographus* L.) in mountains of Poland. I. Chemical composition of volatile oils from healthy trees and those infested with the honey fungus [*Armillaria mellea* (Vahl.) Quel.]. Z Angew Entomol 83:409–415

Makomaska M (1978) Heavy metals contamination of pinewoods in the Niepołomice Forest (Southern Poland). Bull Acad Pol Sci Cl II 26:679–685

Mamajew SA, Skharlat OD (1972) Effects of air and soil pollution by industrial waste on the fructification of Scots pine in the Urals. 7th Intern Arbeitstagung Forstl Rauchschadensachverständiger Essen (1970). Mitt der Forstl Bundes-Versuchsanstalt 97:443–450

Manecki A (1976) Aeromineralogy – mineralogy of atmospheric dust. Miner Polon 7:91–97

Manecki A (1978) The classification and mineralogical composition of atmospheric dusts. Pr Mineral 57:5–17

Manecki A, Skowroński A (1970) Extraterrestrial spherules from Carboniferous montmorillonite clays. Bull Acad Pol Sci S Sci Géol Géogr 18:69–73

Manecki A, Skowroński A (1976) Uranium contents of some industrial dusts emitted to atmosphere. Pr Mineral 50:65–71

Manecki A, Kłapyta Z, Schejbal-Chwastek M, Skowroński A, Tarkowski J, Tokarz M (1981) The effect of industrial pollutants of the atmosphere on the geochemistry of natural environment of the Niepołomice Forest. Pr Mineral 71:1–58

Manecki A, Skowroński A, Schejbal M (1981) Dust separated from ice of Broggi Glacier. Pr Mineral 64:47–65

Margalef R (1960) Idea for a synthetic approach to the ecology of running waters. Int Rev Gesamten Hydrobiol 45:132–153

Markiewicz M (1978) Food of tits in Niepołomice coniferous Forest (in Polish). MSc Thesis – Jagiellon Univ, Kraków; 18 pp

Marvan P (1979) Algal assays – an introduction into the problem. In: Marvan P, Přibil S, Lhotsky O (eds) Algal assays and monitoring eutrophication, Schweizerbart'sche Verlagsbuchhandlung, Stuttgart; 17–22 pp

Materna J (1980) Effects of pollution on the capacity of vegetation to perform such functions as water retention, soil protection, wildlife habitat, ect. In: Papers presented Symp Effects of Air-Borne Pollution on Vegetation, Warsaw; 242–252

Matusiak K, Dusoge I, Kocieszczenko I, Bieszkiewicz E (1976) Studies of purification of waste water from nitrogen fertilizer industry by intensive algal cultures. III. Growth of *Chlorella vulgaris* and utilization of nitrogen in waste from the nitrogen fertilizer industry. Acta microbiol Pol 25:375–382

Matuła H (1977) Influence of snow depth on roe-deer diet in Niepołomice Forest (in Polish). MSc thesis – Jagiellon Univ, Kraków; 22 pp

Mattson WJ, Addy ND (1975) Phytophagous insects as regulators of forest primary production. Science 190:515–522

Mayer R, Ulrich S (1976) Acidity of precipitation as influenced by the filtering of atmospheric sulphur and nitrogen compounds – Its role in the element balance and effect on soil. In: Dochinger LS, Seliga TA (eds) Proc 1st Int Symp Acid Precipitation. Columbus, Ohio: 737–744 pp

Measurement of air pollution. Recommended methods for deposited matter, smoke, and sulphur dioxide (1957). Dep Scient Ind Res, London; 14–20 pp

Medwecka-Kornaś A, Łomnicki A, Bandoła-Ciołczyk E (1974) Energy flow in the oak-hornbeam forest (IBP Project "Ispina"). Bull Acad Pol Sci Ci II 22:563–567

Mejnartowicz L (1981) Application of biochemical genetics in the selection of trees more resistant to the action of sulphur dioxide. Second Ann Report I-MOA-USDA-11; PL-FS-86. Institute of Dendrology, Pol Acad Sci, Kórnik near Poznań; 15–38

Mészárosz E, Várhelyi C (1977) An attempt to estimate the continental sulphur emission on the basis of atmospheric measurements. Atmos Environ 11:169–172

Metcheva R (1979) Parameters of tawny owl bioenergetics (*Strix aluco* L.) (in Bulgarian). Biul 10th Konf Twór Nauk Techn Sofia; 43–47 pp

Mikola P (1960) Comparative experiment on decomposition rates of forest litter in southern and northern Finland. Oikos 11:161–166

Milczanowski Z (1980) Estimates of numbers and total consumption of red fox population in Niepołomice Forest (in Polish). MSc thesis – Jagiellon Univ, Kraków; 27 pp

Molchanov AA (1971) Productivity of organic mass in the forests at different zones. Izd Nauka, Moskva; 274 pp

Mooi J (1976) Report on investigation of the SO_2, O_3 and C_2H_4 content in woody plants performed by means of long-term artificial fumigation in course of 1973–1974–1975 (in Dutch). Institute for Plant Protection, Wageningen; Report R 157:1–43

Moors PJ (1977) Studies of the metabolism, food consumption and assimilation efficiency of a small carnivore, the weasel (*Mustela nivalis* L.). Oecologia (Berl) 27:185–202

Morawska-Horawska M (1970) Investigations of SO_2 spreading caused by meteorological conditions. Ochrona Powietrza 6:18–20

Mrkwa R (1969) Einfluß der Immissionen auf die Saatgutgüte der Kiefer (*Pinus silvestris* L.) im Gebiet des Forstbetriebes Břeclav (Südmähren). Acta Univ Agric Fac Silvic, Brno 38:345–360

Mucha W, Ratajszczak H (1977) Methods of increasing the productivity of poor forest sites on soils formed from the dune sands of the Noteć Forest (in Polish). Postępy techn w leśn 29:49–58

Mucha W, Sienkiewicz A (1966) Problems of fertilization of forests and nature environment protection. Proc of Symp Soil as a site factor for forests of the temperate and cool zones. Zvolen 2:140–150

Mucha W, Sienkiewicz A (1977) Problems of fertilization of forests and nature environment protection. In: Soils as a site factor for forests of temperate and cool zones. Proceedings II. Žilina, DT SVTS, 140–150 pp
Mudd JB, Kozlowski TT (eds) (1975) Responses of plants to air pollution. Academic Press, New York San Francisco London; 383 pp
Munzar J (1977) Anthropogenic influences on the weather and climate Ostravska and Mostecka (in Czech, PhD thesis), Geolog Ustav CSAV, Brno; 175 pp
Myczkowski S (1954) Biological war against vermins causing death of elms. Chrońmy przyr ojczystą 10:3–20
Myczkowski S, Lesiński JA (1976) Ecosystems of Niepołomice Forest as influenced by industrial fumes. 16th Congr IUFRO, Oslo; 09:1–8 pp
Myczkowski S, Lesiński JA (1982) A model of multiple use forestry in the Niepołomice Forest. Studia Ośrodka Dokumentacji Fizjograficznej 10:223–229
Myczkowski S, Lesiński JA (in press) Selected symptoms of anthropogenic succession of Niepołomice Forest. Stud geogr
Myrcha A, Jezierski W (1972) Metabolic rate during the postnatal development of wild boars. Acta theriol 17:443–454
Newton I (1976) Breeding of sparrowhawks (*Accipiter nisus*) in different environments. J anim Ecol 45:831–849
Niedźwiedź (1969) Synoptic weather situations and their effect on some elements of the climate. Zesz nauk Jagiellon Univ, Pr Inst Geogr 25:63–98
Niesterov ND (1960) Draft of sylviculture (in Russian). Gosud Izd Selskochazj Lit. Moskva; 23–27 pp
Niethammer G (1938) Handbuch der deutschen Vogelkunde. Akad Verlagsges, Leipzig 2:545 pp
Nikolajewskij VS (1979) Biological basis of resistance of plants to air pollution (in Russian). Izd Nauka Sibirskoje ot, Novosibirsk; 278 pp
Norton SA (1976) Changes in chemical processes in soils caused by acid precipitation. In: Dochinger LS, Seliga TA (eds) Proc 1st Int Symp Acid Precipitation. Columbus, Ohio; 711–724 pp
Nováková E, Paukert J (1973) Influence of industrial imissions on cation contents in the hair of common hare (*Lepus europaeus* Pall). 11th Int Congr Game Biologists, Stockholm; 423–438 pp
Nowiński M (1967) Polish grass and sedge communities (in Polish). PWRiL, Warszawa; 284 pp
Nunberg M (1946) *Acantholyda nemoralis* C G Thoms in Poland. Inst Bad Leśn, Rozpr i sprawozd s A 46:1–53
Obmiński Z (1977) Forest ecology (in Polish). PWN, Warszawa; 481 pp
Odén S (1976) The acidity problem – an outline of concepts. Water Air Soil Pollut 6:137–166
Odum EP (1969) The strategy of ecosystem development. Science 164:262–270
Odum EP (1971) Fundamentals of ecology. 3rd ed WB Saunders, Philadelphia; 678 pp
OECD (1977) The OECD programme on long range transport of air pollutants. OECD, Paris; 6:1–26 pp
Ogijewski AW (1951) Drought hydrology. Gosud Izd Sielskohozjajst Lit, Moskva; 161–262 pp
Oke TR (1973) City size and the urban heat island. Atmos Environ 7:769–779
Oko Z (1975) The composition of natural food of pheasant (*Phasianus colchicus* L.) in annual cycle. Roczn Akad Rol Poznań; 87:125–169
Olson JS (1963) Energy storage and the balance of producers and decomposers in ecological systems. Ecology 44:322–331
O'Neill RV, Reichle DF (1980) Dimensions of ecosystem theory. In: Waring RH (ed) Forests. Fresh perspective from ecosystem analysis. Proc 40th Ann Biol Confer Oregon State Univ Press, Corvallis OR; 11–26 pp
Overrein LN, Seip HM, Tollan A (1981) Acid precipitation – effects on forest and fish. Final report SNSF-project 1972-1980, Oslo-Ås; 175 pp
Papanek F (1974) Forest function planning. Acta Inst For Zvolensis 4:245–256
Papers presented to the Symposium on the effects of air-borne pollution on vegetation (1980) UN Economic Commission for Europe, Warszawa; 410 pp
Parczewski W (1971) Wind coefficient of dilution of the atmospheric air pollution. Wiad Służ hydrol met 7:17–25
Pasławski T (1977) Game management for foresters and hunters (in Polish). PWRiL, Warszawa; 327 pp

Pasternak D (1974) Primary production of field with winter wheat. Ekol pol 22:369–378

Pasternak K (1976) The influence of the catchment basin on the chemical composition of surface waters (in Polish). In: Starmach K, Wróbel S, Pasternak K (eds) Hydrobiology – limnology. PWN, Warszawa; 59–80 pp

Pasternak K, Reczyńska-Dutka M (1979) The occurrence of mercury in surface waters at different regions of Southern Poland. Acta Hydrobiol 21:331–340

Patterson DT (1975) Nutrient return in the stemflow and throughfall of individual trees in the Piedmont deciduous forest. In: Howell FG, Gentry JB, Smith MH (eds) Mineral Cycling in Southeastern Ecosystems: ERDA Symposium Series, Springfield; 800–812 pp

Pauli FW (1975) Heavy metal humates and their behaviour against hydrogen sulfide. Soil Sci 119:98–105

Pawłowski B (1932) Low birch (*Betula humilis* Ehrh) in the Niepołomice Forest, Ochr Przyr 12:170

Pelikán J, Vačkar J (1978) Densities and fluctuations in numbers of red fox, badger and pine marten in the "Bučin" forest. Folia Zool 27:289–303

Pelz E, Beyer H, Blayer G (1963) A study of diagnostic methods and effects of smoke damage in the neighbourhood of a lead works. Wiss Z Tech Univ Dresden der Sektion Forstwirtschaft der TU, Dresden; 12:209–216

Persson T (ed) (1980) Structure and function of northern coniferous forest – an ecosystem study. Ecol Bull (Stockholm); 32:1–609

Perzanowski K (1978) The effect of winter food composition on roe deer energy budget. Acta theriol 23:451–467

Perzanowski K, Pradel A, Sikorski M, Mydlarz J (1982) Food resources for deer in Niepołomicka Forest. Acta theriol 27:509–519

Petrusewicz K, Grodziński W (1973) The role of herbivorous animals in ecosystems. Ekologiya 4:6–17

Petrusewicz K, Grodziński W (1975) The role of herbivore consumers in various ecosystems. In: Reichle DE, Franklin JF, Goodall DW (eds) Productivity of world ecosystems. Nat Acad Sci, Washington DC; 64–70 pp

Petrusewicz K, Górecki A, Grodziński W, Kozłowski J (1983) Productivity and energetics. In: Petrusewicz K (ed) Ecology of the bank vole. Acta theriol 28, Suppl. 1:173–205

Phillipson J, Putman RJ, Steel J, Woodel SRJ (1975) Litter input, litter decomposition and the evolution of carbon dioxide in a beech woodland, Wytham Woods, Oxford, Oecologia (Berl) 20:203–217

Pieczyńska E (1972) Ecology of the eulittoral zone of lakes. Ekol pol 20:637–732

Piekara A (1975) Forest management and the law of environmental protection (in Polish). In: Law and environment protection. Ossolineum, Wrocław Warszawa Kraków Gdańsk; 171–183 pp

Pietryga Z (1969) Hydrogeological conditions of the Vistula valley area (92–151 km) (in Polish, Ms). IMGW, Kraków; 36 pp

Pillay KNS (1976) Activation analysis and dendrochronology for estimating pollution histories. J Radioanal Chem 32:151–171

Plewczyńska U (1970) Herb layer production and plant fall in the association *Pino-Quercetum*. Kozłowska 1925 in the Pisz Forest. Ekol Pol 18:757–778

Płużański A (1976) The influence of the catchment basin management on the water purity of the Solina impoundment (in Polish). PhD thesis, Politechnika Rzeszów; 112 pp

Prajsnar T (1981) Nutrient concentration in red fox populations in the Niepołomice Forest (in Polish). MSc thesis – Jagiellon Univ, Kraków; 38 pp

Proceedings of Symposium on Effects of air pollutants on Mediterranean and temperate forest ecosystems (1980) Pacific Southwest Forest Range Exp Station Tech Rep PSW-43, Berkeley CA; 256 pp

Przybylski T, Giertych M, Białobok S (1976) Genetics of Scotch pine (*Pinus silvestris* L.). Acad Sci et Artium Slavorum Meridionalium Annls Forestales, Zagreb; 7:59–105

Pucek Z, Bobek B, Łabudzki L, Miłkowski L, Morów K, Tomek A (1975) Estimates of density and numbers of ungulates. Pol Ecol Stud 1:121–135

Puchalski T, Prusinkiewicz Z (1975) Ecological basis for forest site classification (in Polish). PWRiL, Warszawa; 463 pp

Punzet J (1959) Utilization of decrease curve of flows to determine ground outflow of rivers (in Polish). Wiad Służ hydrol met 7:25–44 pp

Punzet J (1975) Average outflow in the basin of the Upper Vistula (in Polish). Gospod Wodna 2:45–49

Raciborski M (1884) Changes of flora in Cracow countryside during the last twenty years (in Polish). Spraw Komis Fizjogr PAU 19:3–24

Rajchel R (1965) Net primary productivity of the herb layer in two forest associations of the Ojców National Park (Southern Poland). Fragm florist geobot 11:121–150

Rams B (1978) The content of some selected elements in the plants of the ground flora and in the leaves of trees in the oak-hornbeam stands of the Niepołomice Forest. Stud Nat Sci A 14:145–164

Ranft H, Dässler HG (1970) Rauchhärtetest an Gehölzen in SO_2-Kabinenversuch. Flora (Jena); 159:573–588

Reczyńska-Dutka M (1984) Some surface waters (mainly reservoirs) in forest-agricultural catchment basin of the River Brynica nearby Upper Silesia Industrial Region. Atmospheric heavy metal pollution – Impoundment Kozłowa Góra (especially bottom sediments). Acta Hydrobiol (in press)

Reichle DE (ed) (1981) Dynamic properties of forest ecosystems. Cambridge Univ Press, Cambridge London New York; 683 pp

Rhode W (1976) Algae in culture and nature. Scr limnol Uppsala Coll 12:427 + 1-22 pp

Rich V (1979) Down in a Polish forest, something is stirring. Nature, 277:9

Rochow JJ (1975) Mineral nutrient and cycling in a Missouri Forest. J Ecol 63:985–994

Rodin LE, Bazilevich NI (1967) Production and mineral cycling in terrestrial vegetation. Oliver and Boyd, Edinburgh London; 288 pp

Rohmeder E, von Schönborn A (1965) Der Einfluß von Umwelt und Erbgut auf die Widerstandsfähigkeit der Waldbäume gegenüber Luftverunreinigung durch Industrieabgase. Forstwiss Centralbl (HAMB) 84:1–13

Rokosz M (in press) Outline of the history of the Niepołomice Forest region (in Polish). Stud hist Philos Sci

Romer E (1949) The climatic regions in Poland (in Polish). Pr Wrocł Tow Nauk S B 16:453–472

Rudnik Z (1980) Estimation of heavy metal and nutrient contents in bank vole food from Niepołomice Forest (in Polish, MSc Thesis – Jagiellon Univ), Kraków; 37 pp

Rühling A, Tyler G (1971) Regional differences in the deposition of heavy metals over Scandinavia. J appl Ecol 8:497–507

Rühling A, Tyler G (1973) Heavy metals pollution and decomposition of spruce needle litter. Oikos 24:402–416

Ryszkowski L (1982) Structure and function of the mammal community in an agricultural landscape. Acta Zool Fenn 169:45–59

Sauerbrey R (1975) Untersuchungen zur Stickstoffdüngung des Weidegrünlands in der Altmärkischen Wische. Arch Acker Pflanzenbau Bodenkde 19:205–214

Sawicka-Kapusta K (1978) Estimation of the concentration of heavy metals in antlers of roe deer from Silesian woods. Arch Ochr Środowiska 1:107–121

Sawicka-Kapusta K (1979) Roe deer antlers as bioindicators of environmental pollution in Southern Poland. Environ Pollut A 19:283–293

Sawicka-Kapusta K, Perzanowski K, Bobek B (1981) Heavy metal intake by roe deer in Silesian forests. Ekol pol 29:353–359

Sawicki M (1977) Estimation of European badger (*Meles meles* L.) population in Niepołomice Forest (in Polish). MSc Thesis – Jagiellon Univ, Kraków; 21 pp

Schauermann I (1977) Energy metabolism of rhizophagous insects and their role in ecosystems. Soil organisms as components of ecosystems. Ecol Bull (Stockholm) 25:310–319

Schmidt-Vogt (1968) Zur Anerkennung von Fichtenbeständen. Allg Forstztg 23:123–127

Schneider Z (1971) Susceptibility to insect feeding of various tree species and their utility for the reconstruction of stands in industrial regions (in Polish). Komisja d/s Leśnej Rekultywacji Ministerstwa Leśnictwa i Przemysłu Drzewnego 4:107–121

Scholz F (1983) Kann Züchtung auf Immissionsresistenz zur Lösung des Problems „Waldsterben" beitragen? Allg. Forst Zeitschrift 12:281–283

Scholz F, Timmann T, Krusche D (1978) Untersuchungen zur Variation der Resistenz gegen HF-Begasung bei *Picea abies* Familien. 10th Intern Arbeitstagung IUFRO Fachgruppe S2.09 Luftverunreinigung, Ljublana; mimeogr 1–10 pp

Scholz F, Timmann T, Krusche D (1980) Genotypic and environmental variance in the response of Norway spruce families to HF-fumigation. Papers presented to the Symposium on the Effects of Air-Borne Pollution on Vegetation, Warsaw; 328–354

Schönbach H, Dässler HG, Polster H, Börtitz S, Enderlein H, Lux H, Ranf H, Stein G, Vogl M (1968) How to warrant productivity of forest influenced by industrial smoke. In: Vincent G (ed) How to increase forest productivity. Statni zemědělske naklad, Praha; 435–484 pp

Schubert R (1977) Angewählte pflanzliche Bioindikatoren zur Erfassung ökologischer Veränderung in terrestrischen Ökosystemen durch anthropogene Beeinflussung unter besonderer Berücksichtigung industrieller Ballungsgebiete. Hercynia NF 14:399–412

Šestak Z, Čatsky J, Jarvis PG (eds) (1971) Plant photosynthetic production. Manual of methods. Junk, The Hague; 818 pp

Serafiński W (1954) Studies on food composition of tawny owl (in Polish). Chrońmy przyr ojcz 10:40–49

Siemińska J (1956) Microbiological and fisheries characteristics of Brynica River (in Polish). Arch hydrobiol 3:69–160

Siemieńska A (1977) Food and feeding habits of *Clethrionomys glareolus* in a coniferous forest in Niepołomice Forest (in Polish). MSc thesis – Jagiellon Univ, Kraków; 23 pp

Sierek B (1977) Food and feeding habits of *Apodemus flavicollis* in a coniferous forest in Niepołomice Forest (in Polish). MSc thesis – Jagiellon Univ, Kraków; 19 pp

Sikora S (1975) Occurrence and composition of food of nuthatch (*Sitta europaea* L.) in forest biotopes. Roczn Acad Rol Poznań; 193–210

Silvicultural norms (in Polish) (1969). PWRiL, Warszawa; 410 pp

Siuda A, Żurowski W, Siuda H (1970) The food of the roe deer. Acta theriol 14:247–262

Skoog P (1970) The food of the Swedish badger. *Meles meles* L. Viltrevy (Stockh) 7:1–120

Skulberg OM (1964) Algal problems related to the eutrophication of European water supplies and bioassay method to assess fertilizing influence of pollution inland waters. In: Jackson DF (ed) Algae and man. Plenum, New York; 262–299

Skuratowicz W (1950) Studies on food composition of tawny owl (*Strix aluco* L.) in 1946/48 (in Polish). Pr Kom biol PTPN 12:226–236

Skye E (1968) Lichens and air pollution: a study of cryptogamic epiphytes and environment in the Stockholm region. Acta phytogeogr suec 52:1–123

Sládek J (1961) Notes on feeding ecology of bussard *Buteo buteo* (L.) (in Czech). Zool Listy 4:331–345

Sławoń J, Woliński Z (1975) Breeding of fox (in Polish). PWRiL, Warszawa; 406 pp

Smith WH (1974) Air pollution – effects on the structure and function of the temperate forest ecosystem. Environ Pollut A 6:111–129

Smith WH (1981) Air pollution and forests – interactions between air contaminants and forest ecosystems. Springer, New York Heidelberg Berlin; 379 pp

Smólski S (1965) Niepołomice Forest. In: Żabko-Potopowicz A (ed) The history of forests, forestry and timber utilization in Poland (in Polish). PWRiL Warszawa; 666–677 pp

Sollins P, Grier CC, McCorison FM, Cromack K Jr, Fogel B, Fredriksen RL (1980) The internal element cycles of an old-growth douglas-fir ecosystem in Western Oregon. Ecol Monogr 50:261–285

Spahova AS, Riazanceva ŁA (1978) Sensitivity of some woody plants to sulfur dioxide (in Russian). Fiziol Rast 28:407–409

Spector WS (ed) (1956) Handbook of biological data. WB Saunders, Philadelphia London; 584 pp

Sporek K (1983) Influence of air pollution on the growth in height of Scots pine. PWN, Warszawa, Wrocław; 99 pp

Starkel L (1972) Sandomierz basin (in Polish). In: Klimaszewski M (ed) Geomorphology of Poland. 1, PWN, Warszawa; 138–163 pp

Starzecka A (1979) Bacteriological characteristics of water in the River Nida and its tributaries. Acta Hydrobiol 24:341–360

Starzecki W (1975) Ecophysiological investigation on photosynthetic productivity of leaves of chosen tree species. Pol ecol Stud 1:51–63

Starzyk JR (1973) The communities of noxious forest insects of the selected tree stands in the Niepołomice Forest (in Polish). Zesz nauk Akad Roln, Kraków; 80:129–130

Starzyk JR (1975) Cerambycial beetles in Niepołomice Forest wood stands and some effects of industrial air pollution. In: Dobrowolski BW, Płotnikov WF (eds) 8th Inter Plant Protect Congr Sect 6, Moskov; 249–255

Starzyk JR (1976) Grouping of cerambycios (*Coleoptera, Cerambycidae*) on the background of types of environment of the Niepołomice Forest near Kraków. Acta Agrar Silvestria Ser Silvestris 16:131–152

Starzyk JR (1977 a) The influence of isolation of the stand upon the appearance of the timber beetles (*Col, Cerambycidae*) in the Niepołomice Forest. Sylwan 120:41–49

Starzyk JR (1977b) Influence of the stand age upon the quality of the composition and the numerousness of the timber beetles (*Col, Cerambycidae*) in the Niepołomice Forest. Acta Agrar Silvestria Ser Silvestris 17:117–135

Starzyk JR, Starzyk K (1981) Cambiophagous, cambio-xylophagous and xylophagous insects in stands of Niepołomice Forest. Studia Ośrodka Dokumentacji Fizjograficznej 9:255–291

Starzyk JR, Witkowski Z (1981) Changes of the parameters describing the cambio and xylophagous insect communities during the secondary succession of the oak-hornbeam association in the Niepołomice Forest near Kraków. Z angew Entomol 91:525–533

Statistical Yearbooks 1971–1983 (in Polish). Vol. 31–43 Główny Urząd Statystyczny, Warszawa

Steinberg Ch (1977) Stable nitrogen – containing dissolved organic substances in Lake Schöhsee and in cultures of algae. Arch hydrobiol 53:48–158

Strojan CL (1978) Forest leaf litter decomposition in the vicinity of a zinc smelter. Oecologia (Berl) 32:203–212

Strojny W (1961) *Acanthoryptus afflictor* (Grav) (*Hym, Ichneumonidae*) parasit *Billaca irrorata* (Meig) (Dipt, *Tachinidae*). Pol Pismo Entomol Ser B, Entomol Stosow 31:79–88

Stubbe M (1973a) Der Fuchs (*Vulpes vulpes* L.). In: Stubbe M (ed) Buch der Hege. Landwirtschaftsverlag, Berlin; 182–212 pp

Stubbe M (1973b) Schutz und Hege des Dachses (*Meles meles* L.). In: Stubbe M (ed) Buch der Hege. Landwirtschaftsverlag, Berlin; 227–249 pp

Sukachev V, Dylis N (1964) Fundamentals of forest biogeocenology. Oliver and Boyd, Edinburgh London; 672 pp

Suliński J (1978) Studies on the movement of ground water level in selected areas of Niepołomice Forest (in Polish). PhD Thesis Acad Agr, Kraków; 1–65 pp

Sunda WG, Lewis JAM (1978) Effect of complexation by natural organic ligands on the toxicity of copper to unicellural alga, *Monochrysis lutheri*. Limnol Oceanogr 23:870–876

Swank WT (1981) Forest ecology and hydrology studies. (MS Lectures presented at the Univ Padova), Coweeta Hydrologic Lab Otto NC; 146 pp

Swank WT, Douglass JE (1977) Nutrient budgets for undisturbed and manipulated hardwood forest ecosystems in the mountains of North Carolina. In: Corell DL (ed) Watershed research in Eastern North America, Smithsonian Inst Edgewater MD; 1:343–363

Swank WT, Waide JB (1980) Interpretation of nutrient cycling research in a management context: evaluating potential effects of alternative management strategies on site productivity. In: Waring RW (ed) Forests: fresh perspectives from ecosystem analysis. Proc 40th Ann Biol Colloq, Oregon State Univ Press, Corvalis OR; 137–158 pp

Świdziński H (1952) General geological map of Poland (in Polish). Sect F 4, Nowy Sącz, PIG, Warszawa

Swoboda AR, Thomas GW (1965) The movement of sulfate salts in soil. Soil Sci Soc Am Proc 29:320–331

Szafer W (1977) Plant cover of lowland Poland (in Polish). In: Szafer W, Zarzycki K (eds) Vegetation of Poland. 2, PWN, Warszawa; 93–103 pp

Szmidt A (1970) Catalase variability in populations of Scots pine under the influence of industrial pollution (in Polish). Summaries papers conference Biological reactions of trees to industrial emissions. Kórnik near Poznań; 13–14 pp

Szponder A (1955) Pollen analysis of the holocen peat bog "Wielkie Błoto" in Niepołomice Forest (in Polish) MSc thesis – Jagiellon Univ, Kraków; 32 pp

Szulakowska G (1974) Nutritive value of the food for deer in pinewoods of Niepołomice Forest (in Polish). MSc thesis – Jagiellon Univ, Kraków; 29 pp

Tappeiner JC, Alm AA (1975) Undergrowth vegetation effect on the nutrient content of litterfall and soils in red pine and birch stands in Northern Minnesota. Ecology 56:1193–1200

Tertil R (1977) Impact of the common vole, *Microtus arvalis* (Pallas) on winter wheat and alfalfa crops. EPPO Bull 7:317–339

Todd RL, Waide JB, Cornaby BW (1975) Significance of biological nitrogen fixation and denitrification in a deciduous forest ecosystem. In: Howell FG, Gentry JB, Smith MH (eds) Mineral Cycling in Southeastern Ecosystems: ERDA Symp Ser Springfield; 729–735 pp

Tomanek J (1963) Meteorology and climatology for foresters (in Polish). PWRiL, Warszawa; 310 pp

Traczyk T (1967) Studies on herb layer production estimate and the size of plant fall. Ekol pol Ser A 15:837–867

Traczyk T (1971) Productivity investigation of two types of meadows in the Vistula valley. I Geobotanical description and primary production. Ekol pol 19:93–196

Traczyk M, Traczyk T (1967) Tentative estimation of the production of herb layer Ekol pol Ser A 15:823–835

Traczyk T, Traczyk M, Moszyńska B (1973) Herb layer production of two pinewood communities in the Kampinos National Park. Ekol pol 21:37–55

Trafas K (1975) Changes of the Vistula river east of Cracow in the light of archival maps and photointerpretation. Zesz nauk Jagiellon Univ, Pr monogr 40, Pr Inst Geogr 62:1–87

Trela J (1938) "Błoto" peat bog in the Niepołomice Forest (in Polish). MSc thesis – Jagiellon Univ, Kraków; 32 pp

Tumidajowicz D (1973) The dynamics of biomass and primary production of herb layer plants in the deciduous *Tilio-Carpinetum* association of the Niepołomice Forest. Bull Acad Pol Sci Cl II 21:101–107

Turunen J, Visapää A (1972) Sulphur in pine needles and birch leaves II. Paperi ja Puu 4:1–6

Tyler G (1972) Heavy metals pollute nature may reduce productivity. Ambio 1:51–59

Uggla H, Uggla Z (1979) Forest pedology (in Polish). PWRiL, Warszawa; 523 pp

Unsworth MH (1977) Dry deposition of gases and particles to vegetation: a review. In: Institute of Terrestrial Ecology Workshop on Methods in Acid Precipitation in Forest Ecosystems. ITE, Edinburgh; 1–11 pp

Uttendörfer O (1939) Die Ernährung der Raubvögel und Eulen. Neumann, Berlin; 412 pp

Vahrenkamp H (1973) Metalle in Lebensprozessen. Chem unserer Zeit 7:97–105

Van Hook RI, Harris WF, Henderson GS (1977) Cadmium, lead, and zinc distributions and cycling in a mixed deciduous forest. Ambio 6:281–286

Vogtsberger LM, Barrett GW (1973) Bioenergetics of capative red foxes. J Wildl Manage 37:495–500

Voisin A (1964) Productivity of pastures (in Polish). PWRiL, Warszawa; 367 pp

Walczewski J (1976) Some characteristics of industrial smokes propagation in the Silesian region, Poland. Artif Earth Satell 11:71–74

Walczewski J (1980) First series of acoustic sounding of the atmosphere in Cracow. Przegl geofiz 25:145–154

Wasylik K (1971) Algal communities in Czarny Dunajec river (Southern Poland) and in some its affluents. Fragm florist geobot 17:257–354

Weaver GT (1975) The quantity and distribution of four nutrient elements in high-elevation forest ecosystems, Balsam Mountains, North Carolina. In: Howell FG, Gentry JB, Smith MH (eds) Mineral Cycling in Southeastern Ecosystems. ERDA Symp Ser, Springfield; 715–728 pp

Weiner J (1973) Dressing percentage, gross body composition and caloric value of the roe deer. Acta theriol 18:209–222

Weiner J (1977 a) Energy metabolism of the roe-deer. Acta theriol 22:3–24

Weiner J (1977 b) Energy requirements of the roe-deer (in Polish). PhD thesis – Jagiellon Univ, Kraków; 56 pp

Weiner J, Głowaciński Z (1975) Energy flow through a bird community in a deciduous forest in Southern Poland. Condor 77:233–242

Weiner J, Górecki A, Zieliński J (1982) The effect of rodents on the rate of matter and energy cycling in ecosystem of arid steppe of central eastern Mongolia. Pol ecol Stud 8:69–86

Wentzel KF (1963) Waldbauliche Maßnahmen gegen Immissionen. Allg ForstZtg 18:101–108

Wentzel KF (1971) Habitus-Änderung der Waldbäume durch Luftverunreinigung. Forstarchiv 41:165–172

Wentzel KF (1975) Preventation and protection against injuries caused by air pollution. In: Buchwald K, Engelhardt W (eds) Landscape planning and protection of nature (in Polish). PWRiL, Warszawa; 567–570 pp

Wentzel KF (1978) Immissionsgrenzwerte für den Wald. Schweiz Z Forstwes 5:368–380

Wentzel KF (1981) Maximale Immissionswerte zum Schutze der Wälder. Mitt Forstl Bundesversuchsanst 137:175–180

Wesely ML, Hicks BB, Dannevik WP, Frisella S, Husar RB (1977) An eddy correlation measurement of particulate deposition from the atmosphere. Atmos Environ 11:561–563

West PW, Gaeke GC (1956) Fixation of sulphur dioxide as disulfidomercurate and subsequent colorimetric estimation. Anal Chem 28:1816–1819

Westlake DF (1966) The biomass and productivity of *Glyceria maxima* L. seasonal changes in biomass. J Ecol 54:745–753
Whelpdale DM, Shaw RW (1974) Sulphur dioxide removal by turbulent transfer over grass, snow and water surface. Tellus 26:196–205
Wiegert RG, Owen DF (1971) Trophic structure, available resources and population density in terrestrial vs. aquatic ecosystems. J theor Biol 30:69–81
Wiener JG (1975) Nutrient cycles, nutrient limitation and vertebrate population. Biologist 57:104–124
Wilczyńska-Michalik W (1981) Mineralogical study of dusts emitted by Lenin's Steel Plant in Kraków. Pr Mineral 68:7–52
Williams RJP (1967) Schwermetalle in biologischen Systemen. Endeavour 26:96–100
Winchester JW, Nifong GD (1971) Water pollution in Lake Michigan by trace elements from pollution aerosol fallout. Water Air Soil Pollut 1:50–64
Witkowski Z (1969) The community of weevils (*Coleoptera, Curculionidae*) on the mown and unmown meadow in the Ojców National Park. Ochr Przyr 34:185–204
Witkowski Z (1975a) The ecology and succession of weevils (*Coleoptera, Curculionidae*) on mowed meadows in the Zabierzów area. Stud Nat Sci A 12:1–81
Witkowski Z (1975b) Environmental regulation of the population size of the oak leaf roller moth (*Tortrix viridana* L.) in the Niepołomice Forest. Bull Acad Pol Sci Cl II 23:513–519
Witkowski Z (1980) The changes of parameters describing plants and weevils (*Coleoptera, Curculionidae*) of the herb stratum during the secondary succession of oak-hornbeam forest. Bull Acad Pol Sci Cl II 27:533–539
Witkowski Z, Kosior A (1974) Energetics of the larval development of the oak leaf roller moth, *Tortrix viridana* L. (*Lepidoptera, Tortricidae*) and an estimate of the energy budget in caterpillar development of other insects feeding on oak leaves. Stud Nat Sci A 9:93–106
Witkowski Z, Langer M (1978) Endangerment of oak stands by mass appearances of the oak leaf roller moth *Tortrix viridana* L. (*Lepidoptera, Tortricidae*) in the Niepołomice Forest. Stud Nat Sci A 14:191–204
Witkowski Z, Mazur M (in press) Plants and weevils (*Coleoptera, Curculionidae*) of the herb stratum during the secondary succession of the oak-hornbeam association in the Niepołomice Forest. Stud Nat Sci A
de Witt T (1976) Flechtenverbreitung in den Niederlanden und ihre Abhängigkeit von der Luftverunreinigung. Schriftenr Vegetationskd 10:169–176
Wittich W (1951) Der Einfluß der Baumart auf den Bodenzustand. AFZ, 2:1–4
Wolak J (1977) Relationship between increase in air-pollution toxicity and elevation above ground. Forest Res Inst (IBL), Warszawa; 247 pp
Wolak J, Harasymowicz M, Sienkiewicz J (1977) Plant sociological studies. In: Wolak J (ed) Relationship between increase in air-pollution toxicity and elevation above ground. IBL, Warszawa; 12–28 pp
Wong SL, Beaver JL (1980) Algal bioassays to determine toxicity of metal mixtures. Hydrobiol 74:199–208
Wong SL, Beaver JL (1981) Metal interactions in algal toxicology: conventional versus in vivo tests. Hydrobiol 85:67–71
Wood TG (1974) Field investigations on the decomposition of leaves of *Eucalyptus delegatiensis* in relation to environmental factors. Pedobiologia 14:343–371
Woodwell GM (1970) The energy cycle of the biosphere. Sci Am 223:64–74
Wójcik Z (1973) Productivity of a randy ryefield. Ekol pol 21:339–357
Yamamura S, Fritz J (1955) Rapid microfiltration of sulfate. Anal Chem 27:1461–1464
Yount JD (1975) Forest-floor nutrient dynamics in Southern Appalachian hardwood and white pine plantation ecosystems. In: Howell FG, Gentry JB, Smith MH (eds) Mineral Cycling in Southeastern Ecosystems: ERDA Symp Ser, Springfield; 598–608 pp
Yurkievich LA, Yaroshevich EP (1974) Biological productivity of types and associations of pine forests. Nauka i Technika Publ, Minsk; 290 pp
Zabielski B (1976) Tables of coefficients K used for the determination of pure and mixed stands increments of the more important species of forest trees in Poland (in Polish). PWRiL, Warszawa; 273 pp
Zawołek T (1978) Heavy metals contamination in the bank vole (*Clethrionomys glareolus* Schr.) tissue (in Polish). Msc thesis – Jagiellon Univ, Kraków; 20 pp

Zdanowska D (1976) Effect of Warsaw metal and steel works on acidification of Kampinos Forest. Fragm faun 20:353–367

Zejda J (1961) Age structure in population of the bank vole (*Clethrionomys glareolus* Schreber 1780). Zool Listy 10:249–264

Zemanek M (1972) Food and feeding habits of rodents in a deciduous forest. Acta theriol 17:315–325

Zieliński J (1980) The effect of nitrogen content on the rate of organic matter decomposition. Pol ecol Stud 6:167–182

Zinke PJ (1980) Influence of chronic air pollution on mineral cycling in forests. Proc Symp Effects air pollutants on Mediterranean and temperate forest ecosystems, Pacific Southwest Forest Range Exp Station, Berkeley CA; 88–89 pp

Zinkiewicz W (1962) Thermoisodromes of Poland. Czas geogr 3:73–81

Zitko V, Carson WV (1972) Release of heavy metals from sediments by nitrilotriacetic acid (NTA). Chemosphere 3:113–118

Zlotin RJ (1975) Estimation of influence of herbivorous animals on the primary production of southern-steppe pasture. In: Zlotin RJ, Isekov JA, Hodašova KS (eds) Rol životnych v funkcjonowanii ecosystem. Izd Nauka MOIP, Moskva; 18–22 pp

Zlotin RJ, Hodašova KS (1974) The role of animals in biological cycles of forest-steppe ecosystem. Izd Nauka, Moskva; 200 pp

Żelawski W, Łotocki A (1974) Photosynthetic capacity of Scots pine (*Pinus silvestris*) needles from a forest subjected to chronic industrial pollution. Bull Acad Pol Sci Cl II 22:431–434

Żmuda A (1920) Rare and new plants of the Cracow flora (in Polish). Spraw Komis Fizjogr PAU 53/54:30–76

Subject Index

Abies alba 240, 241
Abies grandis 99
Abundance (density), of
  bacteria in streams 176–179
  birds 126, 127
  carnivores 135
  raptors 132, 133
  rodents 113–116
  ungulates 122
Acanthocinus aedilis 112
Acantholyda nemoralis 2
Accipiter gentilis 132, 133, 134
Accipiter nisus 132, 133, 134
Accumulation 204
  energy, see Energy accumulation
  nutrients, see Nutrients accumulation
  pollutants, see Pollutants accumulation
Acer platanoides 3, 9, 240, 241
Acer pseudoplatanus 3, 9, 234, 241
Acer spp 244
Achnanthes spp 181
Acid phosphatase (soil enzyme) 152, 153
Acidification, of
  soil 13
  tree bark 28–30
Aegopodium podagraria 15, 103
Aeromineralogy 59
Aerosol 40–44, 46
Agrilus angustulus 106
Agrilus biguttatus 106
Agrostis stolinifera 95
Agrostis tenuis 87
Agrostis vulgaris 88, 152
Air circulation 17, 19
Air humidity 22
Air temperatures 16, 19, 45
  annual 16
  maximum 20
  mean daily 18
  minimum 20
  ten-day average 20
Alces alces 26
Alder, see *Alnus*
Algae 180
  chrysophytes 181
  diatoms 181
  euglenoids 181
  green 181
Alnetalia glutinosae 6, 10
Alnion glutinosae 6
Alno-Padion 6, 8, 10
Alnus glutinosa 3, 6, 7, 9, 10, 29, 71, 73, 103, 237, 240, 241, 244
Alnus incana 9, 241
Alouda arvensis 126
Aluminosilicates 61
Aluminum (Al) in industrial dusts 62, 63, 64, 65
Amino acids, free 179, 180
Ammonia
  in soil 14
  in water 187, 188
  run-off 191
Ammonium ions 57, 58
Ammonium nitrate 238, 239
Amphipyra pyramidea 107
Anaerobiosis (in soil) 11
Andromeda polifolia 4
Anemone nemorosa 9, 103, 117
Anglesite 41
Anhydrite 41
Anthericum ramosum 5
Anthus pratensis 126
Anthus trivialis 125, 126
Apatite 41
Apion flavipes 104
Apodemus agrarius 113, 115
Apodemus flavicollis 113, 115, 117–121, 137, 139
Aquatic communities 92
Arabis halleri 5
Archips crataegena 107
Archips podana 107
Arrhenatheretalia 94
Arthrobacter 152
Ash, see *Fraxinus*
Asio otus 132
Aspen, see *Populus tremula*
Assmann psychrometer 16
ASTM – Designation P-1608-60 38
Atmospheric (thermal) inversion 44, 49
Atomic absorption spectrophotometer (AAS) 56, 59, 65, 97, 139
Augustus II Saxon, king 26
Auroch, see *Bos primigenius*

Bacillus 152
Bacteria, in waters
  ammonifying 179, 180, 202
  decomposing protein 177–179
  denitrifying 177–179, 202
  desulfurizing 177–179, 202
  H_2S releasing 177–179
  iron 177–179, 202
  NH_3 releasing 177–179
Baczków Creek
  water flow 171
  watershed 169
Badger, see *Meles meles*
Bank vole, see *Cletrionomys glareolus*
Barium (Ba), in industrial dusts 64, 65
Batrachium aquatile 170
Bean, see *Phaseolus*
Bear, see *Ursus arctos*
Beech, see *Fagus*
Beryllium (Be), in industrial dusts 64, 65
Betula papyrifera 161

Betula pubescens 6, 10, 241, 244
Betula spp 234, 244
Betula verrucosa 4, 6, 8, 10, 70, 73, 87, 240, 241, 243, 244
B-glucosidase (soil enzyme) 152, 153
Białowieża National Park 15, 29, 30, 31, 32, 97, 140, 150
Bioindication 27–34
Biomass
consumption, see energy flow
of herbs 88, 104
standing aboveground 75–78, 205, 207
of tree bark 205, 206
of tree roots 205
Birch, see *Betula*
Bird, cherry, see *Padus avium*
Bison bonasus 26
Bitumens, in soil 15
Black locust, see *Robinia pseudoacacia*
Black pine, see *Pinus nigra*
Black stork, see *Ciconia nigra*
Blastophagus minor 3, 105, 106, 110, 112
Blastophagus piniperda 3, 105, 106, 110, 112
Blechnum spicant 5
Blueberry, see *Vaccinium myrtillus*
Boron (B), in industrial dusts 64, 65
Bos primigenius 26
Bracken, see *Pteridium aquilinum*
Buckthorn, see *Frangula*
Bupalus pinarus 2
Buprestidae 103, 105, 107
Burkeite 41
Buteo buteo 131, 132, 133, 134, 135
Buteo lagopus 132, 134
Butters-Chenery naphelometric method 97
Buzzard, see *Buteo buteo*

Cadmium (Cd) 151
accumulation 158, 162, 198–202
balance in litter 156
budgets 202, 204, 221, 229
emission in dust 39
input 62, 155
outflow from watersheds 197–199
in soil 238
turnover rate 156, 157
Cadmium concentration, in
homeotherms 139–142
industrial dusts 64, 65
insects 112, 113
litter 155
plants 30, 31, 96, 97, 112, 155
rain water 154
water 195
Calamagrostis arundinacea 9, 87, 96, 152
Calamagrostis canescens 9
Calamagrostis epigeios 7, 87
Calamagrostis lanceolata 5
Calcite 41
Calcium (Ca) 150, 151, 152
accumulation 158
balance in litter 156, 159
budgets 204, 220, 228
in industrial dusts 56, 57, 58, 61, 64
input 155
run-off 191–193
in soil 14, 238
turnover rate 157, 162
Calcium concentration, in
birds 129–131
insects 108–111
litter 155
plants 96–98, 155
rain water 154
rodents 120, 121
ungulates 124
water 186
Calcium oxide 39
Calcium phosphate 40
Callidium aenum 112
Calluna vulgaris 87
Caltha palustris 5
Campbell-Stokes heliograph 16
Capreolus capreolus 26, 121–125, 139
Carbon dioxide 38, 39
Carbon monoxide 38, 39
Carbon (C), organic 62
run-off 191–193
in soil 14, 152
in water 189
Carbonates 62
Cardamine impatiens 10
Carex brizoides 5, 9, 10, 87, 88, 90, 96, 103, 236
Carex elongata 9
Carex montana 5
Carex remota 5
Carex silvatica 9
Carex spp 117, 118
Caricetalia fuscae 94
Caricetum gracilis 93, 95
Carici-Agrostietum 95
Carici elongatae-Alnetum 6, 7, 9, 13, 90, 235, 236
Carici elongatae-Alnetum/Pino-Quercetum 90
Carnivores (mammals) 102
Carpinion betuli 6, 10
Carpinus betulus 4, 7, 9, 10, 70, 71, 73, 79, 103, 241
Casimir the Great, king 26
Cat (vagrant), see *Felis catus*
Centorhynchus scrobicollis 104
Cerambycidae 103, 105, 107
Ceratocystis ulmi 3
Certhia sp. 127
Cervus elaphus 26, 121–125
Cetraria islandica 7, 28
Chelidonium majus 5
Chemical erosion (by streams) 190–192
Chimaphila umbellata 5
Chiolite 41
Chlorella pyrenoidosa 176, 180, 181, 202
Chloride
content in water 186
run-off 191
Chromium (Cr), in industrial dusts 40
Ciconia nigra 132, 133, 134
Cidnorhinchus quadrinaculatus 104
Circaea alpina 5
Circaea intermedia 5
Circaeo-Alnetum 6, 7, 9, 10, 11, 13, 15, 90
Circaeo alnetum/Tilio carpinetum 90
Cirsietum rivularis 95
Cirsio-Polygonetum 95
Cladonia rangiferina 7
Clay minerals, in
dusts 62, 67
soil 15
Clethrionomys glareolus 113–115, 117–121, 135, 137, 139
Climacium dendroides 5
Climate 16–23
Climatological stations 16, 46

Closteridium ehrenbergii 181
Closteridium kutzingii 181
Cobalt (Co)
accumulation 98–202
concentration in industrial dusts 64, 65
concentration in water 195
outflow from watershed 197–199
Coccothraustes coccothraustes 127
Colorimeter Spectronom 97
Comarum palustre 4
Computer simulation 125, 225
Consumption of energy, see energy flow
Convallaria majalis 5
Copper (Cu) 151
accumulation 158, 198–202
balance in litter 156
budgets 202, 204, 229
deposition 66
input 62, 155
outflow from watershed 197–199
turnover rate 156, 157
Copper concentration, in
industrial dusts 61, 64, 65
insects 112–113
plants 30, 31, 96, 97, 112, 155
rain water 154
soil 14, 15
water 195
Corundum 41
Corvus corax 132, 133, 134
Corydalis solida 5
Corylus avellana 9
Corylus cornuta 161
Cosmia trapezina 107
Coturnix coturnix 126
Coweeta basin, USA 219
Cracow 2, 24, 39, 44
Cracow agglomeration (region) 36, 37, 49, 54, 56, 58, 225, 237
Crategus monogyna 8
Crategus oxyacantha 7, 8, 9
Crategus spp 3
Criocephalus rusticus 3, 106, 112
Cryolite 41
Cucubalus baccifer 5
Curculionidae 103, 105
Cynips liquicola 113
Cytisus ratystonensis 5, 7

Damages in forest, by
consumers 143–148
insects 110–113
rodents 121
ungulates 124, 125
Daphne mezereum 5
Decomposition rate 149, 150, 163–165
in coniferous and deciduous forests 165
inhibition of 150, 165, 224
in litter 163–164
Dendrolimus pini 2
Dicrano-Pinion 6
Diffractometric DSA analysis 59
Dolomite 41, 238
Drosera rotundifolia 4
Drwinka River 2, 11, 24, 26, 91, 168, 176, 186
Dryopteris cristata 5
Dryopteris oreopteris 5
Dryopteris thelypteris 9
Dusts (industrial)
chemical composition 64–67
fall 59–68
mineral composition 61–64

East Germany 233, 234
Ecological efficiency, in
birds 125, 127, 128
carnivores 138
consumers 214
rodents 120
Elder, see *Sambucus nigra*
Electron microscopy 59
scanning (SEM) 59, 64, 65
transmission 59
Elm, see *Ulmus*
Elodea canadensis 170
Emberiza citrinella 126
Emission, of
industrial dusts 36–44, 225
sulfur dioxide 36–39, 223
Energetic value of plant material 95, 206
Energy accumulation 206, 211, 213
Energy budget, of
forest ecosystems 226
primary production 205, 207
Energy flow, through
birds 125–129
cambio- and xylophagous insects 105–107
consumers 143, 144, 213–215
ecosystems 222
phyllophagous insects 107, 108
phytophagous insects 110–112
raptors and carnivores 134–138
rodents 113–121
ungulates 122, 123
Energy intake by consumers 213–215
see also Energy flow
Energy partitioning in above-ground biomass 209
Environmental deterioration 27–34
Epilobium montanum 5
Epsomite 41, 61, 62
Equisetum limosum 170
Eriophorum vaginatum 10
Erithacus rubecula 125, 126
Ermine, see *Mustela erminea*
Erzgebirge 240
Euglena acus 182
Euglena granulata 182
Eunotia arcus 181
Eunotia bigibba 181
Eunotia exigus 181
Eunotia praerupta 181
Eunotia spp 181
European bison, see *Bison bonasus* 26
European elk, see *Alces alces*
Evonymus europaea 3, 10
Evonymus verrucosa 5, 7
Exoteleia dodecella 113

Fagetalia silvaticae 6
Fagus silvatica 3, 9, 73, 103
Falconiformes 132
Falco tinnunculus 132, 133, 134
Feldspars 41, 62
Felis catus 132
Festuca ovina 88
Ficario-Ulmetum 10, 11, 13, 90
Ficario-Ulmetum campestris 6
Ficario-Ulmetum stachyetosum 90
Ficedula albicollis 126
Field weed community, see Secali-Violetalia
Fluoride 243, 244

Fluorine (F) 241, 243, 244
Food composition, of
  carnivores 135–137
  raptors 134
  rodents 116–119
  ungulates 124
Food resources available, to
  herbivores 144
  rodents 118
  ungulates 123
Forest communities 6–11
  alder, see Carici elongatae-Alnetum
  coniferous mixed, see Pino-Quercetum
  Douglas-fir 219
  elm 2, 10
  hardwood, northern 219
  humid hornbeam-oak, see Tilio-Carpinetum Stachyetosum
  lowland alder, see Circaeo-Alnetum
  mesic oak-hornbeam, see Tilio-Carpinetum typicum
  mixed oak-pine, see Pino-Quercetum
  moist oak-hornbeam, see Tilio-Carpinetum stachyetosum
  oak-hickory 219
  oak-hornbeam, see Tilio-Carpinetum
  oak-hornbeam, transitional, see Tilio-Carpinetum/Pino-Quercetum
  Ponderosa pine 225
  swamp pine, see Vaccinio uliginosi-Pinetum
Forest, filtering capacity 232, 233
Forest management 24
Forest restructuring 239–244
Fox, see *Vulpes vulpes*
Fragillaria spp 181
Frangula alnus 3, 9, 10, 71, 87, 96, 153
Fraxino-Ulmetum/Tilio-Carpinetum stachyetosum 90
Fraxinus excelsior 2, 3, 4, 70, 71, 210, 241
Fraxinus spp 234, 244
Fringilla coelebs 125, 126
Fungi 119, 120

Gagea lutea 10
Galanthus nivalis 5
Galeobdon luteum 8
Galium schultesi 8
Gallium (Ga), in industrial dusts 64, 65
Gemista spp 117
Germanium (Ge), in industrial dusts 64
Geum urbanum 8
Glauberite 41
Glyceria aquatica 170
Glycerietum maximae 93
Glycine spp 234
Gomphonema angustatum 181
Gomphonema olivaceum 181
Goshawk, see *Accipiter gentilis*
Goslarite 41
Grasslands 92
Great Britain 98
Grobelczyk 2, 11, 12
Grobla 2, 11, 12, 25
Ground water 13, 15, 168, 169, 238, 239
  and drainage systems 175, 176
  levels 171–174
  runoff 168
Growing stock 70–78
  volume 70, 71–75
  increment 70, 74
Gypsum 40, 41, 62

Halite 41, 61, 62
Hare (European), see *Lepus europaeus*
Hawthorn, see *Crategus* spp
Heavy metals 36, 39, 57, 58, 62, 64–68, 193
  accumulation 157–160, 198–201, 204, 212, 224
  balance in litter 155–161
  budgets 202, 204
  concentrations in trophic web 215, 216
  concentration in water 193–196,
  see also Cadmium, Copper, Iron, Lead, Nickel, Pollutants, and Zinc
Heavy minerals 62, 63
Hellmann rain gauge 16
Hematite 41, 62
Hexadrite 61
Hexahydrite 41, 62
Homeotherms 113, 139–148
Homogyne alpina 5
Hornbeam, see *Carpinus*
Hottnia palustris 170
Hubbard Brook, New Hampshire, USA V 151, 219, 220, 221
Humid meadows, see Molinietalia
Humus 14
Hunting 26, 123
Hydrocharo-Stratiotetum 95
Hydrogen fluoride 240, 242, 243
Hysne 47, 48, 49, 54, 55, 57

IBP (International Biological Program) V, 102
Idaho, USA 99
Illesite 41
Illite 41
Impatiens noli-tangere 103
Impatiens parviflora 5
Infrared spectroscopy 59
Insects,
  cambiophagous 102, 107
  phyllophagous 107, 108
  phytophagous 102, 112
  xylophagous 103, 107
Insects, outbreaks 2
Insolation 16–17
Iris pseudoacorus 170
Iron (Fe)
  accumulation 158, 199–202
  balance in litter 156
  budgets 204, 228
  deposition 65
  in dust emissions 39
  input 64–68, 155
  outflow from watersheds 197–199
  run-off 191–193
  turnover rate 156, 157, 162
Iron compounds 180
Iron concentration, in
  homeotherms 139–142
  industrial dusts 56, 57, 61, 62
  insects 112, 113
  litter 151
  plants 30, 31, 96, 97, 112, 155
  water (total) 185, 195
Iron oxides 39
Ispina 16, 20

Jack pine, see *Pinus banksiana*
Jadraös (Sweden) V
Japanese larch, see *Larix leptolepsis*

Juncus caprocarpus 170
Juncus sp 170

Kaolinite 41
Kestrel, see *Falco tinnunculus*
Koło 2, 11, 12

Ladislaus Jagiełło, king 26
Ladislaus IV, king 26
LANDSAT satellite 40–43
Langbeinite 41
Lange-Mensing luxometer 16
Larch, see *Larix*
Larix decidua 9, 241, 243, 244
Larix eurolepsis 243
Larix leptolepsis 241, 243, 244
Larix polonica 243, 244
Larix spp 70, 240, 243
Larnite 41
Lathyrus niger 5
Lathyrus vernus 9
Lead (Pb)
  accumulation 158, 164, 198–202
  balance in litter 156
  budgets 202, 204, 221, 229
  deposition 61
  emission in industrial dusts 39
  input 62, 155
  outflow from watersheds 197–199
  in soil 238
  turnover rate 156, 157
Lead concentration, in
  homeotherms 139–142
  industrial dusts 56, 57, 64
  insects 112, 113
  litter 155
  plants 30, 31, 96, 97, 112
  rain water 154
  water 195
Lead peroxide 49
Leafroller, see *Tortrix*
Lecanora conizaeoides 28
Lecanora varia 28
Ledum palustre 4, 10
Leiopus nebulosus 3, 106, 237
Lemna minor 170
Lepus europaeus 26
Lichens 5, 27, 28, 29
Light, in forests 17
Lilium martagon 5
Lime, see *Tilia*
Linden, see *Tilia*
Liophleone lentus 104
Litter (in forests) 150
  enzymatic activity 153
  half-decay 165
  layers 151, 152
  mor type 151
  turnover rate of organic matter, in 160–162
Litter-bag technique 163
Long-eared owl, see *Asio otus*
Lotus corniculatus 238
Lupinus polyphyllus 238
Lupus lupus 26
Luzula pilosa 9
Lycopodium annotinum 5
Lycopodium clavatinum 5
Lycopodium inundatum 4
Lycopus europaeus 9
Lynx, see *Lynx lynx*
Lynx lynx 26

Macroelement run-off 190–193
Macroelements (see also nutrients)
  in plants 95–98
  in soil 14–16
Magnesium (Mg)
  accumulation 158
  balance in litter 156
  in industrial dusts 61
  input 155
  run-off 191–193
  in soil 14, 238
  turnover rate 156, 157
Magnesium concentration, in
  industrial dusts 64
  litter 155
  plants 96, 155
  rain water 154
  water 186
Magnetite 41, 62
Majanthemum bifolium 8, 9
Manganese (Mn) 56, 66, 67
  accumulation 198–202
  input 62, 68
  outflow from watersheds 197–199
  in soil 14, 15
Manganese concentration
  in industrial dusts 57, 64
  in insects 112, 113
  in plants 30, 112
Maple, see *Acer pseudoplatanus*
Marten, see *Martes martes*
Martes martes 132
Mascagnite 41, 62
Melampyrum vulgatum 6
Melanterite 41
Meles meles 102, 132, 135–138
Melilotus albus 238
Mercury (Hg)
  concentration in water 196
Mesic meadows, see *Arrhenatheretalia*
Micaceous minerals 62
Microelements, in
  plants 95–98
  soil 14–16,
  see also pollutants
Micromys minutus 113
Microtus agrestis 137
Microtus arvalis 113, 134
Mineral fertilization 16, 145, 238, 239
Miocene days 168
Mirabilite 41
Moehringia spp 117, 118
Moehringia trinervia 8, 9, 88
Moist deciduous forest 235
Molinia arundinacea 5, 7, 10
Molinia coerulea 10, 87, 152, 236
Molinietalia 93, 94
Molinio-Arrhenatheretea 9
Molybdenum (Mo), in
  industrial dusts 64, 65
  soils 14, 238
Monochamus galloprovincialis 112
Monotropa hipopitys 5
Monterey pine, see *Pinus radiata*
Montmorillonite 67
Moorgrass, see *Molinia*
Mosses 30
Mullite 39, 41, 62
Mus musculus 137
Muscardinus avellanarius 113
Mustela erminea 132
Mustela nivalis 132
Mustela putorius 132
Mycelis muralis 117

Navicula cryptocephala 181
Needle injury 84, 85, 223, 239
Neodiprion pini 2
Nickel (Ni) 151
  accumulation 158, 198–202
  balance in litter 156
  budgets 202, 204, 228
  input 61, 62, 155
  outflow from watersheds 197–199
  turnover rate 156, 157, 163

Nickel concentration, in
industrial dusts 64, 65
insects 112, 113
litter 156
plants 96, 97, 112, 156
rain water 154
water 195
Nicotiana 234
Niepołomice 12, 24, 47, 57
Niepołomice Forest
history of 24–27
location, description V, 2, 37
Nitrates 62
run-off 191
in soil 14, 15
in water 188
Nitrite (in water) 188, 190
Nitrogen (N)
accumulation in litter 158
balance in litter 156, 158
budgets 204, 219, 220, 227
in decomposition of litter 165
input 155
organic 152
in soil 14–16, 238, 239
turnover rate 156, 157
Nitrogen concentration, in
birds 129–131
industrial dusts 62, 154
insects 108–111
plants 96, 98, 155
rain water 154
rodents 120, 121
ungulates 123, 124
Nitrogen, organic in water 188–190
Nitrogen oxides 38, 39, 240, 241, 243
Nitrogen, total in water 188–190
Nitrokalite 41
Nitronatrite 41
Non-forest communities 91, 92
Norway spruce, see *Picea abies*
Nowa Huta 38
Nutrient contents, in
birds 129–131
carnivores 138
insects 108, 109
litter 155
plants 95–98, 155
rodents 120, 121
soil 13–15
ungulates 123, 124,
see also Calcium, Magnesium, Nitrogen, Phosphorus, Potassium
Nutrients
accumulation in ecosystems 216–218
budgets in ecosystems 204, 219, 220, 227, 228
cycling in ecosystems 222
input 155, 217
intake by consumers 213–215
partitioning in above-ground biomass 209
retention 216–218
uptake of forest vegetation 208–212
Nutrients flow, trough
birds 129–131
carnivores 137, 138
consumers 147–158, 213–215
ecosystems 216–220, 222, 227, 228
insects 108–111
rodents 118–121
ungulates 123–124,
see also Macroelements
Nutrients in excrements, of
birds 145–147
insects 145–147
rodents 121, 145–147
ungulates 125, 145–147

Oak, see *Quercus*
Ocueria monacha 2
Oenanthe aquatica 170
Olivine 41
Ontario Canada 151
Oplosia fennica 106
Organic matter 62, 150
annual production 150
balance in litter 156, 160–170
balance in pine forest 154
content in pine forest 153
decomposition 150, 163–165
in litter 152, 162
mineralization in soil 13
outflow from litter 150
in soil 15
in water 187–190
turnover rate in pine forest 157, 160, 162
Orsat analyser 38
Osmunda regalis 10
Ostrava, Czechoslovakia 23
Oxalis acetosella 117
Oxycoccus quadripetalus 4, 10
Ozone (O_3) 240, 241

Padus avium 3, 10, 73
Panolis flammea 2
Parent rock 13
Parmelia furfuracea 28
Parmeliopsis aleurites 28
Particulate matter 56–58, 61, 62
Parus caeruleus 125 126, 127
Parus major 125, 127, 129, 139
Parus palustris 127
Pedunculate oak, see *Quercus robur*
Pendemis ribeana 107
Penicillium 152
Periclase 41
Pesticides 215
Phaegopteris dryopteris 5
Phaegopteris polyploides 5
Phalaridetum 93
PhAR – photosynthetically active radiation 82, 206, 207
Phaseolus 243
Phosphate
content in water 188, 190
run-off 191
Phosphorus (P)
accumulation 158
balance in litter 156, 159
budgets 204, 227
input 155
run-off 191–193
in soil 14, 15, 238
turnover rate 156, 157
Phosphorus concentration, in
birds 129–131
insects 108–111
litter 155
plants 96, 155
rain water 154
rodents 120, 121
ungulates 123, 124
Phosphorus, organic in water 188–190
Phosphorus, total in water 188–190
Photosynthesis 79–87
net rate 79–87
potential 85
Photosynthetic activity, decline due to SO_2 83, 223

Photosynthetic efficiency 27, 32
Photosynthetic productivity 78–87
in oak-hornbeam forest 81, 82
in pine forest 86, 87
Phragmites communis 92
Phragmitetalia 93, 94, 95
Phragmitetea 9
Phyllobius arborator 104
Phylloscopus collybita 126
Phytoplaneton 176, 181
Picea abies 70, 240, 241, 243
Picea excelsa 3, 4, 9
Picea pungens 241
Pine, see *Pinus*
Pino-Quercetum V, 6, 7, 8, 9, 12, 13, 14, 15, 71, 73, 87, 88, 89, 103, 151, 236
Pino-Quercion 6
Pinus banksiana 161, 210
Pinus contorta 241
Pinus nigra 241, 243
Pinus radiata 161
Pinus silvestris 3, 4, 6, 7, 8, 9, 10, 27, 29, 70, 87, 88, 103, 152, 210, 223, 234, 240, 241, 243, 244, 245
Pinus strobus 240, 241, 243
Pinus taeda 161
Pirola uniflora 5
Pissodes pini 112
Pissodes piniphilus 3, 106, 112
Pitymys subterraneus 113, 137
Pityoagenes bidentatus 106, 112
Pityophthorus pityographus 106, 112
Plagionotus arcuatus 3, 106
Platanthera bifolia 5
Pleistocene sands 168
Pleurozium schreberi 30, 31, 87, 88
Poa annua 5
Poa nemoralis 8
Podzolization 13
Poikilotherms 102, 143–148
Polarizing microscope 66
Polecat, see *Mustela putorius*
Polish larch, see *Larix polonica*
Pollutants
accumulation 36, 216–218
budgets in ecosystems 220, 221, 227, 228, 229
cycling in ecosystems 222
intake by consumers 215, 216
retention 216–218, see also Heavy metals, Sulfur, Fluorine
Pollution index 30, 31
Polygonatum multiflorum 8, 9
Polygonatum odoratum 5
Poplar, see *Populus*
Populus spp 70, 234
Populus tremula 4, 8, 9, 234, 241, 243
Poszyna 26
Potametalia 92
Potassium (K) 56, 151, 152
accumulation 158
balance in litter 156, 159
budgets 204, 220, 228
input 155
run-off 191–193
turnover rate 156, 157
Potassium chloride 238
Potassium concentration, in
birds 129–131
industrial dusts 57, 58, 61, 64
insects 108–111
litter 155
plants 96, 98, 155
rain water 154
rodents 120, 121
water 186
Potentilla alba 5
Potentillo albae-Quercetum 5
Precipitation 46
acidity 46, 54, 55, 153
annual 16, 153
atmospheric 16, 20–22
infiltration rate 21, 22, 23
monthly 21, 172
Pressler equation 73
Production (productivity),
primary 98, 99, 110
average annual 99, 209–211
dry mass 209–211
herb layer 87–91, 119, 152, 211
net efficiency 207
non-forest vegetation 92–95, 99
oak-hornbeam forest 98, 99, 212
pine forest 98, 99, 211
seeds 119, 206
tree stands 75–78
Production, secondary, see Energy flow
Pseudotsuga menziesii 241
Pteridium aquilinum 5, 6, 9, 87
Quartz 39, 40, 41, 62
Querco-Fagetea 6, 9
Quercus aegopodialis 210
Quercus marilandica 210
Quercus petraea 3, 4, 245
Quercus robur 2, 3, 4, 6, 7, 9, 10, 28, 71, 87, 88, 210, 234, 237, 240, 243, 244, 245
chemical elements content 95–98
net photosynthesis 79–82, 103
wood increment 71
wood volume 71–75
Quercus rubra 240, 241, 244
Quercus sessilis (=sessiliflora) 9, 153, 237
Quercus stellata 210

Raba River 2, 24, 26, 176
Radish, see *Raphanus*
Rainfall
elemental concentration 154
interception by canopy and litter 154, see also precipitation
Ranunculus cassubicus 10
Ranunculus lanuginosus 117
Raphanus spp 243
Raspberry, see *Rubus hirtus*
Raven, see *Corvus corax*
Red deer, see *Cervus elaphus*
Red oak, see *Quercus rubra*
Red pine, see *Pinus resinosa*
Reed sedge, see *Phragmites communis*
Reed sedge communities, see Phragmitetalia
Respiration of consumers, see Energy flow
Respiration of leaves 80, 81
Retention 204
of nutrients, see Nutrients retention
of pollutants, see Pollutants retention
Rhagium inquisitir 3, 106, 112
Rhagium mordax 3, 106
Rhagium sycophanta 106
Rhyacionia buoliana 113
Ribes schlechtendalii 10
Robinia pseudoacacia 240, 241, 244
Robitsch-Fuesspyranograph 16

Roe deer, see *Capreolus capreolus*
Rough-legged buzzard, see *Buteo lagopus*
Rowan, see *Sorbus* sp
Rubus hirtus 5, 9
Rumex obtusifolius 10

Sacharase (soil enzyme) 152, 153
Salix aurita 10
Salix caprea
  chemical elements content 96, 97
Salix cinerea 9, 10
Salmiac 41, 61, 62
Sambucus nigra 3, 8, 10, 241
Saperda scalaris 3, 106, 237
Saponaria officinalis 5
Saxifraga hirculus 4
Scandium (Sc)
  in industrial dusts 64
Sciaphilus asperatus 104
Scotch pine, see *Pinus silvestris*
Scotylidae 103, 105, 107
Scotylus intricatus 105, 106, 107, 110
Scotylus spp 3
Scripetum silvatici 95
Scrophularia nodosa 8
Secali-Violetalia 93, 94
Senecio fuchsi 5
Sessile oak, see *Quercus petraea* = *Quercus sessilis*
Sigismundus Augustus, king 27
Sigismundus III Vasa, king 24
Silica glass 39, 41
Silica (in water) 186
Silicates 41, 61, 62, 63
Silicone (Si) 39, 41
  in industrial dusts 64, 65
Silver (Ag)
  in industrial dusts 64, 65
Silver birch, see *Betula verrucosa*
Silver fir, see *Abies alba*
Sitta europaea 127
Smigne Creek 168–202
  watershed 168–170
  water flow 171–174
Sodium (Na)
  concentration in ungulates 123, 124
  in dusts 61, 64
  run-off 191–193
  in water 196
Soil 11–16
  brown 11
  cryptopodzolic 235
  fine sands 11
  gley 11
  gleyed 11, 12, 235
  gleyified loam 169
  grey-brown podzolic 13
  half-bog podzolic 14, 169
  heavy loams 11
  humic-brown gleyed 12
  humic gley 235
  leached brown 235
  loamy-clay 12
  marly 12
  mesotrophic-eutrophic 11, 12, 235, 239
  mesotrophic muck-gley 235
  muck gley 12
  mucky podzolic 235
  old warp 11, 235
  oligotrophic half-bog-peat-gley 169
  oligotrophic peat 13, 14, 235
  peat gley 235
  peaty podzolic 235
  podzolic 13, 14, 151, 235
  podzolized brown 235
  podzolized brown gleyed 13
  podzolized lessive 235
  pseudogley 235
  sandy 12, 169
  sandy loam 12, 13
  stagnogley 13
Soil chemical characteristics 14–16
Soil degradation 13
Soil enzymatic activity 152
Soil exchange capacity 14, 15
Soil layer
  heavy loam 12
  humus 150
  hydromull-humus 13, 14
  less-than-loamy-sand 12, 13, 14
  light loamy sand 13
  loamy sand 12, 13
  loose sand 13, 14
  mor-humus 13, 14
  muck-humus 12, 13, 14
  mull-humus 12
  peat 14
  peaty sand 14
  transitional humus 12, 13,
  very fine clay sand 12
  very fine sand 12
Soil microorganisms 150
  abundance in litter 150, 163
  activity 150, 152
  bacteria 152
  fungi 152
Soil moisture 14
Soil pollution 238
Soil-site relationship 235
Solar energy accumulation by plants 81, 82
Solar radiation 16, 17, 18, 23, 45
Solling, West Germany V
Sorbaria sorbifolia 238
Sorbus aucuparia 88, 241
Sorbus spp 73
Sorption capacity in soil 36
Southern Appalachian Mts, USA 219
Soya, see *Glycine*
Sparrow hawk, see *Accipiter nisus*
Sphagnum cuspidatum 7
Sphagnum formosum 7
Sphagnum fuscum 7
Sphagnum spp 7, 10
Spindle tree, see *Evonymus europaea*
Spinels 41
Spondylis buprestoides 3
Spruce, see *Picea*
Standard-minimum method (rodent sampling) 113
Standig crop, of
  biomass 207–213
  energy 206–213
  nutrients 207–213
  (see also Calcium, Nitrogen, Phosphorus, Potassium)
  pollutants 207–213
  (see also Sulfur, Iron, Zinc, Lead, Copper, Nickel, Cadmium)
Stellaria holostea 8, 10
Stenostola dubia 106
Streptomyces 152
Strigiformes 132
Strix aluco 131, 132, 133, 134, 135
Strix uralensis 132, 133, 134
Strontium (Sr) 64
Stroptohosoma capitatum 104
Succession, secondary 127, 128
Sulfate aerosol 47, 50, 242
Sulfate ions 52, 53, 57
  run-off 191
  in water 186
Sulfur (S) 36
  accumulation 158

balance in litter 156
budgets 36, 47, 204, 222, 227
dry deposition 36, 46, 51, 52, 68
input 56, 155
in soil 14–16, 238
turnover rate 157, 212, 213
wet deposition 36, 46, 51–54, 68
Sulfur compounds 45–58
Sulfur concentration, in
liter 156
plants 28, 29, 32, 33, 96, 155
rain water 154
Sulfur dioxide (SO_2) 38, 39, 45, 47–51, 240, 241, 242, 243, 244
Superphosphate (mineral fertilizer) 238
Sus scrofa 26, 102, 121–125
Sycamore maple, see *Acer platanoides*
Syderite 41
Sylvia atricapilla 129
Sylvite 41, 61, 62
Synedra ulna 181
Syngerite 62
Szarów 16, 20

Tachhydrite 41
Tawny owi, see *Strix aluco*
Thecla quercus 107
Thuja plicata 99, 241
Tilia cordata 3, 7, 9, 10, 28, 71, 241
Tilia spp 70, 73, 103, 234, 244
net photosynthesis 79–82
Tilio-Carpinetum stachyetosum 6, 7, 10, 11, 13, 15, 73, 75, 90, 235
Tilio-Carpinetum typicum 6, 7, 8, 15, 79, 90
Tilio-Carpinetum/Pino-Quercetum 90, 91
Tilio-Carpinetum V 6, 8, 13, 15, 103, 236
Timber 24
exploitation 24, 25
volume and increment 74
Tin (Sn) 64, 65
Titanium (Ti) 63, 65
Tobacco, see *Nicotiana*
Tortix viridana 3, 107, 108, 128, 144, 209, 214
Trachelomonas hispida 182
Trachelomonas volvocina 182
Traczówka Creek 168–202
water flow 171–174, 192, 194, 195
watershed 168–170
Trientalis eurpaea 6, 87, 88, 152
Trona 41
Tsuga heterophylla 99
Turnover rates (Θ), of
nutrients in ecosystems 212, 213
nutrients in litter 156, 157
pollutants in ecosystems 212, 213
pollutants in litter 156, 157

Ulmus campestris 10
Ulmus laevis 10
Ulmus spp 2, 3, 234
United States 98
Upper Silesian Industrial District (Region) 37, 42, 68, 232, 237
Ural owl, see *Strix uralensis*
Urea (mineral fertilizer) 239
Urease (soil enzyme) 152, 153
Ursus arctos 26
Urtica dioica 10, 103
Usnea spp 28
USSR 233, 234

Vaccinio-Piceetalia 6, 9
Vaccinio-uliginosi-Pinetum 6, 7, 10, 13, 14, 90, 235
Vaccinium myrtillus 5, 6, 9, 87, 88, 96, 97, 103, 118, 152
Vaccinium uliginosum 10
Vaccinium vitis-idaea 6, 88
Vanadium (V) 65
Vegetation 2–11
period 16, 19
Veratrum lobelianum 5
Veronica montana 10
Verticillium 152
Viola silvestris 9
Vistula River 2, 6, 10, 11, 24, 26, 28, 169, 176
Vulpes vulpes 26, 102, 132, 135–138, 139

Walker Branch, Tennessee, USA 219, 220, 221
Water
BOD_5 (biological oxygen demand) 184
carbon-dioxide content 184
conductance 183, 184
ioinic composition 185–187
oxydability 188
oxygen content 183, 184
pH value 184, 200
physiochemical characteristics 183, 184
relationships 26
temperature 184
total hardness 186
Watershed
characteristics 168–170
ecosystem studies 204, 219–222
Weasel, see *Mustela nivalis*
Weewils, see *Curculionidae* 103, 105
West-Gaeke method 49
West Germany 98, 233, 234
Western Cascade Mts, USA 219
White pine, see *Pinus strobus*
Whytham Woods, England V
Wild boar, see *Sus scrofa*
Wildlife 24, 26, 27
Wind
directions 19
velocity 45, 49
Wolf, see *Lupus lupus*
Woodwell's method (of measuring plant respiration) 87, 98
Wüstite 41

X-ray methods 59
Xyleborus dispar 106
Xyleborus saxeseni 106

Yellow-necked field mouse, see *Apodemus flavicollis*

Zinc (Zn) 56
accumulation 158, 198–202
balance in litter 156
budgets 204, 221, 229
in dust emissions 39
input 61, 66, 155
outflow from watersheds 197–199
in soil 14, 15, 238, 239
turnover rate 156, 157, 163
Zinc concentration, in
homeotherms 139–142
industrial dusts 57, 62, 64
insects 112, 113
litter 155
plants 30, 31, 96, 97, 112, 155
rain water 154
water 195
Zincite 41
Zirconium (Zr) 64, 65

MIX
Papier aus verantwortungsvollen Quellen
Paper from responsible sources
FSC® C105338

If you have any concerns about our products,
you can contact us on
ProductSafety@springernature.com

In case Publisher is established outside the EU,
the EU authorized representative is:
Springer Nature Customer Service Center GmbH
Europaplatz 3, 69115 Heidelberg, Germany

Printed by Libri Plureos GmbH
in Hamburg, Germany